INTEGRINS:
THE
BIOLOGICAL
PROBLEMS

INTEGRINS: THE BIOLOGICAL PROBLEMS

Edited by

Yoshikazu Takada, M.D., Ph.D.
Assistant Member
Department of Vascular Biology
The Scripps Research Institute
La Jolla, California

CRC Press
Taylor & Francis Group
Boca Raton London New York

CRC Press is an imprint of the
Taylor & Francis Group, an **informa** business

First published 1994 by CRC Press
Taylor & Francis Group
6000 Broken Sound Parkway NW, Suite 300
Boca Raton, FL 33487-2742

Reissued 2018 by CRC Press

© 1994 by Taylor & Francis
CRC Press is an imprint of Taylor & Francis Group, an Informa business

No claim to original U.S. Government works

A Library of Congress record exists under LC control number: 94000833

Publisher's Note
The publisher has gone to great lengths to ensure the quality of this reprint but points out that some imperfections in the original copies may be apparent.

Disclaimer
The publisher has made every effort to trace copyright holders and welcomes correspondence from those they have been unable to contact.

ISBN 13: 978-1-138-10598-0 (hbk)
ISBN 13: 978-1-138-56044-4 (pbk)
ISBN 13: 978-0-203-71164-4 (ebk)

Visit the Taylor & Francis Web site at http://www.taylorandfrancis.com and the
CRC Press Web site at http://www.crcpress.com

PREFACE

A group of cell adhesion receptor heterodimers containing VLA proteins, leukocyte adhesion receptors, platelet fibrinogen receptor GPIIb-IIIa, and vitronectin receptor were named "integrins" in 1986. Since then, integrins have been the subject of intense study. Major topics of the research in the field of integrins include:

* Structures of integrins
* Mechanism of ligand binding
* Role in organization of extracellular matrix
* Role in embryogenesis, wound healing, behavior of neoplastic cells (e.g., tumorigenicity and metastasis)
* Role in pathogenesis of diverse diseases
* Role in signal transduction from outside to inside of cells, and from inside to outside of cells

Now the integrin super gene family of cell-adhesion receptors consists of at least 15 α subunits and 8 β subunits. With so much interest in integrins, more than one research paper per day is published these days!

This book, devoted entirely to integrins, was compiled so that researchers could have a reference book covering many aspects of integrins, while it was still possible. In this volume, Hemler et al. (Chapter 1) describe the structure and functions of integrin cytoplasmic domains; Dana and Arnaout (Chapter 2) detail the role of $\beta2$ integrins in leukocyte adhesion; Loftus (Chapter 3) focuses on the role of platelet membrane fibrinogen receptor glycoprotein IIb-IIIa (αIIb$\beta3$) in thrombosis and hemostasis; Gladson and Cheresh (Chapter 4) summarize the functions of αv integrin family; Shimizu (Chapter 5) and Schwartz (Chapter 6) overview the roles of integrins in signal transduction in immune competent cells and in nonimmune cells, respectively; Carter et al. (Chapter 7) focus on integrins and epiligrin in cell-cell and cell-substrate adhesion in the epidermis; Quaranta et al. (Chapter 8) summarize the functions of integrin $\alpha6\beta4$ in epithelial and carcinoma cells; Isberg (Chapter 9) describes the role of integrins in internalization of microbial pathogens through the binding of *Yersinia pseudotuberculosis* invasin protein; DeSimone (Chapter 10) overviews the integrin function in early vertebrate development using the amphibian embryo system. We notice that integrins are involved in such diverse biological and pathological phenomena.

As with any major undertaking, editing a book had its frustrations and rewards. By far the biggest reward was the pleasure of working with and learning from the many contributors.

Yoshikazu Takada, M.D., Ph.D.

THE EDITOR

Yoshikazu Takada, M.D., Ph.D., is Assistant Member of the Department of Vascular Biology at The Scripps Research Institute, La Jolla, California.

Dr. Takada received his M.D. at Wakayama Medical School, Japan in 1976 and Ph.D. (Biochemistry) from Kyushu Dental College, Japan in 1981 and from Wakayama Medical School in 1988. He served as Research Associate in 1978 and then Assistant Professor of Biochemistry in 1981 at Kyushu Dental College until 1988. He spent 3 years as Visiting Assistant Professor of Pathology at Dana-Farber Cancer Institute, Boston, MA from 1986 to 1989. He served as Senior Investigator at the National Cancer Center Research Institute, Tokyo, Japan from 1989 to 1990. He returned to the U.S. as Assistant Member in the Department of Vascular Biology, The Scripps Research Institute, La Jolla, California in 1990.

Dr. Takada is a Member of the American Society of Cell Biology. He has received grants from the National Institutes of Health and Sandoz Pharmaceutical Company. He encountered β1 integrins (VLA proteins) in 1986 when he worked with Dr. Martin Hemler in Boston. His present interests include mechanisms of integrin/ligand interaction, their regulation at the molecular level, and the roles of integrins in tumor metastasis.

CONTRIBUTORS

M. Amin Arnaout
Renal Unit
Department of Medicine
Massachusetts General Hospital
Charlestown, Massachusetts

F. B. Berdichevsky
Dana-Farber Cancer Institute
Harvard Medical School
Boston, Massachusetts

Tod A. Brown
Department of Cell Biology
Fred Hutchinson Cancer Research
    Center
Seattle, Washington

William G. Carter
Department of Cell Biology
Fred Hutchinson Cancer Research
    Center
Seattle, Washington

David A. Cheresh
Department of Immunology
The Scripps Research Institute
La Jolla, California

Ginetta Collo
Department of Cell Biology
The Scripps Research Institute
La Jolla, California

Nava Dana
Renal Unit
Department of Medicine
Massachusetts General Hospital
Charlestown, Massachusetts

Douglas W. DeSimone
Department of Anatomy
    and Cell Biology
University of Virginia
Charlottesville, Virginia

Guido Gaietta
Department of Cell Biology
The Scripps Research Institute
La Jolla, California

Susana G. Gil
Department of Cell Biology
Fred Hutchison Cancer Research Center
Seattle, Washington

Candece L. Gladson
Department of Pathology
    and Neuropathology
University of Alabama at Birmingham
Birmingham, Alabama

Shunji Hattori
Department of Cell Biology
Fred Hutchison Cancer Research Center
Seattle, Washington

Martin E. Hemler
Dana-Farber Cancer Institute
Harvard Medical School
Boston, Massachusetts

Ralph R. Isberg
Howard Hughes Medical Institute
Tufts University
Department of Molecular Biology
Boston, Massachusetts

Paul D. Kassner
Dana-Farber Cancer Institute
Harvard Medical School
Boston, Massachusetts

Satoshi Kawaguchi
Dana-Farber Cancer Institute
Harvard Medical School
Boston, Massachusetts

Joseph C. Loftus
Department of Vascular Biology
The Scripps Research Institute
La Jolla, California

Renata Pasqualini
Dana-Farber Cancer Institute
Harvard Medical School
Boston, Massachusetts

Vito Quaranta
Department of Cell Biology
The Scripps Research Institute
La Jolla, California

Carla Rozzo
Department of Cell Biology
The Scripps Research Institute
La Jolla, California

Maureen C. Ryan
Department of Cell Biology
Fred Hutchinson Cancer Center
Seattle, Washington

Martin A. Schwartz
Department of Vascular Biology
The Scripps Research Institute
La Jolla, California

Yoji Shimizu
Department of Microbiology
    and Immunology
University of Michigan
Medical School
Ann Arbor, Michigan

Lisa Starr
Department of Cell Biology
The Scripps Research Institute
La Jolla, California

Banu E. Symington
Department of Cell Biology
Fred Hutchinson Cancer Research
    Center
Seattle, Washington

Yoshikazu Takada
Department of Vascular Biology
The Scripps Research Institute
La Jolla, California

Richard N. Tamura
Department of Cell Biology
The Scripps Research Institute
La Jolla, California

Jonathan B. Weitzman
Dana-Farber Cancer Institute
Harvard Medical School
Boston, Massachusetts

CONTENTS

Chapter 1

Structure, Biochemical Properties, and Biological Functions of Integrin Cytoplasmic Domains

Martin E. Hemler, Jonathan B. Weitzman, Renata Pasqualini, Satoshi Kawaguchi, Paul D. Kassner, and Feodor B. Berdichevsky

CONTENTS

0-8493-4711-4/94/$0.00+$.50

1

2

I. INTRODUCTION

When the integrin family of adhesion receptors was first described,[1-4] initial emphasis focused on identifying α and β subunits, and on elucidating the ligand-binding specificities for the various heterodimers. More recently, there has been a major focus on signaling through integrins, and several reviews have been written, concentrating on general aspects,[5] specific pathways,[6] extracellular matrix involvement,[7] regulation of lymphocyte adhesion,[8] and the involvement of focal adhesion kinase (FAK).[9]

Any consideration of signaling through integrins must address the roles of their relatively short cytoplasmic domains. Because each of the 14 known α subunits and 8 known β subunits has a distinct cytoplasmic domain sequence, their potential for interaction with diverse cytoskeletal proteins and signaling pathways is obviously rather large. This review will address the structural features of integrin cytoplasmic domains and attempt to relate this information to the emerging knowledge regarding their biochemical and biological functions.

II. CYTOPLASMIC DOMAIN STRUCTURES
A. INTEGRIN α SUBUNIT CYTOPLASMIC DOMAINS
1. Comparison of α Sequences

The active molecular biology studies of recent years have provided primary sequences for 13 different integrin α subunits, in many cases with corresponding homologues from several species (Table 1). There are varying degrees of homology between the different α subunit sequences, allowing division into a number of groups depending on shared structural features.[5,10] Compared to the extracellular and transmembrane regions, the C-terminal cytoplasmic domains of the α subunits generally show more sequence variability, including differences in length. A current hypothesis is that these differences reflect subunit-specific cytoskeletal interactions, regulating specific cellular events subsequent to interaction with common extracellular ligands.

In addition, within the cytoplasmic domains there is a high degree of cross-species conservation, further implying a functional importance for these sequences. For example, human and hamster α3, and human and mouse α5, are each 100% conserved, and α4, α6, and αV are each >90% conserved among human, rodent, and/or chicken. Indeed, antisera to C-terminal peptides of α3 and α5 subunits cross-react with the equivalent subunits from other species.[11] The greater sequence divergence between subunits of some species (e.g., αL or αM in human and mouse) may indicate that there was less functional pressure for conservation. The comparison between cytoplasmic domain sequences of different subunits and species will serve to highlight key residues of interest for mutational studies to define functionally important regions or motifs.

The most striking shared feature of the α subunit cytoplasmic domains is the highly conserved GFFKR sequence near the membrane, the function of which remains unclear (see below). As seen for other transmembrane proteins, the dibasic "KR" residues may serve to anchor the protein on the cytoplasmic side of the membrane. It is noteworthy that the α3, α6, and α8 cytoplasmic domains contain a cysteine residue immediately preceding the GFFKR motif.

2. Alternative Splicing of α Subunits

Both the α3 and α6 subunits have been reported to exist in alternatively spliced forms.[12-14] The alternative forms (referred to as α3B and α6B) also contain a cysteine residue and

Table 1   Integrin α subunit cytoplasmic domains

Subunit[a]	Sequence[b]	Reference
α1	WKIGFFKRPLKKKMEK	116
α1 rat	WKIGFFKRPLKKKMEK	220
α2	WKLGFFKRKYEKMTKNPDEIDETTELSS	221
α3A	WKCGFFKRARTRALYEAKRQKAEMKSQPSETERLTDDY	222,223
α3A rat	WKCGFFKRARTRALYEAKRQKAEMKSQPSETERLTDDY	224
α3A chi	WKCGFFKRASTGAMYEAKGQKAEMRIQPSETERLTDDY	11
α3B ham	""CDFFKPTRYRIMPKYHAVRIREEERYPPPGSTLPTKKHWVTSWQIRDRYY	12
α3B mou	""CDFFKPTRYRIMPKYHAVRIREEERYPPPGSTLPTKKH.....	12
α4	WKAGFFKRQYKSILQEENRRDSWSYINSKSNDD	225
α4 mou	WKAGFFKRQYRSILQEENRRDSWSYVNSKSNDD	226
α5	YKLGFFKRSLPYGTAMEKAQLKPPATSDA	227,228
α5 mou	YKLGFFKRSLPYGTAMEKAQLKPPATSDA	27
α6A	WKCGFFKRNKKDHYDATYHKAEIHAQPSDKERLTSDA	14,43
α6A mou	WKCGFFKRNKKDHYDATYHKAEIHTQPSDKERLTSDA	13
α6A chi	WKCGFFKRSKKDHYDATYHKAEIHAQPSDKERLTSDA	229
α6B	""CGFFKRSRYDDSVPRYHAVRIRKEEREIKDEKYIDNLEKKQWITKWNRNESYS	12,14
α6B mou	""CGFFKRSRYDDSIPRYHAVRIRKEERE...	13
α7 rat	WKLGFFKRAKHPEATVPQYHAVKILREDRQQFKEEKTIQRSNWGNSQWEGSDAHPILAADWHPELGPDGHPVSTA	230
α8 chi	YKCGFFDRARPPQDDMADREQLTNNKTTDA	231
αV	YRMGFFKRVRPPQEEQEREQLQPHENGEGNSET	232
αV chi	YRMGFFKRVRPPQEEQEREQLQPHENGEGTSEA	11
αIIb	WKVGFFKRNRPPLEEDDEEGE	233
αIIb rat	WKVGFFKRNRPPLEEEEEE	234
αL	YKVGFFKRNLKEKMEAGRGVPNGIPAEDSEQLA-SGQEAGDPDCLKPLHEKDSESG	235
αL mou	YKVGFFKRNLKEKMEADGGVPNGSPPEDTDPLAVPGEETKDMGCLEPSGRVTRTKA	236
αM	YKLGFFKRQYKDMMSEGGPPGAEPQ	237–239
αM mou	YKLGFFKRQYKDMMNEAAPQDAPPQ	236
αX	YKVGFFKRQYKEMMEEANGQIAPENGTQTPSPPSEK	240
αPS2 dro	YKCGFFNRNRPTDHSQERQPLRNGYHGDEHL	241

[a]Sequences are of human origin except for those marked "chi", "rat", "mou", or "dro" which are from chicken, rat, mouse, hamster, or drosophila sources, respectively; [b]For the α$_{3B}$ and α$_{6B}$ sequences, only the alternatively spliced residues are shown. The highly conserved GFFKR motifs are underlined.

share a stretch of identical residues that are also similar to sequence in the α7 cytoplasmic domain (Table 1). It will be interesting to see if α7 is also alternatively spliced. The variant forms of α6 and α3 are expressed in a cell-type specific manner[12-15] and offer another layer of complexity to integrin-specific cellular responses. So far, no functional differences have yet been ascribed to the different isoforms, but phosphorylation differences have been noted (see below). The position of the divergence of the two isoforms corresponds to the position of introns (i.e., immediately preceding the GFFKR motif) in all of the α subunit genes that have yet been characterized.[16-19] However, there is not yet evidence that similar alternative splicing events occur for other integrin α subunits besides α3 and α6.

B. INTEGRIN β SUBUNIT CYTOPLASMIC DOMAINS
1. Comparison of β Sequences
There are 8 different integrin β subunits (β1 to β8) that have been identified and sequenced (Table 2). All of these β subunits have similar topographic features, including a large extracellular domain followed by a single putative transmembrane region (23 to 28 residues) and a short cytoplasmic tail (39 to 60 residues). An exception is the cytoplasmic tail of a β4 subunit, which exceeds 1000 amino acids. Studies of genomic organization for the coding region of β1, β2, β3, β6, and β7 subunits[20-25] suggest that all β subunits have a common origin in evolution.[25,26]

Homologues of human β1, β2, β7, and β8 subunits cloned and sequenced from other species have revealed that the cytoplasmic tail represents one of the most conserved parts of the molecule.[27-33] In fact, human, chicken, and frog β1 cytoplasmic domains are identical, and human β2 and β8 differ by only one to two residues from their mouse and rabbit counterparts, respectively (Table 2). Whereas the overall cross-species homology for these molecules is 82 to 87%, there is 96 to 100% conservation of cytoplasmic domains. This high level of amino acid conservation is consistent with these domains being of critical functional importance. Comparison of human, mouse, and rat β7 sequences provides an interesting exception to the high cross-species conservation generally observed. Although 85 to 92% are similar through the first 40 amino acids of the cytoplasmic domain, the last 7 to 16 amino acids are quite distinct (Table 2), possibly implying that this region of the molecule is less functionally important (see also Figure 1).

In contrast to α subunit cytoplasmic domains which are rather distinct (Table 1), the human β1, β2, β3, β5, β6, and β7 subunit cytoplasmic domains are ~25 to 60% similar to each other (Table 3). This roughly correlates with the level of similarity obtained when comparing the sequences of the entire molecules. Thus, we expect that certain β subunit cytoplasmic domain functions may be shared (see below). In this regard, the cytoplasmic domains of β1 and β3 have the highest degree of similarity (~60%) and are functionally interchangeable.[34] In contrast to the others, the cytoplasmic domains of β4 (with three variant forms) and β8 are totally distinct, suggesting that these may mediate unique functions. For example, in contrast to the β1, β2, and β3 integrins which connect with a cellular network of microfilaments,[5,35] the α6β4 integrin was found in hemidesmosomes, in association with intermediate filaments.[36-38]

2. Alternative Splicing of β Subunits
The cytoplasmic domains of the β1,[20,39] β3,[40] and β4[41-43] subunits have alternatively spliced forms, which potentially may provide additional functional diversity for these integrins. An alternatively spliced variant of the β1 subunit (called β1s) has been recently described[39] that diverges from β1 after the first 20 amino acids of the cytoplasmic domain (Table 2). At the protein level, the β1s form is present in minor amounts

Table 2 Integrin β subunit cytoplasmic domains

Subunit[a]	Sequence[b]	Reference
βps dro	HDRREFARFEKERMNAKWDTGENPIYKQATSTFKNPMYAGK	242
β1	HDRREFAKFEKEKMNAKWDTGENPIYKSAVTTVVNPKYEGK	227
β1 mou	HDRREFAKFEKEKMNAKWDTGENPIYKSAVTTVVNPIYEGK	27,28
β1 chi	HDRREFAKFEKEKMNAKWDTGENPIYKSAVTTVVNPKYEGK	29
β1 xen	HDRREFAKFEKEKMNAKWDTGENPIYKSAVTTVVNPKYEGK	30
β1	HDRREFAKFEKEKMNAKWDTGENPIYKSAVTTVVNPKYEGK	227
β1(3'v)	HDRREFAKFEKEKMNAKWDTVSYKTSKKQSGL	20
β1(s)	HDRREFAKFEKEKMNAKWDTSLSVAQPGVQWCDISSLQPLTSRFQQFSCLSLPSTWDYRVKILFIRVP	39
β2	SDLREYRRFEKEKLKSQWNNDNPLFKSATTTVMNPKFAES	243,244
β2 mou	TDLREYRRFEKEKLKSQWNNDNPLFKSATTTVMNPKFAES	31
β3	HDRKEFAKFEEERARAKWDTANNPLYKEATSTFTNITYRG	245–247
β3'	HDRKEFAKFEEERARAKWDTVRDGAGRFLKSLV	40
β4	KACLALLPCCNRGHMVGFKEDHYMLRENLMAS	41–43
	DHLDTPMLRSGNLKGRDVVRWKVT............	
β5	HDRREFAKFQSERSRARYEMASNPLYRKPISTHTVDFTFNKFNKSYNGTVD	45–47
β6	HDRKEVAKFEAERSKAKWQTGTNPLYRGSTSTFKNVTYKHREKQKVDLSTDC	248
β7	YDRREYSRFEKEQQQLNWKQDSNPLYKSAITTINPRFQEAD—SPTL	249,250
β7 mou	YDRREYRRFEKEQQQLNWKQDNNPLYKSAITTVNPRFQGTNGRSPSLSLTREAD	32,251,252
β7 rat	YDRLEYSRFEKERQQLNWKQDSNPLYKSAVTTVVNPRFQGGNKQSLSLPLTQEAD	252
β8	QWNSNKIKSSSDYRVSASKKDKLILQSVCTRAVTYREKEPEEIKMDISKLNAHETFRCNF	33
β8 rab	QWNSSKIKSSSDYRVSASKKDKLILQSVCTRAVTYRREKEPEEIKLDISKLNAHETFRCNF	33

[a]Sequences are of human origin, except for those marked "mou", "chi", "xen", "dro", or "rab", which are from mouse, chicken, xenopus, drosophila, or rabbit, respectively. Both β1 and β3 have alternatively spliced forms [β1(s), β1(3'v), β3'] that are indicated by underlining; [b]The β4 sequence continues for nearly 1000 additional residues.

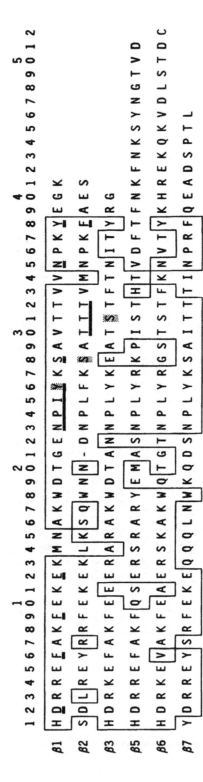

Figure 1 Alignment of similar integrin β subunit cytoplasmic domains. Double-underlined residues in the β₁ sequence have been shown by site-directed mutagenesis to contribute to integrin localization of focal adhesions.⁹⁹ The shaded residue (Y) has been suggested to be the site of phosphorylation by tyrosine kinase. ²⁹,²⁵³ Double-underlined residues in the β₂ sequence are important for cell adhesion mediated by LFA-1.¹⁰¹ The shaded serine is the major residue phosphorylated in response to PMA-stimulation. The shaded serine residue in the β₃ sequence is critical for activiation of α₁₁bβ₃ in platelets.¹⁵⁰

Table 3   Comparison of β subunit cytoplasmic domain sequences

	β1	β2	β3	β4	β5	β6	β7	β8
β1	100							
β2	44	100						
β3	61	34	100					
β4	<10	<10	<10	100				
β5	33	21	38	<10	100			
β6	38	25	50	<10	38	100		
β7	38	49	32	<10	31	26	100	
β8	<10	<10	<10	<10	<10	<10	<10	100

compared to β1.[39] The presence of a domain in β1s with some similarity to the SH2 motif suggests that it could have a specific signaling function.[39]

Besides β1s, an additional β1 transcript (β13'v) and an alternative β3 transcript (β3') with divergent 3' ends were found at a low level in placenta and in several established cell lines.[20,40,44] Since the divergent portions of β13'v and β3' are identical to the 5' ends of the last introns of each gene, both forms may be generated either as a result of premature termination of transcription or due to a lack of splicing of this intron. The β13'v form (renamed β1B) shows a distinct subcellular localization pattern and different tissue distribution when compared to β1 (also called β1A).[44] Notably, the β1, β2, β3, β6, and β7 subunits[20–25] each have an intron/exon boundary at the same location in which β1 and β3 are alternatively spliced (see Figure 1 below), but alternative forms within the cytoplasmic domains of β2, β6, and β7 have not been reported.

Alternative splicing appears to yield three different forms of the long cytoplasmic domain of β4. Analysis by PCR[43] revealed that mRNA coding for the shortest cytoplasmic domain (1019 amino acids) was present in placenta and in 2 carcinoma cell lines. In addition these cells expressed a transcript with an in-frame insertion of 210 bp. A third form of the β4 transcript (originally cloned from a keratinocyte cDNA library) had another in-frame insertion of 159 bp[41] and was also found in placenta.[43]

Transcripts for the integrin β5 subunit can encode either a single "FNK" sequence or a tandem "FNKFNK" sequence near the end of the cytoplasmic domain.[45–47] The genomic organization of β5 has not yet been investigated, and at present it is not clear if these represent alternatively spliced forms.

III. BIOCHEMICAL PROPERTIES OF INTEGRIN CYTOPLASMIC DOMAINS

A. ASSOCIATED MOLECULES

As it becomes increasingly clear that integrins may participate in multiple signaling events, it is hypothesized that the diversity of cytoplasmic domain sequences might allow association with a number of distinct other molecules. However, unlike cadherins which are tightly bound to intracellular proteins called catenins,[48,49] most integrins do not appear to associate tightly with cytoplasmic or cytoskeletal proteins. In one attempt to identify proteins associated with the β1 cytoplasmic domain, a protein called fibulin was isolated from a β1 peptide affinity column,[50] but later it was discovered that this binding was adventitious.[51] Despite the technical difficulties, a few candidate integrin-associated molecules have been identified and are discussed below.

1. Talin

Using equilibrium gel filtration, it was found that a 225-kDa cytoskeletal protein called talin could bind to an integrin complex purified from chicken fibroblasts.[52] Subsequent studies showed that this binding could be blocked with a high concentration (1 mg/ml) of the synthetic decapeptide (WDTGENPIYK) from the cytoplasmic domain of the β1 subunit.[53,54] Moreover, the ability of β1 integrin to interact with talin was decreased when β1 was phosphorylated on Tyr[788] (as isolated from RSV-transformed chick embryonic fibroblasts).[54] However, subsequent mutation of Tyr[788] to Phe did not alter β1 subcellular localization properties.[55] Thus, the significance of Tyr[788] phosphorylation for integrin association with talin or other cytoskeletal proteins remains in doubt. Since the alternatively spliced β1s[39] and β13′v[40] forms lack the sequence that may be responsible for the β1-talin interaction,[54] these are predicted not to interact with talin.

Because β1,[56,57] β2,[58,59] and β3[60] integrins can localize into talin-containing subcellular complexes, it is predicted that a site for direct talin binding might be shared by all of these integrin subunits. In this regard, compared to the WDTGENPIYK sequence implicated in β1-talin binding, β2 and β3 express partly similar WNNDNPLFK and WDTANNPLYK sequences, respectively (see Figure 1 below).

2. α-Actinin

In affinity chromatography experiments, several proteins were retained by Sepharose, coupled with the peptide corresponding to the cytoplasmic domain of the β1 subunit.[61] One of these proteins was identified as the 100-kDa focal adhesion protein α-actinin. Upon further examination in a solid phase binding assay, α-actinin bound to an immobilized β1 cytoplasmic domain peptide with a high affinity. In addition, purified β1 integrin heterodimer, or the platelet αIIbβ3 integrin, also bound to α-actinin, although with a lower affinity compared to the peptide binding. One interpretation of these results is that integrin α subunit domains might regulate the interaction of α-actinin with the β subunit[61] (see discussion of focal adhesions below).

A direct interaction between α-actinin and a β2 integrin (LFA-1) has also been proposed. From a coimmunoprecipitation assay it was found that activation of T cells with PMA or with anti-CD3 antibodies resulted in an apparent association of LFA-1 with α-actinin and vinculin.[62] Information regarding a specific β subunit cytoplasmic domain residues responsible for α-actinin interaction is not yet available.

3. IAP-50

During characterization of a potentially novel integrin called LRI, for "leukocyte response integrin",[63] a monoclonal antibody was selected that blocked the functions of LRI. However, rather than directly recognizing LRI subunits, this antibody instead recognized a 50-kDa protein noncovalently associated with LRI, called integrin associated protein (IAP).[64] The IAP protein sequence suggests that this protein may span the plasma membrane three or five times, with one major extracellular loop containing most of the glycosylation sites.[65] Besides associating with LRI, which is immunologically cross-reactive with β3, the IAP protein can also associate with β3 integrins themselves.[64] The transmembrane nature of IAP, and its specificity for β3 or β3-like integrins, suggests a possible interaction with the β3 transmembrane or cytoplasmic domains, although this has not yet been demonstrated. At present, there is no evidence for any other integrins associating with either IAP-like proteins or other types of membrane channels. Also, it is unclear why cells such as erythrocytes would express IAP when they do not express integrins.[66]

4. Possible Lipid Associations

The integrin β3 subunit can become radiolabeled upon incubation of intact platelets with [³H]palmitate.[67] Because there are no cysteine, serine, or threonine residues in the transmembrane domain, serines or threonines in the β3 cytoplasmic domain are likely candidates for this covalent linkage, of unknown functional importance.

Integrin function can be modulated by altering the membrane lipid composition,[68] by incubation with phosphatidic acid or lysophosphatidic acid,[69] or by adding an unidentified nonphospholipid.[70] At present it is not clear whether any of these interactions is specific for only a limited subset of integrins, or whether cytoplasmic domains might be involved.

5. Other Molecules

Many of the integrin α and β chain cytoplasmic domains are phosphorylated *in vivo* (see next section), thus implying a temporal interaction of the cytoplasmic tails with protein kinase(s) and, perhaps, also protein phosphatase(s). However, the direct association of such molecules with integrins has thus far not been found.

Antibody-induced clustering of integrins, cell spreading, or platelet aggregation all correlate with apparent integrin-dependent tyrosine phosphorylation of various proteins such as the 125-kDa focal adhesion kinase (FAK),[71-76] the 70-kDa cytoskeletal protein paxillin,[74] and the unnamed 105 to 110-kDa protein(s) on lymphocytes[77,78] and other leukocytes.[79] However, at present there is no evidence for direct association of any of these proteins with integrins.

In astrocytes, functionally active α1β1 integrins colocalized with clathrin rather than talin or α-actinin[80] in a subcellular structure called a "point contact", which differed from the commonly observed focal contact. This result implies that integrin cytoplasmic domains might directly interact with other cytoskeletal proteins besides α-actinin and talin.

Immobilized KLGFFKR-peptide has been used in an affinity chromatography experiment to isolate a 60-kDa protein from a Triton-soluble cell extract prepared from human osteosarcoma cells. This 60-kDa protein, which appears to be the human equivalent of rabbit calreticulin,[81] bound to purified α3 subunit, but not β1, consistent with the presence of a KXGFFKR sequence at the transmembrane/cytoplasm interface of α3 (as in most other α subunits). However, the significance of this observation remains to be determined, since *in vivo* p60/calreticulin is predominantly localized in the endoplasmic reticulum (ER). Further experiments will be required to verify the hypothesis that interaction of p60 with integrin α subunits might be important for the assembly of heterodimers in the ER.[81]

With its uniquely long cytoplasmic domain, the integrin β4 subunit is likely to participate in specific biochemical interactions. In this regard, several groups have shown that the α6β4 integrin localizes to hemidesmosomes.[36-38] Specifically, it was found that antibodies directed to the extracellular domain of the β4 subunit prevented hemidesmosomal plaque assembly in explant culture and perturbed the distribution of anchoring filaments.[82] Further studies are required to determine how the long cytoplasmic tail of β4 might participate in the anchoring of intermediate filaments to the membrane.[83]

B. PHOSPHORYLATION OF INTEGRIN CYTOPLASMIC DOMAINS

The presence of tyrosine, serine, and threonine residues in the cytoplasmic domains of integrin α and β subunits (see Tables 1, 2) suggests that they may be substrates for both tyrosine and serine/threonine kinases. Variations in integrin phosphorylation have been found to correlate with changes in the adhesive function of integrins and with changes in their associations with cytoskeletal proteins. This has focused interest on integrin phosphorylation as a candidate regulatory mechanism for controlling integrin function and distribution.

1. α Subunit Phosphorylation

Phorbol ester-stimulation of macrophages can induce a PKC-dependent increase in α6β1-mediated adhesion to laminin without increased expression of α6β1(VLA-6) at the cell surface.[84] This macrophage activation is accompanied by an increase in phosphorylation of the cytoplasmic domain of the α6 (but not β1) subunit, a dramatic increase in the anchoring of α6β1 to the cytoskeleton, and rearrangement of the actin cytoskeleton. Similar results indicating a correlation between α6 phosphorylation and increased laminin adhesion have been obtained in studies of NK cells, activated either by phorbol esters or by cross-linking of CD16.[85] In response to phorbol ester treatment, α6A, but not α6B, showed increased phosphorylation in various cell lines,[15] and this phosphorylation occurred mostly on serine, and, to a lesser extent, on tyrosine.[85] Notably, α6A in platelets and α6B in a carcinoma cell line appeared to adhere actively to laminin even though not phosphorylated.[85] Thus, there is not an absolute correlation between α6 phosphorylation and function, and the significance of the phosphorylation remains to be established.

The α3 subunit in a melanoma cell line was constitutively phosphorylated on serine,[86] and on other cell lines the level of phosphorylation of α3 (α3A in particular) increased upon phorbol ester treatment.[15] There is one serine conserved between α6A and α3A, suggesting that this could be the major site of constitutive and/or phorbol ester-stimulated phosphorylation. The significance of α3 phosphorylation also remains to be established.

In several studies, phosphorylation of uncharacterized mixtures of β1-associated α subunits have been examined. In one case, a marked increase in serine and tyrosine phosphorylation was observed for α subunit(s) from v-src-containing RSV-transformed CEF cells compared to untransformed CEF cells.[87] Consistent with these findings, serine and tyrosine phosphorylation of an unknown β1-associated α subunit was diminished in a muscle cell line, coordinate with cell differentiation when a temperature-sensitive mutant v-src was inactivated at the nonpermissive temperature.[88] In another example, retinoic acid-induced differentiation of F9 teratocarcinoma cells caused a loss of serine phosphorylation from a mixture of β1 and unknown α subunits.[89] Together these studies suggest phosphorylation of unknown β1-associated integrin α subunit(s) may block normal integrin-cytoskeletal organization and cell differentiation.

Constitutive phosphorylation of the integrin α4 subunit has also been observed, but this was not correlated with any obvious integrin function (M. Elices and M. E. Hemler, unpublished).

The β2-associated α subunits (αL, αM, αX) are constitutively phosphorylated in neutrophils, monocytes, and lymphocytes,[62,90-92] with the majority of this phosphorylation occurring on serine.[91] In some studies, levels of constitutive α chain phosphorylation did not change upon cell stimulation with FMLP or PMA.[90-92] In contrast, a transient (15 to 45 min) PKC-dependent hyperphosphorylation of the LFA-1 α subunit was observed upon antigen receptor stimulation (anti-CD3 cross-linking) of T cells, which correlates with rearrangement of the actin cytoskeleton and physical association of LFA-1 with cytoskeletal proteins α-actinin and vinculin.[62]

Thus far, all of the studies of α subunit phosphorylation have been strictly correlative and do not provide definitive evidence for an essential role in integrin regulation.

2. β1 Subunit Phosphorylation

The β1 cytoplasmic domain contains two tyrosines, one serine, and several threonine residues which may act as potential phosphorylation sites[29] and are highly conserved across species (see Table 1). Phosphoamino acid analysis has shown that both the Tyr[788] and Ser[790] residues (positions 26 and 28 in Figure 1) can be phosphorylated in normal and virally transformed avian fibroblast cells.[87,93]

The β1 integrin has a higher specific phosphorylation when localized in adhesion plaques,[93] and decreased phosphorylation in migratory cells that lack focal contacts.[94] These results suggest that integrin phosphorylation may exert a positive controlling effect on cell adhesion. However, it has also been suggested that β1 phosphorylation may be associated with receptor inactivation. Oncogenic transformation by the Rous sarcoma virus resulted in increased integrin phosphorylation (pp60v-src kinase phosphorylation of β1 at Tyr[788], a redistribution of integrin dimers,[93,95] and decreases in binding to fibronectin and talin as assayed by gel filtration experiments.[54] These changes, namely loss of cell-associated fibronectin, disorganization of the cytoskeleton, and reduced cell adhesion, mimic those of the transformed phenotype.[96,97] This correlation is further supported by results showing that cells transformed with a virus containing a mutant form of pp60v-src that lacks the transformed phenotype also fail to phosphorylate β1.[95]

Furthermore, the retinoic acid-induced differentiation of F9 teratocarcinoma stem cells is accompanied by integrin redistribution and fibrous fibronectin deposits and a highly structured actin cytoskeleton. These events coincide temporally with loss of phosphorylation (predominantly on serine) of the β1 integrin subunit (and associated α subunit).[89] The differentiation of avian myoblasts into myotubules is also associated with integrin, fibronectin, and actin redistribution. These correlate with a decrease (three- to fourfold) in tyrosine and serine phosphorylation of β1 (and also α; see above) that could be blocked by the v-src kinase.[88]

Despite these correlative findings, an extensive study involving site-directed mutagenesis of phosphorylated residues is necessary to establish firmly a causal relationship between phosphorylation and regulation of integrin function. Several groups have reported experiments expressing the chicken β1 subunit and truncated or mutated variants in mouse NIH 3T3 cells.[55,98,99] Truncated molecules that contained the tyrosine and serine residues failed to localize to focal contacts.[98] Site-directed mutants of the Tyr[788] (Y788F) or Ser[790] (S790T or S790M) residues localized to focal contacts and were indistinguishable from wild-type transfectants.[55,99] Substitutions of threonine residues also appeared to have no effect on focal contact localization.[99] In addition, mutation of Tyr[788] (Y788F) failed to influence activation of focal adhesion kinase (FAK).[71] Hence, these residues, which can act as phosphorylation sites, appear not to be essential for efficient dimer formation, export or localization to focal contacts, and possibly adhesive function. Acidic substitutions of these residues (Y788E, S790D) resulted in reduced focal contact localization,[99] suggesting that a negative charge at these positions might influence integrin distribution, thereby negatively regulating integrin function.

A role for β1 phosphorylation cannot be ruled out until more extensive studies address the effects on integrin functions beyond focal contact localization, such as signaling events, interaction with cytoplasmic or membrane proteins, and the differences between alternatively spliced forms in which tyrosine, threonine, and serine residues are not conserved.[20,39]

3. β2 Subunit Phosphorylation

Phorbol esters, which activate protein kinase C (PKC), have been demonstrated to induce rapid and sustained phosphorylation of β2 on serine in monocytes, neutrophils, lymphocytes, or β2-transfected COS cells.[90–92,100,101] This induced phosphorylation correlates with the time scale for phorbol ester-induced neutrophil[90] and mononuclear leukocyte[92] aggregation. Phorbol ester stimulation induced only a small amount of β2 phosphorylation on threonine and no detectable tyrosine phosphorylation.[101] Notably, the cytoplasmic domain of the β2 subunit lacks the tyrosine phosphorylation site present in most other β subunits (Figure 1, position 26).

Mutational analysis[101] showed that the serine at position 27 (Figure 1) is the major residue phosphorylated upon phorbol ester stimulation of β2-transfected COS cells or of β2-transfected B lymphoblastoid cells from an LAD patient. A neighboring threonine (position 30), as well as a few residues on the N-terminal side of the serine, appears to contribute to the domain recognized by the kinase responsible for phosphorylation of the serine at position 27.[101] Notably, that serine is conserved in β1 and β7 subunits.

However, substitution of each individual serine, or tyrosine residue, in the β2 cytoplasmic domain had no effect on constitutive adhesion of COS cells to the LFA-1 ligand ICAM-1. Furthermore, substitution of each serine, tyrosine, or threonine residue did not alter PMA-stimulated LFA-1 function in lymphocytes.[101] These studies clearly show that phosphorylation of the β2 cytoplasmic domain is not a critical part of the mechanism of regulation of the avidity of LFA-1 for ICAM-1. Consistent with this conclusion, stimulation of T lymphocytes with anti-CD3 MAb is known to induce a high avidity state for αLβ2,[102,103] but CD3 MAb did not induce β2 phosphorylation.[62]

Simultaneous mutation of the three contiguous threonine residues (at positions 28 to 31) significantly reduced constitutive αLβ2-mediated ICAM-1 binding.[101] However, this effect is probably not due to loss of threonine phosphorylation since there is only minor phosphorylation of threonine residues.

4. Phosphorylation of Other β Subunits (β3, β4, β5)

Tyrosine, serine, and threonine residues are located in the β3 (GPIIIa) cytoplasmic domain at positions conserved in other β subunits (see Figure 1, positions 20, 26, 30 to 32, 38). The phosphorylation of β3 was shown to occur predominantly on threonine, and the levels increased upon platelet activation.[104] However, the stoichiometry is low (0.01 mol of ^{32}P/mol of β3 unstimulated; 0.03 mol of ^{32}P/mol of activated β3) and not likely to affect ligand binding.[105] This suggests that β3 phosphorylation is not functionally relevant, but studies with recombinant mutated variants have yet to address this issue.

There are reports of phosphorylation of other integrin β subunits. For example, β5 is phosphorylated mainly on serine in response to phorbol ester in cells and by PKC in a cell-free assay.[106] Notably, β5 lacks a serine at position 28 (Figure 1) which is phosphorylated in β2, but contains other serines not found in β2 (e.g., at positions 11, 14, 22, 31, 45). The unusually long β4 cytoplasmic domain can also be phosphorylated on serine and tyrosine residues near the cleaved C terminus.[107,108] However, as with the β1, β2, and β3 subunits, the role of β5 and β4 subunit phosphorylation in regulating integrin function remains unclear.

In summary, although many studies have observed correlations between phosphorylation states and changes in cellular adhesion, a clear causal link has yet to be defined. It is also possible that the cellular phenomena studied so far (i.e., adhesion, localization to focal contacts, etc.) may not represent the phosphorylation-dependent events that are important in a physiological context.

C. INTEGRIN SYNTHESIS, EXPRESSION, AND TURNOVER

The role of cytoplasmic domains of the integrin α or β subunits during the biosynthetic process has been examined in several studies employing mutant constructs with deleted or exchanged cytoplasmic domains. The avian β1 subunit was expressed efficiently on the surface of mouse 3T3 cells and formed heterodimers with endogenous mouse α subunits despite serial truncations of the cytoplasmic domain[55,98] or removal of intermediate portions.[99] Surface expression was even observed for a mutant lacking the entire β1 cytoplasmic domain.[109] Similarly, cytoplasmic deletion mutants of β2,[110] β3,[111] or β4 subunit[112] also retained the capacity for expression on COS or CHO cell surfaces in association with appropriate α subunits. The αIIbβ3 integrin was exported to the cell surface despite deletion of both β3 cytoplasmic and transmembrane domains.[113,114] Similarly, both αMβ2[115] and α1β1[116] were secreted in functionally active forms when transmembrane and cytoplasmic domains from α and β subunits were simultaneously deleted. When expressed in the absence of α chain, wild type β3 is retained in the ER and then degraded.[117] However, when β3 lacking transmembrane and cytoplasmic regions was expressed in the absence of an α chain, it was secreted, suggesting that a retention signal may reside either in the transmembrane or cytoplasmic domain.[114] When a chimeric construct consisting of the extracellular domain of avian β1 and the transmembrane and cytoplasmic domains of human β3 was transfected into mouse 3T3 cells, it associated with α subunits appropriate for β1 but not β3.[34] Together these results suggest that the cytoplasmic and transmembrane domains of integrin β subunits are not a determining factor for α subunit selection, assembly of αβ heterodimers, or for cell surface expression. There are naturally occurring mutations which can result in the absence of integrin expression, such as in leukocyte adhesion deficiency (deficiency of integrin β2 family on leukocytes) and Glanzmann thrombasthenia (deficiency of αIIbβ3 integrin on plate-lets). Thus far, none of these have mapped to the cytoplasmic domains of the β2 or β3 subunits,[118–121] providing further evidence that these domains are not essential for chain association and surface expression.

The role of α subunit cytoplasmic domains during integrin synthesis and expression has been less well studied. Deletion of α2 or α4 cytoplasmic domains just after the highly conserved GFFKR sequence (see Table 1) still allowed association with β1 and cell surface expression. However, if cytoplasmic domain deletion included the GFFKR sequence of either α2 or α4, then surface expression was not observed.[122,123] This result is consistent with the "KR" residues perhaps serving a critical anchoring function. Deletion of the αL cytoplasmic domain two residues beyond the GFFKR sequence did not allow cell surface expression, suggesting that additional residues may be required to stabilize αL in the membrane and/or allow association with β2.[110] In contrast to the results with α chains associating with β1 or β2 integrins, a cytoplasmic deletion mutant of the αIIb subunit was expressed on the surface of CHO cells as an αIIbβ3 heterodimer even though it lacked the GFFKR motif.[111] It remains to be determined whether that was an aberrant finding, or an indication that β3 integrins have different rules for chain association and expression.

A low level of endocytosis and recycling of β1,[124–127] β2,[127,128] and β4[127] integrins has been observed. The cytoplasmic domain sequences involved have not been elucidated, but cytoplasmic domains of many integrin β subunits (β1, β2, β3, β5, β6, β7) contain one or two NPXY motifs (Figure 1, positions 23 to 26, 35 to 38) known to be utilized for coated pit-mediated internalization of many cell surface proteins.[129] Supporting a role for these NPXY motifs, their deletion resulted in loss of β2 integrin endocytosis.[128] Since the alternatively spliced forms of β1 and β3 (see Table 2 above) lack these motifs, it will be interesting to see if they are deficient in endocytosis. Following binding to various β1 integrins, invasin-bearing bacteria are internalized into an endocytic vacu-ole,[130] accompanied by cytoskeletal reorganization.[130] It is likely that specific endocytic properties of the β1 cytoplasmic domain may also play a role in this process. The recycling behavior of α5β1 differed from α3β1 and α4β1, and, similarly, αMβ2 differed from αLβ2, despite sharing the same β subunits.[127] Thus, it is possible that α chain cytoplasmic domains also may play a role in integrin recycling.

IV. BIOLOGICAL/FUNCTIONAL ROLES OF CYTOPLASMIC DOMAINS

It has become increasingly well established that integrins participate in both "inside out" and "outside-in" signaling events.[5–9,131,132] Studies aimed at understanding the detailed mechanisms of bidirectional signaling through integrins have begun to focus on integrin cytoplasmic domains. Emerging data are beginning to suggest a critical connection between extracellular ligand-binding domains and intracellular cytoplasmic domains, such that induced conformational changes in either region can have critical consequences in the other.

A. INSIDE-OUT SIGNALING: REGULATION OF INTEGRIN ADHESIVE ACTIVITY

A dominant feature of integrins is that their adhesive functions can be regulated by mechanisms that do not involve changes in integrin expression. A wide variety of cellular agonists, acting through their respective receptors, cause the function of $\beta 1$,[133-136] $\beta 2$,[102,103,137-141] and $\beta 3$[142] integrins to be rapidly upregulated (within seconds or minutes). In addition, integrins can undergo negative regulation. For example, loss of adhesive function, but retention of expression, has been documented for $\beta 1$ integrins.[143-145] This process of "inside-out" integrin regulation is poorly understood, although G proteins, protein kinase C, cAMP, calcineurin, arachidonic acid, and other traditional signaling pathway components have been implicated.[102,146-148] Recent studies involving site-directed mutagenesis, deletion, and cytoplasmic domain swapping indicate that the cytoplasmic domains of both the integrin α and β chains play pivotal roles in the control of functional activity.

1. α Subunit Cytoplasmic Domains

Studies of α-chain cytoplasmic domains in regulation of integrin adhesive function have yielded a range of results, thus making it difficult to discern general principles. Deletion of the αIIb cytoplasmic tail or exchange with the $\alpha 5$ cytoplasmic domain both resulted in an increased affinity of the αIIb$\beta 3$ heterodimer for fibrinogen.[111] That result suggested that in CHO cells, the αIIb cytoplasmic tail may play a negative regulatory role, since its removal resulted in upregulation of function.

In contrast, the cytoplasmic domains of $\alpha 4$, $\alpha 2$, and $\alpha 5$ appear to play a positive role in the regulation of cell adhesion mediated by $\beta 1$ integrins, because deletion of the cytoplasmic domains of either $\alpha 4$[122] or $\alpha 2$[123] resulted in loss of constitutive adhesive activity of the integrins $\alpha 4\beta 1$ and $\alpha 2\beta 1$, respectively. Furthermore, these integrins with deleted cytoplasmic domains could no longer be functionally upregulated when cells were treated with phorbol ester.[122,123] The $\alpha 2$, $\alpha 4$, and $\alpha 5$ cytoplasmic were all equally able to support cell adhesion. Together these results suggest that multiple interchangeable α-chain cytoplasmic domains exert a positive control on adhesion through $\beta 1$ integrins.

It is not yet clear why the cytoplasmic domain of αIIb might play a negative role, whereas those of $\alpha 2$, $\alpha 4$, and $\alpha 5$ would have apparent positive roles in supporting integrin adhesive function. One possibility is that α chains associated with $\beta 3$ and $\beta 1$ integrins have different regulation mechanisms. Alternatively, it should be noted that the αIIb cytoplasmic domain was truncated prior to the GFFKR motif,[111] whereas $\alpha 4$ and $\alpha 2$ were truncated after the GFFKR motif.[122,123] Thus, resolution of this discrepancy awaits analysis of the function of αIIb truncated after the GFFKR motif.

A series of deletions of the αL cytoplasmic tail (up to 39 amino acids) had no effect on LFA-1 mediated adhesion to ICAM-1.[110] However, the longest of these deletions still contained 12 amino acids beyond the conserved GFFKR motif, which could potentially be critical for adhesion. Thus, the αL deletion results are not necessarily at variance with the results obtained using $\alpha 4$ and $\alpha 2$ deletions. Deletion of αL just after the conserved GFFKR motif resulted in loss of expression,[110] thus preventing analysis of functional regulation.

Although the cytoplasmic domains of $\alpha 2$, $\alpha 4$, and $\alpha 5$ appear to play interchangeable positive roles in supporting $\beta 1$ integrin-mediated adhesion, it is not certain that all $\beta 1$ integrins will show a similar pattern. For example, the integrin $\alpha 3\beta 1$(VLA-3) was constitutively active in cells in which $\alpha 2\beta 1$ and $\alpha 4\beta 1$ were inactive, but was inactive in cells in which $\alpha 2\beta 1$ and $\alpha 4\beta 1$ were active.[149] Since it is already established that α chain cytoplasmic domains can influence the constitutive activity of integrins, we hypothesize that the α3 cytoplasmic domain may differ from those of α2, α4, and α5 in this respect.

2. β Subunit Cytoplasmic Domains

The cytoplasmic domains of β subunits also play a critical role in the regulation of integrin-mediated ligand binding and cell adhesion. Deletion of the C-terminal 39 or 25 amino acids, but not 16 or 5 amino acids, eliminated adhesion through β1-containing integrins (Figure 2A, column Ad, lines 3, 5, 6, 9). Interestingly, the maintenance or loss of adhesive capability did not correlate with ability of the truncated β1 integrins to localize to focal contacts (column FA). For example, deletion of 16 amino acids caused loss of the latter, but not the former (Figure 2A, line 5).

Deletion of 5, 10, 13, or 26 amino acids from the C terminus of the β2 cytoplasmic domain resulted in loss of both constitutive adhesion and phorbol ester-stimulated function of αLβ2 expressed in COS cells, or B cells (Figure 2B, lines 2, 3, 4, 6). Surprisingly, with the deletion of more amino acids (39 total), constitutive adhesive function was restored in COS cells (line 8), although not in B lymphoblastoid cells.[110] Deletions of the β2 cytoplasmic domain also altered the ligand binding function of αMβ2 (Mac-1).[150] Again, shorter deletions had a greater adverse effect than did removal of the entire cytoplasmic tail (Figure 2, bottom, lines 5, 7, 9). Similarly, a point mutation in a β3 integrin (Ser[752] to Pro; position 31, Figure 1) completely eliminated function,[150] and yet removal of nearly all of the β3 cytoplasmic domain did not affect ligand binding by αIIbβ3.[111] Together these results possibly suggest that a positive regulatory region may reside closer to the C terminus, and that a negative regulatory region may be present closer to the membrane. Alternatively, shorter deletions (or disruptive point mutations) may have a disturbing effect on the secondary structure that is not seen with longer deletions.

The finding that β2 with 39 amino acids deleted was active in COS cells, but inactive in B lymphoid cells, suggests cell type-specific differences in integrin regulation, and emphasizes the difficulty in elucidating general rules for integrin regulation.

Deletion of N-terminal regions of the β2 cytoplasmic domain (Figure 2B, lines 10, 11) had minimal effect on adhesive function. However, deletion of five residues (ATTTV) near the C terminus resulted in a nearly complete loss of adhesion (line 12). Mutation of the three consecutive threonines (positions 30–32, Figure 1) also abolished LFA-1 adhesive function, thus confirming that this region is essential. Notably, if each threonine was mutated individually, adhesion was not affected. Within β2, Phe[766] (position 38, Figure 1) was also found to be important for LFA-1 adhesion to ICAM-1.[101] When changed to lysine or alanine, adhesion was eliminated. However, a conservative substitution of tyrosine permitted adhesion, suggesting that an aromatic amino acid is required to maintain structural integrity.

At least two of the three threonines in β2 (Figure 1, #30 to 32) are conserved as a threonine or serine in β1, β3, β5, β6, and β7, and the tyrosine at position #38, also essential for β2 adhesion function, is conserved as either a phenylalanine or tyrosine in β1, β3, β6, and β7. This conservation suggests that these amino acids also may be essential for the adhesive function of these other integrins. In this regard, a naturally occurring mutation of serine at position #31 in β3 resulted in a functionally inactive αIIbβ3 integrin.[150] Also, when the cytoplasmic domain of β1 was exchanged with that of β3,[34] or β5 (R. Pasqualini and M. Hemler, unpublished), adhesive function was not altered, thus indicating that some of these cytoplasmic domains are indeed interchangeable with respect to the regulation of adhesive functions. Notably, β5 lacks a phenylalanine or tyrosine at position 38 (Figure 1), but does contain a phenylalanine at position 39.

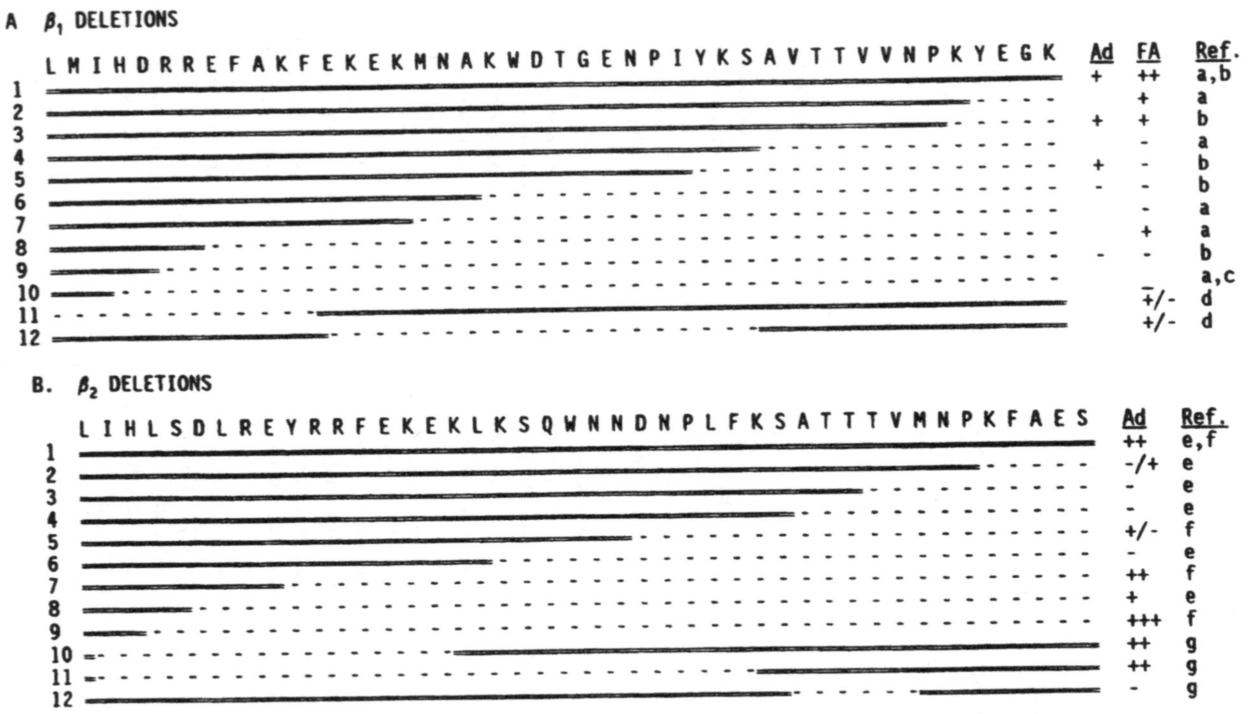

Figure 2 Schematic summary of results from integrin β subunit deletion experiments. For β₁ deletion experiments (part A), cell adhesion (abbreviated Ad) was analyzed in a few cases and localization to focal adhesions was analyzed (abbreviated FA). For β₂ deletion experiments (part B), either adhesion to ICAM-1 mediated by LFA-1,[110] or binding to iC3b mediated by Mac-1[128] was measured (Ad). The undeleted sequence is indicated by a double line, and deleted regions are indicated by a dashed line. References are as follows: a,[96] b,[55] c,[109] d,[99] e,[110] f,[128] g.[101]

B. OUTSIDE-IN SIGNALING: REGULATION OF FOCAL ADHESION FORMATION

1. Characteristics of Focal Adhesions

Focal adhesions are discrete subcellular regions where the plasma membrane is in tight association with the underlying substrate. These subcellular complexes, which are specialized for anchoring microfilament bundles known as stress fibers,[97] contain several cytoskeletal proteins, including α-actinin, vinculin, and talin. In a substrate-dependent manner, integrins also localize to focal adhesions, thus providing a transmembrane linkage between the extracellular matrix and the cytoskeleton.[56,57,60,97,151] Protein kinase C[152] and the majority of tyrosine-phosphorylated proteins[71,153] also appear to concentrate within focal adhesions, suggesting that these complexes are important sites of cellular signaling.

2. Coordinate Roles of α and β Cytoplasmic Domains

Several careful studies have focused attention on integrin cytoplasmic domains as critical for controlling localization into focal adhesions.[34,55,98,99,109,154,155] An emerging model is that the integrin β chain (β1 in particular), but not the α chain, contains sufficient information for integrins to form focal adhesions.[154,155] The α chain is hypothesized to act as a negative regulator, preventing nonspecific localization to focal adhesions containing other integrins. Presumably, ligand binding causes release of this negative regulation.[154] Interestingly, a form of β1, mutated so that it was unable to bind to a ligand, still was recruited into focal adhesions.[156] Thus, in this case, the α chain cytoplasmic domain (α5) did not prevent the integrin from joining with wild-type α5β1 focal adhesions.

3. Insights from β1 Mutations

Deletions of the C-terminal 39 to 40 residues from the β1 cytoplasmic domain eliminated focal adhesion formation, but retention of the R–R residues was sufficient to restore this activity (Figure 2A, lines 9 to 11). Consistent with the importance of this region, an internal deletion spanning the R–R sequence also caused loss of focal adhesions (line 11). Surprisingly, shorter deletions (13 to 28 residues) prevented integrin localization to focal adhesions (lines 4 to 7). An interpretation of these results is that partial sequences may disrupt the overall structure of the integrin.[98] The same interpretation potentially may apply to another internal deletion which mostly eliminated focal adhesions (line 12). The C-terminal 4 to 5 residues appeared not to be required for focal adhesion formation (Figure 2A, lines 1 to 3).

Different conclusions regarding specific β1 residues were obtained in another study involving point mutagenesis of essentially every residue in the β1 cytoplasmic domain.[99] Shown in Figure 1 are the ten most critical residues for focal adhesion localization (double-underlined), scattered throughout the cytoplasmic domain of β1. Notably these do not include the R–R residues (positions 2 and 3) implicated from the deletion studies, but they do include a tyrosine (at position #38) not implicated in the deletion studies. The difficulty in reconciling deletion and mutagenesis results suggests that there may be complex folding of the integrin β1 cytoplasmic domain critical for focal adhesion localization.

The simultaneous mutation of both threonines at positions 30 and 31 had no effect on focal adhesions, despite their importance for the adhesive function of β2 integrins (see discussion above). This provides further evidence that regulation of integrin localization to focal adhesions may be distinct from regulation of adhesion.

4. The Role of Other β Subunits

Comparison with other β subunits indicates that most of the ten critical residues implicated in the point mutagenesis study are highly conserved. Thus it is expected that β2, β3, β6, and β7 would show similar capabilities with respect to subcellular localization. In this regard, the β1 and β3 cytoplasmic domains were shown to be functionally interchangeable as far as their ability to localize to focal adhesions.[34] Notably, the β5 cytoplasmic domain does not contain three of the critical ten residues apparently required for focal adhesion localization (see positions 27, 34, 37). Consistent with this, αVβ3 but not αVβ5 was found in focal adhesions when cells were spread on vitronectin, although both integrins were equally able to adhere to vitronectin.[157] Similarly, the replacement of the β1 cytoplasmic domain with that of β5 correlated with a more diffuse distribution in cells spread on fibronectin (R. Pasqualini and M. Hemler, unpublished). Alternative splicing of both β1 and β3 results in removal of six out of the ten residues critical for focal adhesion formation.[20,39,40] Thus, it would be predicted that these alternatively spliced forms of β1 and β3 should not localize to focal adhesions. In agreement with this prediction, the β13'v alternatively spliced form (now called β1B) did not localize to focal adhesions.[44]

Integrins may not only localize to focal adhesions, but also to other types of subcellular structures such as podosomes[158,159] or point contacts.[80] At present, there is no information regarding specific roles for cytoplasmic domains for localization to these other structures.

C. OUTSIDE-IN SIGNALING: INTEGRIN-DEPENDENT ACTIVATION OF p125[FAK]

Integrin-mediated cell adhesion and spreading,[71,74–76] antibody-induced clustering of integrins,[72] and integrin-dependent platelet aggregation[73] all induce the tyrosine phosphorylation of a novel protein called focal adhesion kinase,[9,160,161] or p125[FAK], named for its property of localizing to focal adhesions. Not surprisingly, integrin β1 deletion mutants that prevented localization to focal adhesions also abolished anti-β1 antibody-induced activation of p125[FAK].[71] However, induction of p125[FAK] tyrosine phosphorylation was also lost for two β1 deletions (lines 2 and 8, Figure 2A) that retained the ability to localize to focal adhesions,[71] suggesting that activation of p125[FAK] may require more than just integrin localization to focal adhesions. Notably, Tyr[788] (position 25, Figure 1) was not required for β1-integrin dependent p125[FAK] activation.[71] Thus far, activation of p125[FAK] has been seen as a result of triggering through β1 and β3 integrins. It remains to be determined whether any of the other β subunits with highly conserved cytoplasmic domains also can participate in activation of p125[FAK].

In addition to p125[FAK], other proteins, such as paxillin in fibroblasts,[74] and various uncharacterized proteins in platelets,[162,163] in T cells,[77] in B cells,[78] and in a basophilic leukemia line,[79] undergo tyrosine phosphorylation in an integrin-dependent manner. However, at present there is no information available regarding possible specific roles for integrin cytoplasmic domains during the phosphorylation of these other proteins.

D. OUTSIDE-IN SIGNALING: CALCIUM AND pH CHANGES

Adhesion and spreading mediated by β1,[164] β2,[165–167] and β3[76,168] integrins can result in calcium mobilization, sometimes observed as oscillations.[76,166,167,169] Such calcium oscillations may be a prerequisite for subsequent phosphorylation of p125[FAK] and cell spreading,[76] or may be a result of cell spreading.[164] In another report, binding of osteopontin and related peptides to αVβ3 in osteoclasts caused a prompt reduction in cytosolic calcium by a calmodulin-dependent mechanism.[170] In a recent study, it was noted that triggering through the αVβ3, but not the α2β1, integrin led to a rise in internal calcium in endothelial cells. Also, cell migration on vitronectin, but not on collagen, was dependent on extracellular calcium.[171] Both of these observations point to an interesting difference between specific integrins with respect to calcium signaling, thus implying that the respective cytoplasmic domains may play distinct roles.

Integrin-dependent cell spreading or clustering of integrins by antibodies leads to an elevation of intracellular pH, due to activation of the Na–H antiporter.[172-174] However, since all integrins tested could induce elevation of intracellular pH, there is not yet any evidence for specialized roles. In fact, integrins ($\alpha2\beta1$ and $\alpha V\beta3$) that showed differences with respect to calcium signaling were both able to induce a similar elevation in intracellular pH.[171]

E. INTEGRIN-DEPENDENT CELL MIGRATION

Several different integrins have been shown to play a central role during both random cell migration and haptotactic migration *in vitro*, and during cell migration through or across tissues *in vivo*.[147,171,175-188] The dynamic process of cell migration likely involves simultaneous outside-in and inside-out signaling through integrins. In one scenario, initial adhesion may trigger a rise in calcium,[171] which then could lead to activation of the calcium-dependent phosphatase calcineurin, which is required to feed back and cause de-adhesion of the integrin to allow cell migration.[147] By this model, the occurrence of calcium transients would correspond to cycles of adhesion and de-adhesion. Notably, endothelial cells utilized $\alpha2\beta1$ and $\alpha V\beta3$ to migrate on collagen and vitronectin, respectively, but only the latter integrin required the presence of extracellular calcium.[171] Consistent with this, neutrophils showed calcineurin-dependent migration on vitronectin but not fibronectin.[147] These results suggest that different models for migration may apply to different integrins, and strongly suggest that cytoplasmic domains may play important roles in coordinating the distinct intracellular signaling events associated with different integrins.

Because integrin cytoplasmic domains play a fundamental role in subcellular localization (see above), and because integrins tend to be localized in a more diffuse pattern during cell migration,[184] it would then be expected that integrin cytoplasmic domains that do not localize to focal adhesion sites would perhaps favor cell migration. Indeed, compared to the $\beta1$ cytoplasmic domain, the $\beta5$ cytoplasmic domain (in a $\beta1/5$ chimeric molecule) has been found to favor migration of CHO cells on fibronectin. At the same time, the $\beta1$ cytoplasmic domain, but not that of $\beta5$, was found in focal adhesions (R. Pasqualini and M. Hemler, unpublished). However, a comparison of $\alpha V\beta3$ and $\alpha V\beta5$ in carcinoma cells found that the former was better able to localize to focal adhesions and also was better able to promote cell migration.[157,186] Obviously, the specific roles of the respective cytoplasmic domains and the relationships between migration and subcellular localization need to be further clarified.

The cytoplasmic domains of integrin α chains can play a key role during cell migration. Mutant forms of the integrin $\alpha2\beta1$ (VLA-2) were prepared in which the cytoplasmic domain of $\alpha2$ was replaced with that of $\alpha4$ or $\alpha5$. When tested in random cell migration assays, the $\alpha4$ cytoplasmic domain supported migration to a markedly greater extent than did the other cytoplasmic domains, whereas cell adhesion was not altered.[188] Although it is not clear whether the $\alpha4$ cytoplasmic domain may exert a positive effect in support of cell migration, or whether the $\alpha2$ and $\alpha5$ cytoplasmic domains are suppressing migration, the α chains obviously do play an important and specific role.

We hypothesize that although multiple receptors might recognize the same ligand, the different cytoplasmic domains are likely to associate with distinct sets of cytoskeletal or cytoplasmic proteins, thus allowing a single ligand to stimulate a variety of post-ligand binding signals involved in cell migration and other events.

F. MATRIX REMODELING—COLLAGEN GEL CONTRACTION

During wound healing and embryogenesis, cells can exert a physical force on the surrounding tissue structure. A good *in vitro* model for this process is the contraction of a three-dimensional collagen lattice by cells distributed within.[189-191] Collagen gel contraction is now known to be integrin dependent,[192] with α2β1 (VLA-2) playing a particularly important role.[188,193,194] A study using chimeric forms of α2 showed that the cytoplasmic domain of the α2 or α5 subunits supported collagen gel contraction, whereas that from α4 did not.[188] This result was exactly the opposite from results obtained in cell migration studies, and confirms that α chain cytoplasmic domains can have distinct postligand-binding functions. The strength of cell adhesion is much greater during collagen gel contraction than cell migration,[190,191] and the organization of the actin cytoskeleton is totally different in the two processes.[189,191] Thus, we hypothesize that different integrin cytoplasmic domains are required to provide optimal interaction with the relevant cytoskeletal-signaling components involved.

G. OTHER OUTSIDE-IN INTEGRIN-MEDIATED SIGNALING EVENTS
1. Cell Proliferation
A role for integrins in cell proliferation is well established based on many studies showing costimulatory[195–200] or direct stimulatory[201–203] effects of various integrin ligands or anti-integrin antibodies. The mechanism by which this occurs is not clear, although integrin-mediated alterations in calcium and intracellular pH are likely to be relevant, since these pathways contribute to the regulation of cell proliferation.[204,205]

Thus far, there is little published data indicating specific roles for integrin cytoplasmic domains during cell proliferation. However, in our comparisons of the β1 subunit with a β1/5 chimera in CHO cells, we found that the latter lost the ability to transmit a proliferative signal when exposed to fibronectin plus an anti-β1 antibody (R. Pasqualini and M. Hemler, unpublished). Further studies will be required to determine the critical pathways selectively supported by the β1 cytoplasmic domain and to relate this difference to differences in the structures of these domains.

2. Gene Induction
The interaction of various integrins with either ligands or specific antibodies has been shown to cause the induction of genes for metalloproteinases[206] and cytokines,[207] and, in another study, fibronectin acting through α5β1 contributed to the production of the AP-1 transcription factor.[208] It is likely that specific gene induction is somewhat downstream of the initial adhesion and signaling events mediated by integrins; so, it remains a challenging goal to link specific gene induction with the unique cytoplasmic domain of a particular integrin. The finding that different ECM proteins have distinct effects on β-casein production by mammary epithelial cells[209] supports the concept that variations in integrin signaling could contribute to specific gene induction.

3. Integrin-Dependent Phagocytosis, Respiratory Burst, and cAMP Changes
It has been well established that ECM proteins, acting through integrins, can upregulate the phagocytic functions of neutrophils and macrophages.[63,210–212] Interestingly, it has been suggested that αMβ2 (Mac-1) may have a signaling role during phagocytosis that is distinct from its involvement in adhesion phenomena.[213,214]

Extracellular matrix proteins also potentiate the respiratory burst response of neutrophils to cytokines such as TNF.[215] Members of the β2 integrin family are required for this stimulation of the respiratory burst even though some stimulatory ECM proteins (such as thrombospondin and laminin) are not direct ligands for β2 integrins.[215] Independent of their adhesive functions, β2 integrins appear to stimulate the respiratory burst by transmitting a signal leading to a drop in neutrophil cAMP levels.[216] Importantly, αLβ2 and αXβ2, but not αMβ2, could deliver the signals needed to stimulate respiratory burst, suggesting that functional differences may reside within the α chain cytoplasmic domains.[217] As shown in Table 1, the cytoplasmic domains of αX and αL are both longer than that of αM, but there is no other obvious similarity specific for αX and αL.

Whereas stimulation of β2 integrins leads to a drop in cAMP in neutrophils,[216,217] triggering of β1 caused cAMP to increase in T lymphocytes.[218] Conversely, elevation of cAMP caused a decrease in αLβ2-mediated adhesiveness in T lymphocytes,[102] but did not diminish the adhesive function of β1 and β3 integrins in endothelial cells, although in the latter case, focal contact formation was severely impaired and cell motility was inhibited.[219] These results emphasize that cAMP may be a critical mediator of integrin signaling. However, it remains to be determined whether the variety of cAMP effects are due to different cell types and assay systems or are due to specific differences in integrin cytoplasmic domains.

V. CONCLUSIONS AND FUTURE DIRECTIONS

The recent discoveries of the role of integrins in a wide variety of signaling events have provided a range of different systems in which to investigate the functional variability among integrin cytoplasmic domains. Accordingly, evidence for functional diversity of integrin cytoplasmic domains has begun to emerge, consistent with their structural diversity. Initial work carried out in this regard suggests that this will continue to be a fruitful endeavor. An additional challenge now will be to sort out which signaling events directly intersect with integrin functions, and to understand the precise biochemical mechanisms for the involvement of specific cytoplasmic domains.

ACKNOWLEDGMENTS

We thank Dr. Qian Yuan for his critical reading of this manuscript. Also, we acknowledge grant support (to M.E.H.) from the National Institutes of Health (CA42368, GM46526), support from the DFCI/Sandoz Drug Discovery Program, and fellowship support from the Lady Tata Memorial Trust (to J.B.W.) and from the Conselho Nacional de Desenvolvimento Cientifico e Technológico, Brazil (to R.P.)

REFERENCES

1. Hynes, R. O., Integrins: a family of cell surface receptors, *Cell*, 48, 549, 1987.
2. Hemler, M. E., Adhesive protein receptors on hematopoietic cells, *Immunol. Today*, 41, 109, 1988.
3. Ginsberg, M. H., Loftus, J. C., and Plow, E. F., Cytoadhesins, integrins and platelets, *Thromb. Hemostas.*, 59, 1, 1988.
4. Anderson, D. C. and Springer, T. A., Leukocyte adhesion deficiency: an inherited defect in the Mac-1, LFA-1, and p150,95 glycoproteins, *Annu. Rev. Med.*, 38, 175, 1987.
5 Hynes, R. O., Integrins: versatility, modulation and signaling in cell adhesion, *Cell*, 69, 11, 1992.
6. Schwartz, M. A., Transmembrane signaling by integrins, *Trends Cell Biol.*, 2, 304, 1992.
7. Damsky, C. H. and Werb, Z., Signal transduction by integrin receptors for extracellular matrix: cooperative processing of extracellular information, *Curr. Op. Cell Biol.*, 4, 772, 1992.
8. Pardi, R., Inverardi, L., and Bender, J. R., Regulatory mechanisms in lymphocyte adhesion: flexible receptors for sophisticated travelers, *Immunol. Today*, 13, 224, 1992.
9. Zachary, I. and Rozengurt, E., Focal adhesion kinase (p125^FAK): a point of convergence in the action of neuropeptides, integrins, and oncogenes, *Cell*, 71, 891, 1993.

10. Hemler, M. E., Structures and functions of VLA proteins and related integrins, in *Receptors for Extracellular Matrix Proteins (Series: Biology of Extracellular Matrix)*, Mecham, R. P. and McDonald, J. A., Eds., Academic Press, San Diego, CA, 1991, 255.

11. Hynes, R. O., Marcantonio, E. E., Stepp, M. A., Urry, L. A., and Yee, G. H., Integrin heterodimer and receptor complexity in avian and mammalian cells, *J. Cell Biol.*, 109, 409, 1989.

12. Tamura, R. N., Cooper, H. M., Collo, G., and Quaranta, V., Cell type-specific integrin variants with alternative α chain cytoplasmic domains, *Proc. Natl. Acad. Sci. U.S.A.*, 88, 10183, 1991.

13. Cooper, H. M., Tamura, R. N., and Quaranta, V., The major laminin receptor of mouse embryonic stem cells is a novel isoform of the α6β1 integrin, *J. Cell Biol.*, 115, 843, 1991.

14. Hogervorst, F., Kuikman, I., Van Kessel, A. G., and Sonnenberg, A., Molecular cloning of the human α6 integrin subunit. Alternative splicing of α6 mRNA and chromosomal localization of the α6 and β4 genes, *Eur. J. Biochem.*, 199, 425, 1991.

15. Hogervorst, F., Admiraal, L. G., Niessen, C., Kuikman, I., Janssen, H., Daams, H., and Sonnenberg, A., Biochemical characterization and tissue distribution of the A and B variants of the integrin α6 subunit, *J. Cell Biol.*, 121, 179, 1993.

16. Corbi, A. L., Garcia-Aguilar, J., and Springer, T. A., Genomic structure of an integrin alpha subunit, the leukocyte p150,95 molecule, *J. Biol. Chem.*, 265, 2782, 1990.

17. Heidenreich, R., Eisman, R., Surrey, S., Delgrosso, K., Bennett, J. S., Schwartz, E., and Poncz, M., Organization of the gene for platelet glycoprotein IIb, *Biochemistry*, 29, 1232, 1990.

18. Brown, N. H., King, D. L., Wilcox, M., and Kafatos, F. C., Developmentally regulated alternative splicing of drosophila integrin PS2 alpha transcripts, *Cell*, 59, 185, 1989.

19. Fleming, J. C., Pahl, H. L., Gonzalez, D. A., Smith, T. F., and Tenen, D. G., Structural analysis of the CD11b gene and phylogenetic analysis of the α integrin gene family demonstrate remarkable conservation of genomic organization and suggest early diversification during evolution, *J. Immunol.*, 150, 480, 1993.

20. Altruda, F., Cervella, P., Tarone, G., Botta, C., Balzac, F., Stefanuto, G., and Silengo, L., A human integrin β1 subunit with a unique cytoplasmic domain generated by alternative mRNA processing, *Gene*, 95, 261, 1990.

21. Weitzman, J. B., Wells, C. E., Wright, A. H., Clark, P. A., and Law, S. K. A., The gene organization of the human β2 integrin subunit (CD18), *FEBS Lett.*, 294, 97, 1991.

22. Lanza, F., Kieffer, N., Phillips, D. R., and Fitzgerald, L. A., Characterization of the human platelet glycoprotein IIIa gene, *J. Biol. Chem.*, 265, 18098, 1990.

23. Zimrin, A. B., Gidwitz, S., Lord, S., Schwartz, E., Bennett, J. S., White, G. C., and Poncz, M., The genomic organization of platelet glycoprotein IIIa, *J. Biol. Chem.*, 265, 8590, 1990.

24. Krissansen, G. W., Yuan, Q., Jenkins, D., Jiang, W.-M., Rooke, L., Spurr, N. K., Eccles, M., Leung, E., and Watson, J. D., Chromosomal locations of the genes coding for the integrin β6 and β7 subunits, *Immunogenetics*, 35, 58, 1992.

25. Jiang, W.-M., Jenkins, D., Yuan, Q., Leung, E., Choo, A. K. H., Watson, J. D., and Krissansen, G. W., The gene organization of the human β7 subunit, the common β subunit of the leukocyte integrins HML-1 and LPAM-1, *Int. Immunol.*, 4, 1031, 1992.

26. Hughes, A. L., Coevolution of the vertebrate integrin alpha- and beta-chain genes. *Mol. Biol. Evol.*, 9, 216, 1992.

27. Holers, V. M., Ruff, T. G., Parks, D. L., McDonald, J. A., Ballard, L. L., and Brown, E. J., Molecular cloning of a murine fibronectin receptor and its expression during inflammation, *J. Exp. Med.*, 169, 1589, 1989.

28. Tominaga, S., Murine mRNA for the β-subunit of integrin is increased in BALB/c-3T3 cells entering the G1 phase from the Go state, *FEBS Lett.*, 238, 315, 1988.

29. Tamkun, J. W., DeSimone, D. W., Fonda, D., Patel, R. S., Buck, C., Horwitz, A. F., and Hynes, R. O., Structure of integrin, a glycoprotein involved in the transmembrane linkage between fibronectin and actin, *Cell*, 46, 271, 1986.

30. DeSimone, D. W. and Hynes, R. O., Xenopus laevis integrins: structural conservation and evolutionary divergence of integrin beta subunits, *J. Biol. Chem.*, 263, 5333, 1988.

31. Wilson, R., O, Brien, W., and Beaudet, A., Nucleotide sequence of the cDNA from the mouse leukocyte adhesion protein CD18, *Nucl. Acids Res.*, 17, 5397, 1992.

32. Yuan, Q., Jiang, W.-M., Leung, E., Hollander, D., Watson, J. D., and Krissansen, G. W., Molecular cloning of the mouse integrin β7 subunit, *J. Biol. Chem.*, 267, 7352, 1992.

33. Moyle, M., Napier, M. A., and McLean, J. W., Cloning and expression of a divergent integrin subunit β8, *J. Biol. Chem.*, 266, 19650, 1991.

34. Solowska, J., Edelman, J. M., Albelda, S. M., and Buck, C. A., Cytoplasmic and transmembrane domains of integrin β_1 and β_3 subunits are functionally interchangeable, *J. Cell Biol.*, 114, 1079, 1991.

35. Burridge, K., Fath, K., Kelly, T., Nuckolls, G., and Turner, C., Focal adhesions: transmembrane junctions between the extracellular matrix and the cytoskeleton, *Annu. Rev. Cell Biol.*, 4, 487, 1988.

36. Stepp, M. A., Spurr-Michaud, S., Tisdale, A., Elwell, J., and Gipson, I. K., $\alpha_6\beta_4$ integrin heterodimer is a component of hemidesmosomes, *Proc. Natl. Acad. Sci. U.S.A.*, 87, 8970, 1990.

37. Sonnenberg, A., Calafat, J., Janssen, H., Daams, H., van der Raaij-Helmer, L. M. H., Falcioni, R., Kennel, S. J., Aplin, J. D., Baker, J., Loizidou, M., and Garrod, D., Integrin α6β4 complex is located in hemidesmosomes, suggesting a major role in epidermal cell-basement membrane adhesion, *J. Cell Biol.*, 113, 907, 1991.

38. Jones, J. C. R., Kurpakus, M. A., Cooper, H. M., and Quaranta, V., A function for the integrin alpha6beta4 in the hemidesmosome, *Cell Regul.*, 2, 427, 1991.

39. Languino, L. R. and Ruoslahti, E., An alternative form of the integrin β_1 subunit with a variant cytoplasmic domain, *J. Biol. Chem.*, 267, 7116, 1992.

40. Van Kuppevelt, T. H., Languino, L. R., Gailit, J. O., Suzuki, S., and Ruoslahti, E., An alternative cytoplasmic domain of the integrin β3 subunit, *Proc. Natl. Acad. Sci. U.S.A.*, 86, 5415, 1989.

41. Hogervorst, F., Kuikman, I., von dem Borne, A. E. G. Kr., and Sonnenberg, A., Cloning and sequence analysis of beta-4 cDNA: an integrin subunit that contains a unique 118 kd cytoplasmic domain, *EMBO J.*, 9, 765, 1990.

42. Suzuki, S. and Naitoh, Y., Amino acid sequence of a novel integrin β4 subunit and primary expression of the mRNA in epithelial cells, *EMBO J.*, 9, 757, 1990.

43. Tamura, R. N., Rozzo, C., Starr, L., Chambers, J., Reichardt, L. F., Cooper, H. M., and Quaranta, V., Epiththelial integrin α6β4: complete primary structure of α6 and variant forms of β4, *J. Cell Biol.*, 111, 1593, 1990.

44. Balzac, F., Belkin, A. M., Koteliansky, V. E., Balabanov, Y. V., and Altruda, F., Expression and functional analysis of a cytoplasmic domain variant of the β1 integrin subunit, *J. Cell Biol.*, 121, 171, 1993.

45. Ramaswamy, H. and Hemler, M. E., Cloning, primary structure, and properties of a novel human integrin β subunit, *EMBO J.*, 9, 1561, 1990.

46. Suzuki, S., Huang, Z.-S. and Tanihara, H., Cloning of an integrin β subunit exhibiting high homology with integrin β3 subunit, *Proc. Natl. Acad. Sci. U.S.A.*, 87, 5354, 1990.

47. McLean, J. W., Vestal, D. J., Cheresh, D. A., and Bodary, S. C., cDNA sequence of the human integrin β5 subunit, *J. Biol. Chem.*, 265, 17126, 1990.

48. Takeichi, M., Cadherins: a molecular family important in selective cell-cell adhesion, *Annu. Rev. Biochem.*, 59, 237, 1990.

49. Geiger, B. and Ayalon, O., Cadherins, *Annu. Rev. Cell Biol.*, 8, 307, 1992.

50. Argraves, W. S., Dickerson, K., Burgess, W. H., and Ruoslahti, E., Fibulin, a novel protein that interacts with the fibronectin receptor β subunit cytoplasmic domain, *Cell*, 58, 623, 1989.

51. Argraves, W. S., Tran, H., Burgess, W. H., and Dickerson, K., Fibulin is an extracellular matrix and plasma glycoprotein with repeated domain structure, *J. Cell Biol.*, 111, 3155, 1990.

52. Horwitz, A., Duggan, K., Buck, C., Beckerle, M. C., and Burridge, K., Interaction of plasma membrane fibronectin receptor with talin-a transmembrane linkage, *Nature*, 320, 531, 1986.

53. Miller, L. J., Wiebe, M., and Springer, T. A., Purification and α subunit N-terminal sequences of human Mac-1 and p150,95 leukocyte adhesion proteins, *J. Immunol.*, 138, 2381, 1987.

54. Tapley, P., Horwitz, A., Buck, C., Duggan, K., and Rohrschneider, L., Integrins isolated from Rous sarcoma virus-transformed chicken embryo fibroblasts, *Oncogene*, 4, 325, 1989.

55. Hayashi, Y., Haimovich, B., Reszka, A., Boettiger, D., and Horwitz, A., Expression and function of chicken integrin beta-1 subunit and its cytoplasmic domain mutants in mouse NIH 3T3 cells, *J. Cell Biol.*, 110, 175, 1990.

56. Chen, W.-T., Hasegawa, E., Hasegawa, T., Weinstock, C., and Yamada, K. M., Development of cell surface linkage complexes in cultured fibroblasts, *J. Cell Biol.*, 100, 1103, 1985.

57. Damsky, C. H., Knudsen, K. A., Bradley, D., Buck, C. A., and Horwitz, A. F., Distribution of the CSAT cell-matrix antigen on myogenic and fibroblastic cells in culture, *J. Cell Biol.*, 100, 1528, 1985.

58. Burn, P., Kupfer, A., and Singer, S. J., Dynamic membrane-cytoskeletal interactions: specific association of integrin and talin arises in vivo after phorbol ester treatment of peripheral blood lymphocytes, *Proc. Natl. Acad. Sci. U.S.A.*, 85, 497, 1988.

59. Kupfer, A. and Singer, S. J., Cell biology of cytotoxic and helper T cell functions: immunofluorescence microscopic studies of single cells and cell couples, *Annu. Rev. Immunol.*, 7, 309, 1989.

60. Singer, I. I., Scott, S., Kawka, D. W., Kazazis, D. M., Gailit, J. and Ruoslahti, E., Cell surface distribution of fibronectin and vitronectin receptors depends on substrate composition and extracellular matrix accumulation, *J. Cell Biol.*, 106, 2171, 1988.

61. Otey, C. A., Pavalko, F. M., and Burridge, K., An interaction between α-actinin and the β1 integrin subunit in vitro, *J. Cell Biol.*, 111, 721, 1990.

62. Pardi, R., Inverardi, L., Rugarli, C., and Bender, J. R., Antigen-receptor complex stimulation triggers protein kinase C-dependent CD11a/CD18-cytoskeleton association in T lymphocytes, *J. Cell Biol.*, 116, 1211, 1992.

63. Gresham, H. D., Goodwin, J. L., Allen, P. M., Anderson, D. C., and Brown, E. J., A novel member of the integrin receptor family mediates Arg-Gly-Asp-stimulated neutrophil phagocytosis, *J. Cell Biol.*, 108, 1935, 1989.

64. Brown, E., Hooper, L., Ho, T., and Gresham, H., Integrin-associated protein: a 50 kD plasma membrane antigen physically and functionally associated with integrins, *J. Cell Biol.*, 111, 2785, 1990.

65. Brown, E. J., Signal transduction through leukocyte integrins, in *Cell Adhesion Molecules*, Mihich, E. and Hemler, M. E., Eds., Plenum, New York, 1994.

66. Rosales, C., Gresham, H. D., and Brown, E. J., Expression of the 50-kDa integrin-associated protein on myeloid cells and erythrocytes, *J. Immunol.*, 149, 2759, 1992.

67. Cierniewski, C. S., Krzeslowska, J., Pawlowska, Z., Witas, H., and Meyer, M., Palmitoylation of the glycoprotein IIb-IIIa complex in human blood platelets, *J. Biol. Chem.*, 264, 12158, 1989.

68. Conforti, G., Zanetti, A., Pasquali-Ronchetti, I., Quaglino, D., Jr., Neyroz, P., and

Dejana, E., Modulation of vitronectin receptor binding by membrane lipid composition, *J. Biol. Chem.*, 265, 4011, 1990.

69. Smyth, S. S., Hillery, C. A., and Parise, L. V., Fibrinogen binding to purified platelet glycoprotein IIb-IIIa (Integrin $\alpha_{IIb}\beta_3$) is modulated by lipids, *J. Biol. Chem.*, 267, 15568, 1992.

70. Hermanowski-Vosatka, A., Van Strijp, J. A. G., Swiggard, W. J., and Wright, S. D., Integrin modulating factor-1: a lipid that alters the function of leukocyte integrins, *Cell*, 68, 341, 1992.

71. Guan, J.-L., Trevithick, J. E., and Hynes, R. O., Fibronectin/integrin interaction induces tyrosine phosphorylation of a 120 kDa protein, *Cell Regul.*, 2, 951, 1991.

72. Kornberg, L. J., Earp, H. S., Turner, C. E., Prockop, C., and Juliano, R. L., Signal transduction by integrins: increased protein tyrosine phosphorylation caused by clustering of $\beta 1$ integrins, *Proc. Natl. Acad. Sci. U.S.A.*, 88, 8392, 1991.

73. Lipfert, L., Haimovich, B., Schaller, M. D., Cobb, B. S., Parsons, J. T., and Brugge, J. S., Integrin-dependent phosphorylation and activation of the protein tyrosine kinase pp125^{FAK} in platelets, *J. Cell Biol.*, 119, 905, 1993.

74. Burridge, K., Turner, C. E., and Romer, L. H., Tyrosine phosphorylation of paxillin and pp125^{FAK} accompanies cell adhesion to extracellular matrix: a role in cytoskeletal assembly, *J. Cell Biol.*, 119, 893, 1992.

75. Kornberg, L., Earp, H. S., Parsons, J. T., Schaller, M., and Juliano, R. L., Cell adhesion or integrin clustering increases phorphorylation of a focal adhesion-associated tyrosine kinase, *J. Biol. Chem.*, 267, 23439, 1992.

76. Pelletier, A. J., Bodary, S. C., and Levinson, A. D., Signal transduction by the platelet integrin $\alpha_{IIb}\beta 3$: induction of calcium oscillations required for protein-tyrosine phosphorylation and ligand-induced spreading of stably transfected cells, *Mol. Biol. Cell*, 3, 989, 1992.

77. Nojima, Y., Rothstein, D. M., Sugita, K., Schlossman, S. F., and Morimoto, C., Ligation of VLA-4 on T cells stimulates tyrosine phosphorylation of a 105-kD protein, *J. Exp. Med.*, 175, 1045, 1992.

78. Freedman, A. S., Rhynhart, K., Nojima, Y., Svahn, J., Eliseo, L., Benjamin, C. D., Morimoto, C., and Vivier, E., Stimulation of protein tyrosine phosphorylation in human B cells after ligation of the $\beta 1$ integrin VLA-4, *J. Immunol.*, 150, 1645, 1993.

79. Hamawy, M. M., Mergenhagen, S. E., and Siraganian, R. P., Cell adherence to fibronectin and the aggregation of the high affinity immunoblobulin E receptor synergistically regulate tyrosine phosphorylation of 105-115-kDa proteins, *J. Biol. Chem.*, 268, 5227, 1993.

80. Tawil, N., Wilson, P., and Carbonetto, S., Integrins in point contacts mediate cell spreading: factors that regulate integrin accumulation in point contacts vs. focal contacts, *J. Cell Biol.*, 120, 261, 1993.

81. Rojiani, M. V., Finlay, B. B., Gray, V., and Dedhar, S., In vitro interaction of a polypeptide homologous to human Ro/SS-A antigen (calreticulin) with a highly conserved amino acid sequence in the cytoplasmic domain of integrin α subunits, *Biochemistry*, 30, 9859, 1991.

82. Kurpakus, M. A., Quaranta, V., and Jones, J. C. R., Surface relocation of alpha6beta4 integrins and assembly of hemidesmosomes in an in vitro model of wound healing, *J. Cell Biol.*, 115, 1737, 1991.

83. Legan, P. K., Collins, J. E., and Garrod, D. R., The molecular biology of desmosomes and hemidesmosomes: "What's in a name?", *BioEssay*, 14, 385, 1992.

84. Shaw, L. M., Messier, J. M., and Mercurio, A. M., The activation dependent adhesion of macrophages to laminin involves cytoskeleton anchoring and phosphorylation of the alpha-6 beta-1 integrin, *J. Cell Biol.*, 110, 2167, 1990.

85. Gismondi, A., Mainiero, F., Morrone, S., Palmieri, G., Piccoli, M., Frati, L., and

Santoni, A., Triggering through CD16 or phorbol esters enhances adhesion of NK cells to laminin via very late antigen 6, *J. Exp. Med.*, 176, 1251, 1992.

86. Kantor, R. R. S., Mattes, M. J., Lloyd, K. O., Old, L. J., and Albino, A. P., Biochemical analysis of two cell surface glycoprotein complexes: VCA-1 and VCA-2. Relationship to VLA T cell antigens, *J. Biol. Chem.*, 262, 15158, 1987.

87. Hemler, M. E., Elices, M. J., Parker, C., and Takada, Y., Structure of the integrin VLA-4 and its cell-cell and cell-matrix adhesion functions, *Immunol. Rev.*, 114, 45, 1990.

88. Aneskievich, B. J., Haimovich, B., and Boettiger, D., Phosphorylation of integrin in differentiating ts-Rous sarcoma virus-infected myogenic cells, *Oncogene*, 6, 1381, 1991.

89. Savill, J., Dransfield, I., Hogg, N., and Haslett, C., Vitronectin receptor-mediated phagocytosis of cells undergoing apoptosis, *Nature*, 343, 170, 1990.

90. Buyon, J. P., Slade, S., Reibman, J., Abramson, S. B., Philips, M. R., Weissmann, G., and Winchester, R., Constitutive and induced phosphorylation of the α- and β-chains of the CD11/CD18 leukocyte integrin family, *J. Immunol.*, 144, 191, 1990.

91. Chatila, T. A., Geha, R. S., and Arnaout, M. A., Constitutive and stimulus-induced phosphorylation of CD11/CD18 leukocyte adhesion molecules, *J. Cell Biol.*, 109, 3435, 1989.

92. Valmu, L., Autero, M., Siljander, P., Patarroyo, M., and Gahmberg, C. G., Phosphorylation of the β-subunit of CD11/CD18 integrins by protein kinase C correlates with leukocyte adhesion, *Eur. J. Immunol.*, 21, 2857, 1991.

93. Haimovich, B., Aneskievich, B. J., and Boettiger, D., Cellular partitioning of β1 integrins and their phosphorylated forms is altered after transformation by Rous sarcoma virus or treatment with cytochalasin D, *Cell Regul.*, 2, 271, 1991.

94. Duband, J.-L., Dufour, S., Yamada, K. M., and Thiery, J. P., The migratory behavior of avian embryonic cells does not require phosphorylation of the fibronectin-receptor complex, *FEBS Lett.*, 230, 181, 1988.

95. Horvath, A. R., Elmore, A. R., and Kellie, S., Differential tyrosine-specific phosphorylation of integrin in Rous sarcoma virus transformed cells with differing transformed phenotypes, *Oncogene*, 5, 1349, 1990.

96. Burridge, K., Substrate adhesions in normal and transformed fibroblasts: organization and regulation of cytskeletal, membrane and extracellular matrix components at focal contacts, *Cancer Rev.*, 4, 18, 1986.

97. Kramer, R. H., McDonald, K. A., Crowley, E., Ramos, D. M., and Damsky, C. H., Melanoma cell adhesion to basement membrane mediated by integrin-related complexes, *Cancer Res.*, 49, 393, 1989.

98. Marcantonio, E. E., Guan, J., Trevithick, J. E., and Hynes, R. O., Mapping of the functional determinants of the integrin beta-1 cytoplasmic domain by site-directed mutagenesis, *Cell Regul.*, 1, 597, 1990.

99. Reszka, A. A., Hayashi, Y., and Horwitz, A. F., Identification of amino acid sequences in the integrin β₁ cytoplasmic domain implicated in cytoskeletal association, *J. Cell Biol.*, 117, 1321, 1992.

100. Merrill, J. T., Slade, S. G., Weissmann, G., Winchester, R., and Buyon, J. P., Two pathways of CD11b/CD18-mediated neutrophil aggregation with different involvement of protein kinase C-dependent phosphorylation, *J. Immunol.*, 145, 2608, 1990.

101. Hibbs, M. L., Jakes, S., Stacker, S. A., Wallace, R. W., and Springer, T. A., The cytoplasmic domain of the integrin lymphocyte function-associated antigen 1 β subunit: sites required for binding to intercellular adhesion molecule 1 and the phorbol ester-stimulated phosphorylation site, *J. Exp. Med.*, 174, 1227, 1991.

102. Dustin, M. L. and Springer, T. A., T-cell receptor cross-linking transiently stimulates adhesiveness through LFA-1, *Nature*, 341, 619, 1989.

103. Van Kooyk, Y., Van DeWiel-Van Kemenade, P., Weder, P., Kuijpers, T. W., and Figdor, C. G., Enhancement of LFA-1-mediated cell adhesion by triggering through CD2 or CD3 on T lymphocytes, *Nature*, 342, 811, 1989.

104. Parise, L. V., Criss, A. B., Nannizzi, L., and Wardell, M. R., Glycoprotein IIIa is phosphorylated in intact human platelets, *Blood*, 75, 2363, 1990.

105. Hillery, C. A., Smyth, S. S., and Parise, L. V., Phosphorylation of human platelet glycoprotein IIIa (GPIIIa): dissociation from fibrinogen receptor activation and phosphorylation of GPIIIa *in vitro*, *J. Biol. Chem.*, 266, 14663, 1991.

106. Freed, E., Gailit, J., van der Geer, P., Ruoslahti, E., and Hunter, T., A novel integrin β subunit is associated with the vitronectin receptor α subunit (αv) in a human osteosarcoma cell line and is a substrate for protein kinase C, *EMBO J.*, 8, 2955, 1989.

107. Sacchi, A., Falcioni, R., Piaggio, G., Gianfelice, M. A., Perrotti, N., and Kennel, S. J., Ligand-induced phosphorylation of a murine tumor surface protein (TSP-180) associated with metastatic phenotype, *Cancer Res.*, 49, 2615, 1989.

108. Falcioni, R., Perrotti, N., Piaggio, G., Kennel, S. J., and Sacchi, A., Insulin-induced phosphorylation of the beta-4 integrin subunit expressed on murine metastatic carcinoma cells, *Mol. Carcin.*, 2, 361, 1989.

109. Solowska, J., Guan, J-L., Marcantonio, E. E., Trevithick, J. E., Buck, C. A., and Hynes, R. O., Expression of normal and mutant avian integrin subunits in rodent cells, *J. Cell Biol.*, 109, 853, 1989.

110. Hibbs, M. L., Xu, H., Stacker, S. A., and Springer, T. A., Regulation of adhesion to ICAM-1 by the cytoplasmic domain of LFA-1 integrin beta subunit, *Science*, 251, 1611, 1991.

111. O'Toole, T. E., Mandelman, D., Forsyth, J., Shattil, S. J., Plow, E. F., and Ginsberg, M. H., Modulation of the affinity of integrin αIIbβ3 (GPIIb-IIIa) by the cytoplasmic domain of αIIb, *Science*, 254, 845, 1991.

112. Giancotti, F. G., Stepp, M. A., Suzuki, S., Engvall, E., and Ruoslahti, E., Proteolytic processing of endogenous and recombinant β₄ integrin subunit, *J. Cell Biol.*, 118, 951, 1992.

113. Frachet, P., Duperray, A., Delachanal, E., and Marguerie, G., Role of the transmembrane and cytoplasmic domains in the assembly and surface exposure of the platelet integrin GPIIb/IIIa, *Biochemistry*, 31, 2408, 1992.

114. Bennett, J. S., Kolodziej, M. A., Vilaire, G., and Poncz, M., Determinants of the intracellular fate of truncated forms of the platelet glycoproteins IIb and IIIa, *J. Biol. Chem.*, 268, 3580, 1993.

115. Dana, N., Fathallah, D. M., and Arnaout, M. A., Expression of a soluble and functional form of the human β2 integrin CD11b/CD18, *Proc. Natl. Acad. Sci. U.S.A.*, 88, 3106, 1991.

116. Briesewitz, R., Epstein, M. R. and Marcantonio, E. E., Expression of native and truncated forms of the human integin α1 subunit, *J. Biol. Chem.*, 268, 2989, 1993.

117. Kolodziej, M. A., Vilaire, G., Rifat, S., Poncz, M., and Bennett, J. S., Effect of deletion of glycoprotein IIb Exon 28 on the expression of the platelet glycoprotein IIb/IIIa complex, *Blood*, 78, 2344, 1991.

118. Arnaout, M. A., Dana, N., Gupta, S. K., Tenen, D. G., and Fathallah, D. M., Point mutations impairing cell surface expression of the common β subunit (CD18) in a patient with leukocyte adhesion molecule (Leu-CAM) deficiency, *J. Clin. Invest.*, 85, 977, 1990.

119. Wardlaw, A. J., Hibbs, M. L., Stacker, S. A., and Springer, T. A., Distinct mutations in two patients with leukocyte adhesion deficiency and their functional correlates, *J. Exp. Med.*, 172, 335, 1990.

120. Back, A. L., Kwok, W. W., and Hickstein, D. D., Identification of two molecular defects in a child with leukocyte adherence deficiency, *J. Biol. Chem.*, 267, 5482, 1992.

121. Newman, P. J., Seligsohn, U., Lyman, S., and Coller, B. S., The molecular genetic basis of Glanzmann thrombasthenia in the Iraqi-Jewish and Arab populations in Israel, *Proc. Natl. Acad. Sci. U.S.A.*, 88, 3160, 1991.

122. Kassner, P. D. and Hemler, M. E., Interchangeable alpha chain cytoplasmic domains play a positive role in control of cell adhesion mediated by VLA-4, a β_1-integrin, *J. Exp. Med.*, 178, 649, 1993.

123. Kawaguchi, S. and Hemler, M. E., Role of the α subunit cytoplasmic domain in regulation of adhesive activity mediated by the integrin VLA-2, *J. Biol. Chem.*, 268, 1629, 1993.

124. Lotz, M. M., Burdsal, C. A., Erickson, H. P., and McClay, D. R., Cell adhesion to fibronectin and tenascin: quantitative measurements of initial binding and subsequent strengthening response, *J. Cell Biol.*, 109, 1795, 1989.

125. Raub, T. J. and Kuentzel, S. L., Kinetic and morphological evidence for endocytosis of mammalian cell integrin receptors by using an anti-fibronectin receptor beta subunit monoclonal antibody, *Exp. Cell Res.*, 184, 407, 1989.

126. Sczekan, M. and Juliano, R. L., Internalization of the fibronectin receptor is a constitutive process, *J. Cell Physiol.*, 142, 574, 1990.

127. Bretscher, M. S., Circulating integrins: $\alpha_5\beta_1$, $\alpha_6\beta_1$, and Mac-1, but not $\alpha_3\beta_1$ or LFA-1, *EMBO J.*, 11, 405, 1992.

128. Arnaout, M. A., Michishita, M., and Sharma, C. P., On the regulation of $\beta2$ integrins, in *Mechanisms of Lymphocyte Activation and Immune Regulation IV: Cellular Communications*, Gupta, S. and Waldmann, T. A., Eds., Plenum, New York, 1992, 171.

129. Chen, W.-J., Goldstein, J. L., and Brown, M. S., NPXY, a sequence often found in cytoplasmic tails, is required for coated pit-mediated internalization of the low density lipoprotein receptor, *J. Biol. Chem.*, 265, 3116, 1990.

130. Young, V. B., Falkow, S., and Schoolnik, G. K., The invasin protein of *Yersinia enterocolitica*: internalization of invasin-bearing bacteria by eukaryotic cells is associated with reorganization of the cytoskeleton, *J. Cell Biol.*, 116, 197, 1992.

131. Gudewicz, P. W., Molnar, J., Lai, M. Z., Beezhold, D. W., Siefring, G. E., Jr., Credo, R. B., and Lorand, L., Fibronectin-mediated uptake of gelatin-coated latex particles by peritoneal macrophages, *J. Cell Biol.*, 87, 427, 1980.

132. Ginsberg, M. H., Du, X., and Plow, E. F., Inside-out integrin signaling, *Curr. Op. Cell Biol.*, 4, 766, 1992.

133. Danilov, Y. N. and Juliano, R. L., Phorbol ester modulation of integrin-mediated cell adhesion: a postreceptor event, *J. Cell Biol.*, 108, 1925, 1989.

134. Shimizu, Y., Van Seventer, G. A., Horgan, K. J., and Shaw, S., Regulated expression and binding of three VLA ($\beta1$) integrin receptors on T cells, *Nature*, 345, 250, 1990.

135. Chan, B. M. C., Wong, J., Rao, A., and Hemler, M. E., T cell receptor dependent, antigen specific stimulation of a murine T cell clone induces a transient VLA protein-mediated binding to extracellular matrix, *J. Immunol.*, 147, 398, 1991.

136. Wilkins, J. A., Stupack, D., Stewart, S., and Caixia, S., β_1 integrin-mediated lymphocyte adherence to extracellular matrix is enhanced by phorbol ester treatment, *Eur. J. Immunol.*, 21, 517, 1991.

137. Altieri, D. C., Bader, R., Mannucci, P. M., and Edgington, T. S., Oligospecificity of the cellular adhesion receptor MAC-1 encompasses an inducible recognition specificity for fibrinogen, *J. Cell Biol.*, 107, 1893, 1988.

138. Altieri, D. C. and Edgington, T. S., The saturable high affinity association of factor X to ADP-stimulated monocytes defines a novel function of the Mac-1 receptor, *J. Biol. Chem.*, 263, 7007, 1988.

139. Buyon, J. P., Abramson, S. B., Philips, M. R., Slade, S. G., Ross, G. D., Weissman, G., and Winchester, R. J., Dissociation between increased surface expression of Gp165/95 and homotypic neutrophil aggregation, *J. Immunol.*, 140, 3156, 1988.

140. Detmers, P. A., Lo, S. K., Olsen-Egbert, E., Walz, A., Bagglioni, M., and Cohn, Z. A., Neutrophil-activating protein 1/interleukin 8 stimulates the binding activity of the leukocyte adhesion receptor CD11b/CD18 on human neutrophils, *J. Exp. Med.*, 171, 1155, 1990.

141. Wright, S. D. and Meyer, C. B., Fibronectin receptor of human macrophages recognizes the sequence arg-gly-asp-ser, *J. Exp. Med.*, 162, 762, 1985.

142. Bennett, J. S. and Vilaire, G., Exposure of platelet fibrinogen receptors by ADP and epinephrine, *J. Clin. Invest.*, 64, 1393, 1979.

143. Dahl, S. C. and Grabel, L. B., Integrin phosphorylation is modulated during the differentiation of F-9 teratocarcinoma stem cells, *J. Cell Biol.*, 108, 183, 1989.

144. Adams, J. C. and Watt, F. M., Changes in keratinocyte adhesion during terminal differentiation: reduction in fibronectin binding precedes $\alpha 5\beta 1$ integrin loss from the cell surface, *Cell*, 63, 425, 1990.

145. Neugebauer, K. M. and Reichardt, L. F., Cell-surface regulation of β_1-integrin activity on developing retinal neurons, *Nature*, 350, 68, 1991.

146. Shattil, S. J. and Brass, L. F., Induction of the fibrinogen receptor on human platelets by intracellular mediators, *J. Biol. Chem.*, 262, 992, 1987.

147. Hendey, B., Klee, C. B., and Maxfield, F. R., Inhibition of neutrophil chemokinesis on vitronectin by inhibitors of calcineurin, *Science*, 258, 296, 1992.

148. Lefkowith, J. B., Rogers, M., Lennartz, M. R., and Brown, E. J., Essential fatty acid deficiency impairs macrophage spreading and adherence: role of arachidonate in cell adhesion, *J. Biol. Chem.*, 266, 1071, 1991.

149. Weitzman, J. B., Pasqualini, R., Takada, Y., and Hemler, M. E., The function and distinctive regulation of the integrin VLA-3 in cell adhesion, spreading and homotypic cell aggregation, *J. Biol. Chem.*, 268, 8651, 1993.

150. Chen, Y. P., Djaffar, I., Pidard, D., Steiner, B., Cieutat, A. M., Caen, J. P., and Rosa, J. P., Ser-752 > Pro mutation in the cytoplasmic domain of integrin β_3 subunit and defective activation of platelet integrin $\alpha_{IIb}\beta_3$ (glycoprotein IIb-IIIa) in a variant of Glanzmann thrombasthenia, *Proc. Natl. Acad. Sci. U.S.A.*, 89, 10169, 1992.

151. Dejana, E., Colella, S., Conforti, G., Abbadini, M., Gaboli, M., and Marchisio, P. C., Fibronectin and vitronectin regulate the organization of their respective Arg-Gly-Asp adhesion receptors in cultured human endothelial cells, *J. Cell Biol.*, 107, 1215, 1988.

152. Jaken, S., Leach, K., and Klauck, T., Association of type 3 protein kinase C with focal contacts in rat embryo fibroblasts, *J. Cell Biol.*, 109, 697, 1989.

153. Maher, P. A., Pasquale, E. B., Wang, J. Y. J., and Singer, S. J., Phosphotyrosine-containing proteins are concentrated in focal adhesions and intercellular junctions in normal cells, *Proc. Natl. Acad. Sci. U.S.A.*, 82, 6576, 1985.

154. LaFlamme, S. E., Akiyama, S. K., and Yamada, K. M., Regulation of fibronectin receptor distribution, *J. Cell Biol.*, 117, 437, 1992.

155. Geiger, B., Salomon, D., Takeichi, M., and Hynes, R. O., A chimeric N-cadherin/β_1-integrin receptor which localizes to both cell-cell and cell-matrix adhesions, *J. Cell Sci.*, 103, 943, 1992.

156. Takada, Y., Ylänne, Y., Mandelman, D., Puzon, W., and Ginsberg, M. H., A point mutation of integrin β_1 subunit blocks binding of $\alpha_5\beta_1$ to fibronectin and invasin but not recruitment to adhesion plaques, *J. Cell Biol.*, 119, 913, 1992.

157. Wayner, E. A., Orlando, R. A., and Cheresh, D. A., Integrins $\alpha v\beta 3$ and $\alpha v\beta 5$ contribute to cell attachment to vitronectin but differentially distribute on the cell surface, *J. Cell Biol.*, 113, 919, 1991.

158. Marchisio, P. C., Bergui, L., Corbascio, G. C., Cremona, O., D'Urso, N., Schena, M., Tesio, L., and Cappio, F. C., Vinculin, talin, and integrins are localized at specific adhesion sites of malignant B lymphocytes, *Blood*, 72, 830, 1988.

159. Zambonin-Zallone, A., Teti, A., Grano, M., Rubinacci, A., Abbadini, M., Gaboli, M., and Marchisio, P. C., Immunocytochemical distribution of extracellular matrix receptors in human osteoclasts: a beta-3 integrin is colocalized with vinculin and talin in the podosomes of osteoclastoma giant cells, *Exp. Cell Res.*, 182, 645, 1989.

160. Schaller, M. D., Borgman, C. A., Cobb, B. S., Vines, R. R., Reynolds, A. B., and Parsons, J. T., pp125[FAK], a structurally distinctive protein-tyrosine kinase associated with focal adhesions, *Proc. Natl. Acad. Sci. U.S.A.*, 89, 5192, 1992.

161. Hanks, S. K., Calalb, M. B., Harper, M. C., and Patel, S. K., Focal adhesion protein-tyrosine kinase phosphorylated in response to cell attachment to fibronectin, *Proc. Natl. Acad. Sci. U.S.A.*, 89, 8487, 1992.

162. Ferrell, J. E. and Martin, G. S., Tyrosine-specific protein phosphorylation is regulated by glycoprotein IIb-IIIa in platelets, *Proc. Natl. Acad. Sci. U.S.A.*, 86, 2234, 1989.

163. Golden, A., Brugge, J. S., and Shattil, S. J., Role of platelet membrane glycoprotein IIb-IIIa in agonist-induced tyrosine phosphorylation of platelet proteins, *J. Cell Biol.*, 111, 3117, 1990.

164. Schwartz, M. A., Spreading of human endothelial cells on fibronectin or vitronectin triggers elevation of intracellular free calcium, *J. Cell Biol.*, 120, 1003, 1993.

165. Pardi, R., Bender, J. R., Dettori, C., Giannazza, E., and Engleman, E. G., Heterogeneous distribution and transmembrane signaling properties of lymphocyte function-associated antigen (LFA-1) in human lymphocyte subsets, *J. Immunol.*, 143, 3157, 1989.

166. Richter, J., Ng-Sikorski, J., Olsson, I., and Andersson, T., Tumor necrosis factor-induced degranulation in adherent human neutrophils is dependent on CD11b/CD18-integrin-triggered oscillations of cytosolic free Ca^{2+}, *Proc. Natl. Acad. Sci. U.S.A.*, 87, 9472, 1991.

167. Jaconi, M. E. E., Theler, J. M., Schlegel, W., Appel, R. D., Wright, S. D., and Lew, P. D., Multiple elevations of cytosolic-free Ca^{2+} in human neutrophils: initiation by adherence receptors of the integrin family, *J. Cell Biol.*, 112, 1249, 1991.

168. Yamiguchi, A., Tanoue, K., and Yaazaki, H., Secondary signals mediated by GPIIbIIIa in thrombin-activated platelets, *Biochim. Biophys. Acta*, 1054, 8, 1990.

169. Ng-Sikorski, J., Andersson, R., Patarroyo, M., and Andersson, T., Calcium signaling capacity of the CD11b/CD18 integrin on human neutrophils, *Exp. Cell Res.*, 195, 504, 1991.

170. Miyauchi, A., Alvarez, J., Greenfield, E. M., Teti, A., Grano, M., Colucci, S., Zambonin-Zallone, A., Ross, F. P., Teitelbaum, S. L., Cheresh, D., and Hruska, K. A., Recognition of osteopontin and related peptides by an $\alpha_v\beta_3$ integrin stimulates immediate cell signals in osteoclasts, *J. Biol. Chem.*, 266, 20369, 1991.

171. Leavesley, D. I., Schwartz, M. A., Rosenfeld, M., and Cheresh, D. A., Integrin $\beta1$- and $\beta3$-mediated endothelial cell migration is triggered through distinct signaling mechanisms, *J. Cell Biol.*, 121, 163, 1993.

172. Schwartz, M. A., Lechene, C., and Ingber, D. E., Insoluble fibronectin activates the Na/H antiporter by clustering and immobilizing integrin $\alpha_5\beta_1$, independent of cell shape, *Proc. Natl. Acad. Sci. U.S.A.*, 88, 7849, 1991.

173. Schwartz, M. A., Ingber, D. E., Lawrence, M., Springer, T. A., and Lechene, C., Multiple integrins share the ability to induce elevation of intracellular pH, *Exp. Cell Res.*, 195, 533, 1991.

174. Ingber, D. E., Prusty, D., Frangioni, J. J., Cragoe, Jr., E. J., Lechene, C. P., and Schwartz, M. A., Control of intracellular pH and growth by fibronectin in capillary endothelial cells, *J. Cell Biol.*, 110, 1803, 1990.

175. Akiyama, S. K., Yamada, S. S., Chen, W.-T., and Yamada, K. M., Analysis of fibronectin receptor function with monoclonal antibodies: roles in cell adhesion, migration, matrix assembly, and cytoskeletal organization, *J. Cell Biol.*, 109, 863, 1989.

176. Straus, A. H., Carter, W. G., Wayner, E. A., and Hakomori, S. I., Mechanism of fibronectin-mediated cell migration: dependence or independence of cell migration susceptibility on RGDS-directed receptor (integrin), *Exp. Cell Res.*, 183, 126, 1989.

177. Jaffredo, T., Horwitz, A. F., Buck, C. A., Rong, P. M., and Dieterlen-Lievre, F., Myoblast migration specifically inhibited in the chick embryo by grafted CSAT hybridoma cells secreting an anti-integrin antibody, *Development*, 103, 431, 1988.

178. Letourneau, P. C., Pech, I. V., Rogers, S. L., Palm, S. L., McCarthy, J. B., and Furcht, L. T., Growth cone migration across extracellular matrix components depends on integrin, but migration across glioma cells does not, *J. Neur. Res.*, 21, 286, 1988.

179. Perris, R., Paulsson, M., and Bronner-Fraser, M., Molecular mechanisms of avian neural crest cell migration on fibronectin and laminin, *Dev. Biol.*, 136, 222, 1989.

180. Smith, C. W., Marlin, S. D., Rothlein, R., Toman, C., and Anderson, D. C., Cooperative interactions of LFA-1 and Mac-1 with intercellular adhesion molecule-1 in facilitating adherence and transendothelial migration of human neutrophils in vitro, *J. Clin. Invest.*, 83, 2008, 1989.

181. Yamada, K. M., Kennedy, D. W., Yamada, S. S., Gralnick, H., Chen, W.-T., and Akiyama, S. K., Monoclonal antibody and synthetic peptide inhibitors of human tumor cell migration, *Cancer Res.*, 50, 4485, 1990.

182. McCarthy, J. B. and Furcht, L. T., Laminin and fibronectin promote the haptotactic migration of B16 mouse melanoma cells in vitro, *J. Cell Biol.*, 98, 1474, 1984.

183. Kavanaugh, A. F., Lightfoot, E., Lipsky, P. E., and Oppenheimer-Marks, N., Role of CD11/CD18 in adhesion and transendothelial migration of T cells: analysis utilizing CD18-deficient T cell clones, *J. Immunol.*, 146, 4149, 1991.

184. Duband, J.-L., Nuckolls, G. H., Ishihara, A., Hasegawa, T., Yamada, K. M., Thiery, J. P., and Jacobson, K., Fibronectin receptor exhibits high lateral motility in embryonic locomoting cells but is immobile in focal contacts and fibrillar streaks in stationary cells, *J. Cell Biol.*, 107, 1385, 1988.

185. Grzesiak, J. J., Davis, G. E., Kirchhofer, D., and Pierschbacher, M. D., Regulation of $\alpha_2\beta_1$-mediated fibroblast migration on type I collagen by shifts in the concentrations of extracellular Mg^{2+} and Ca^{2+}, *J. Cell Biol.*, 117, 1109, 1992.

186. Leavesley, D. J., Ferguson, G. D., Wayner, E. A., and Cheresh, D. A., Requirement of the integrin $\beta3$ subunit for carcinoma cell spreading or migration on vitronectin and fibrinogen, *J. Cell Biol.*, 117, 1101, 1992.

187. Miyake, K., Hasunuma, Y., Yagita, H., and Kimoto, M., Requirement for VLA-4 and VLA-5 integrins in lymphoma cells binding to and migration beneath stromal cells in culture, *J. Cell Biol.*, 119, 653, 1992.

188. Chan, B. M. C., Kassner, P. D., Schiro, J. A., Byers, H. R., Kupper, T. S., and Hemler, M. E., Distinct cellular functions mediated by different VLA integrin α subunit cytoplasmic domains, *Cell*, 68, 1051, 1992.

189. Bellows, C. G., Melcher, A. H., and Aubin, J. E., Association between tension and orientation of periodontal ligament fibroblasts and exogenous collagen fibres in collagen gels *in vitro*, *J. Cell Sci.*, 58, 125, 1982.

190. Harris, A. K., Stopak, D., and Wild, P., Fibroblast traction as a mechanism for collagen morphogenesis, *Nature*, 290, 249, 1981.

191. Tucker, R. P., Edwards, B. F., and Erickson, C. A., Tension in the culture dish: microfilament organization and migratory behavior of quail neural crest cells, *Cell Motil.*, 5, 225, 1985.

192. Gullberg, D., Tingstrom, A., Thuresson, A. C., Olsson, L., Terracio, L., Borg, T. K., and Rubin, K., Beta-1 integrin-mediated collagen gel contraction is stimulated by PDGF, *Exp. Cell Res.*, 186, 264, 1990.

193. Schiro, J., Chan, B. M. C., Roswit, W. T., Kassner, P. D., Pentland, A., Hemler, M. E., Eisen, A. Z., and Kupper, T. S., Integrin α2β1 (VLA-2) mediates reorganization and contraction of collagen matrices by human cells, *Cell*, 67, 403, 1991.

194. Klein, C. E., Dressel, D., Steinmayer, T., Mauch, C., Eckes, B., Kreig, T., Bankert, R. B., and Weber, L., Integrin α2β1 is upregulated in fibroblasts and highly aggressive melanoma cells in three-dimensional collagen lattices and mediates the reorganization of collagen I fibrils, *J. Cell Biol.*, 115, 1427, 1991.

195. Matsuyama, T., Yamada, A., Kay, J., Yamada, K. M., Akiyama, S. K., Schlossman, S. F., and Morimoto, C., Activation of CD4 cells by fibronectin and anti-CD3 antibody: a synergistic effect mediated by the VLA-5 fibronectin receptor complex, *J. Exp. Med.*, 170, 1133, 1989.

196. Davis, L. S., Oppenheimer-Marks, N., Bednarczyk, J. L., McIntyre, B. W., and Lipsky, P. E., Fibronectin promotes proliferation of naive and memory T cells by signaling through both the VLA-4 and VLA-5 integrin molecules, *J. Immunol.*, 145, 785, 1990.

197. Shimizu, Y., Van Seventer, G. A., Horgan, K. J., and Shaw, S., Costimulation of proliferative responses of resting CD4+ T cells by the interaction of VLA-4 and VLA-5 with fibronectin or VLA-6 with laminin, *J. Immunol.*, 145, 59, 1990.

198. Yamada, A., Nojima, Y., Sugita, K., Dang, N. H., Schlossman, S. F., and Morimoto, C., Cross-linking of VLA/CD29 molecule has a co-mitogenic effect with anti-CD3 on CD4 cell activation in serum-free culture system, *Eur. J. Immunol.*, 21, 319, 1991.

199. Roberts, K., Yokoyama, W. M., Kehn, P. J., and Shevach, E. M., The vitronectin receptor serves as an accessory molecule for the activation of a subset of τ/δ T cells, *J. Exp. Med.*, 173, 231, 1991.

200. Nojima, Y., Humphries, M. J., Mould, A. P., Komoriya, A., Yamada, K. M., Schlossman, S. F., and Morimoto, C., VLA-4 mediates CD3-dependent CD4+ T cell activation via the CS1 alternatively spliced domain of fibronectin, *J. Exp. Med.*, 172, 1185, 1990.

201. Takahashi, K., Nakamura, T., Adachi, H., Yagita, H., and Okumura, K., Antigen-independent T cell activation mediated by a very late activation antigen-like extracellular matrix receptor, *Eur. J. Immunol.*, 21, 1559, 1991.

202. Mortarini, R., Gismondi, A., Santoni, A., Parmiani, G., and Anichini, A., Role of the α5β1 integrin receptor in the proliferative response of quiescent human melanoma cells to fibronectin, *Cancer Res.*, 52, 4499, 1992.

203. Symington, B. E., Fibronectin receptor modulates cyclin-dependent kinase activity, *J. Biol. Chem.*, 267, 25744, 1992.

204. Poenie, M., Alderton, J., Tsein, R. Y., and Steinhardt, R. A., Changes of free calcium levels with stages of the cell division cycle, *Nature*, 315, 147, 1985.

205. Grandin, N. and Charbonneau, M., Cycling of intracellular pH during cell division of *xenopus* embryos is a cytoplasmic activity depending on protein synthesis and phosphorylation, *J. Cell Biol.*, 111, 523, 1990.

206. Cadroy, Y., Houghten, R. A., and Hanson, S. R., RGDV peptide selectively inhibits platelet-dependent thrombus formation in vivo, *J. Clin. Invest.*, 84, 939, 1989.

207. Yurochko, A. D., Liu, D. Y., Eierman, D., and Haskill, S., Integrins as a primary signal transduction molecule regulating monocyte immediate-early gene induction, *Proc. Natl. Acad. Sci. U.S.A.*, 89, 9034, 1992.

208. Yamada, A., Nikaido, T., Nojima, Y., Schlossman, S. F., and Morimoto, C., Activation of human CD4 T lymphocytes: interaction of fibronectin with VLA-5 receptor on CD4 cells induces the AP-1 transcription factor, *J. Immunol.*, 146, 53, 1991.

209. Streuli, C. H., Bailey, N., and Bissell, M. J., Control of mammary epithelial differentia-

tion: basement membrane induces tissue-specific gene expression in the absence of cell-cell interaction and morphological polarity, *J. Cell Biol.*, 115, 1383, 1991.

210. Bohnsack, J. F., Kleinman, H. K., Takahashi, T., O'Shea, J. J., and Brown, E. J., Connective tissue proteins and phagocytic cell function: laminin enhances complement and Fc-mediated phagocytosis by cultured human macrophages, *J. Exp. Med.*, 161, 912, 1985.

211. Pommier, C. G., Inada, S., Fries, L. F., Takahashi, T., Frank, M. M., and Brown, E. J., Plasma fibronectin enhances phagotyctosis of opsonized particles by human peripheral blood monocytes, *J. Exp. Med.*, 157, 1844, 1983.

212. Wright, S. D., Craigmyle, L. S., and Silverstein, S. C., Fibronectin and serum amyloid p component stimulate C3b- and C3bi-mediated phagocytosis in cultured human monocytes, *J. Exp. Med.*, 158, 1338, 1983.

213. Graham, I. L., Gresham, H. D., and Brown, E. J., An immobile subset of plasma membrane CD11b/CD18 is involved in phagocytosis of targets recognized by multiple receptors, *J. Immunol.*, 142, 2352, 1989.

214. Graham, I. L. and Brown, E. J., Extracellular calcium results in a conformational change in Mac-1 (CD11b/CD18) on neutrophils: differentiation of adhesion and phagocytosis functions of Mac-1, *J. Immunol.*, 146, 685, 1991.

215. Nathan, C., Srimal, S., Farber, C., Sanchez, E., Kabbash, L., Asch, A., Gailit, J., and Wright, S. D., Cytokine-induced respiratory burst of human neutrophils: dependence on interaction of CD11/CD18 integrins with extracellular matrix proteins, *J. Cell Biol.*, 109, 1341, 1989.

216. Nathan, C. and Sanchez, E., Tumor necrosis factor and CD11/CD18 (β2) integrins act synergistically to lower cAMP in human neutrophils, *J. Cell Biol.*, 111, 2171, 1990.

217. Berton, G., Laudanna, C., Sorio, C., and Rossi, F., Generation of signals activating neutrophil functions by leukocyte integrins: LFA-1 and gp150/95, but not CR3, are able to stimulate the respiratory burst of human neutrophils, *J. Cell Biol.*, 116, 1007, 1992.

218. Groux, H., Huet, S., Valentin, H., Pham, D., and Bernard, A., Suppressor effects and cyclic AMP accumulation by the CD29 molecule of CD4+ lymphocytes, *Nature*, 339, 152, 1989.

219. Lampugnani, M. G., Giorgi, M., Gaboli, M., Dejana, E., and Marchisio, P. C., Endothelial cell motility, integrin receptor clustering, and microfilament organization are inhibited by agents that increase intracellular cAMP, *Lab. Invest.*, 63, 521, 1990.

220. Ignatius, M. J., Large, T. H., Houde, M., Tawil, J. W., Barton, A., Esch, F., Carbonetto, S., and Reichardt, L. F., Molecular cloning of the rat integrin α1 subunit: a receptor for laminin and collagen, *J. Cell Biol.*, 111, 709, 1990.

221. Takada, Y. and Hemler, M. E., The primary structure of the VLA-2/collagen receptor α2 subunit (platelet GP Ia): homology to other integrins and the presence of a possible collagen-binding domain, *J. Cell Biol.*, 109, 397, 1989.

222. Takada, Y., Murphy, E., Pil, P., Chen, C., Ginsberg, M. H., and Hemler, M. E., Molelcular cloning and expression of the cDNA for the α3 chain of human α3β1 (VLA-3), an integrin receptor for fibronectin, laminin and collagen, *J. Cell Biol.*, 115, 257, 1991.

223. Tsuji, T., Hakomori, S., and Osawa, T., Identification of human galactoprotein b3, an oncogenic transformation-induced membrane glycoprotein, as VLA-3 α subunit: the primary structure of human integrin α3, *J. Biochem.*, 109, 659, 1991.

224. Tsuji, T., Yamamoto, F., Miura, Y., Takio, K., Titani, K., Pawar, S., Osawa, T., and Hakomori, S., Characterization through cDNA cloning of galactoprotein b3 (Gap b3), a cell surface membrane glycoprotein showing enhanced expression on oncogenic transformation: identification of Gap b3 as a member of the integrin superfamily, *J. Biol. Chem.*, 265, 7016, 1990.

225. Takada, Y., Elices, M. J., Crouse, C., and Hemler, M. E., The primary structure of the α4 subunit of VLA-4: homology to other integrins and a possible cell-cell adhesion function, *EMBO J.*, 8, 1361, 1989.

226. Neuhaus, H., Hu, M. C-T., Hemler, M. E., Takada, Y., Holzmann, B., and Weissman, I. L., Cloning and expression of cDNAs for the α subunit of the murine lymphocyute-peyer's patch homing receptor, *J. Cell Biol.*, 115, 1149, 1991.

227. Argraves, W. S., Suzuki, S., Arai, H., Thompson, K., Pierschbacher, M. D., and Ruoslahti, E., Amino acid sequence of the human fibronectin receptor, *J. Cell Biol.*, 105, 1183, 1987.

228. Fitzgerald, L. A., Poncz, M., Steiner, B., Rall, S. C., Jr., Bennett, J. S., and Phillips, D. R., Comparison of cDNA-derived protein sequences of the human fibronectin and vitronectin receptor alpha-subunits and platelet glycoprotein IIb, *Biochemistry*, 26, 8158, 1987.

229. de Curtis, I., Quaranta, V., Tamura, R. N., and Reichardt, L. F., Laminin receptors in the retina: sequence analysis of the chick integrin alpha-6 subunit, *J. Cell Biol.*, 113, 405, 1991.

230. Song, W. K., Wang, W., Forster, R. F., Bielser, D. A., and Kaufman, S. J., H36-α7 is a novel integrin alpha chain that is developmentally regulated during skeletal myogenesis, *J. Cell Biol.*, 117, 643, 1992.

231. Bossy, B., Bossy-Wetzel, E., and Reichardt, L. F., Characterization of the integrin α8 subunit: a new integrin β1-associated subunit, which is prominently expressed on axons and on cells in contact with basal laminae in chick embryos, *EMBO J.*, 10, 2375, 1991.

232. Suzuki, S., Argraves, W. S., Arai, H., Languino, L. R., Pierschbacher, M., and Ruoslahti, E., Amino acid sequence of the vitronectin receptor alpha subunit and comparative expression of adhesion receptor mRNAs, *J. Biol. Chem.*, 262, 14080, 1987.

233. Poncz, M., Eisman, R., Heidenreich, R., Silver, S. M., Vilaire, G., Surrey, S., Schwartz, E., and Bennett, J. S., Structure of the platelet membrane glycoprotein IIb: homology to the alpha subunits of the vitronectin and fibronectin membrane receptors, *J. Biol. Chem.*, 262, 8476, 1987.

234. Poncz, M. and Newman, P. J., Analysis of rodent platelet glycoprotein IIb: evidence for evolutionarily conserved domains and alternative proteolytic processing, *Blood*, 75, 1282, 1990.

235. Larson, R. S., Corbi, A. L., Berman, L., and Springer, T. A., Primary structure of the LFA-1 a subunit: an Integrin with an embedded domain defining a protein superfamily, *J. Cell Biol.*, 108, 703, 1989.

236. Kaufmann, Y., Tseng, E., and Springer, T. A., Cloning of the murine lymphocyte function-associated molecule-1 α subunit and its expression in COS cells, *J. Immunol.*, 147, 369, 1992.

237. Corbi, A. L., Kishimoto, T. K., Miller, L. J., and Springer, T. A., The human leukocyte adhesion glycoprotein Mac-1 (complement receptor type 3, CD11b) α subunit: cloning, primary structure, and relation to the integrins, von Willebrand factor and factor B, *J. Biol. Chem.*, 263, 12403, 1988.

238. Arnaout, M. A., Gupta, S. K., Pierce, M. W., and Tenen, D. G., Amino acid sequence of the alpha subunit of human leukocyte adhesion receptor Mo1 (complement receptor type 3), *J. Cell Biol.*, 106, 2153, 1988.

239. Hickstein, D. D., Hickey, M. J., Ozols, J., Baker, D. M., Back, A. L., and Roth, G. J., cDNA sequence for the alpha-M subunit of the human neutrophil adherence receptor indicates homology to integrin alpha subunits, *Proc. Natl. Acad. Sci. U.S.A.*, 86, 257, 1989.

240. Corbi, A. L., Miller, L. J., O'Connor, K., Larson, R. S., and Springer, T. A., cDNA

cloning and complete primary structure of the alpha subunit of a leukocyte adhesion glycoprotein, p150,95, *EMBO J.*, 6, 4023, 1987.

241. Bogaert, T., Brown, N., and Wilcox, M., The Drosophila PS2 antigen is an invertebrate integrin which like the fibronectin receptor becomes localized to muscle attachments, *Cell*, 51, 929, 1988.

242. MacKrell, A. J., Blumberg, B., Haynes, S. R., and Fessler, J. H., The lethal myospheroid gene of Drosophila encodes a membrane protein homologous to vertebrate integrin B subunits, *Proc. Natl. Acad. Sci. U.S.A.*, 85, 2633, 1988.

243. Kishimoto, T. K., O'Connor, K., Lee, A., Roberts, T. M., and Springer, T. A., Cloning of the beta subunit of the leukocyte adhesion proteins: homology to an extracellular matrix receptor defines a novel supergene family, *Cell*, 48, 681, 1987.

244. Law, S. K. A., Gagnon, J., Hildreth, J. E. K., Wells, C. E., Willis, A. C., and Wong, A. J., The primary structure of the beta subunit of the cell surface adhesion glycoproteins LFA-1, CR3, and P150,95 and its relationship to the fibronectin receptor, *EMBO J.*, 6, 915, 1987.

245. Fitzgerald, L. A., Steiner, B., Rall, S. C., Jr., Lo, S., and Phillips, D. R., Protein sequence of endothelial glycoprotein IIIa derived from a cDNA clone. Identity with platelet glycoprotein IIIa and similarity to "integrin", *J. Biol. Chem.*, 262, 3936, 1987.

246. Rosa, J. P., Bray, P. F., Gayet, O., Johnston, G. I., Cook, R. G., Jackson, K. W., Shuman, M. A., and McEver, R. P., Cloning of glycoprotein IIIa cDNA from human erythroleukemia cells and localization of the gene to chromosome 17, *Blood*, 72, 593, 1988.

247. Zimrin, A. B., Eisman, R., Vilaire, G., Schwartz, E., and Bennett, J. S., Structure of platelet Glycoprotein IIIa: a common subunit for two different membrane receptors, *J. Clin. Invest.*, 81, 1470, 1988.

248. Sheppard, D., Rozzo, C., Starr, L., Quaranta, V., Erle, D. J., and Pytela, R., Complete amino acid sequence of a novel integrin β subunit (β6) identified in epithelial cells using the polymerase chain reaction, *J. Biol. Chem.*, 265, 11502, 1990.

249. Yuan, Q., Jiang, W.-M., Krissansen, G. W., and Watson, J. D., Cloning and sequence analysis of a novel β2-related integrin transcript from T lymphocytes: homology of integrin cysteine-rich repeats to domain III of laminin B chains, *Int. Immunol.*, 2, 1097, 1990.

250. Erle, D. J., Rüegg, C., Sheppard, D., and Pytela, R., Complete amino acid sequence of an integrin β subunit (β7) identified in leukocytes, *J. Biol. Chem.*, 266, 11009, 1991.

251. Hu, M. C.-T., Crowe, D. T., Weissman, I. L., and Holzmann, B., Cloning and expression of mouse integrin βp(β7): a functional role in Peyer's patch-specific lymphocyte homing, *Proc. Natl. Acad. Sci. U.S.A.*, 89, 8254, 1992.

252. Gurish, M. F., Bell, A. F., Smith, T. J., Ducharme, L. A., Wang, R. K., and Weis, J. H., Expression of murine β7, α4, and β1 integrin genes by rodent mast cells, *J. Immunol.*, 149, 1964, 1992.

253. Hirst, R., Horwitz, A., Buck, C., and Rohrschneider, L., Phosphorylation of the fibronectin receptor complex in cells transformed by oncogenes that encode tyrosine kinases, *Proc. Natl. Acad. Sci. U.S.A.*, 83, 6470, 1986.

Chapter 2

The Role of β2 Integrins in Leukocyte Adhesion

Nava Dana and M. Amin Arnaout

CONTENTS

An intact immune system is critical for protecting the organism from threats both from the outside environment, such as various pathogens, and threats from within, such as malignant transformation, cell injury, or death. It is necessary for the functioning immune system to react quickly and specifically to such targets; first, by discriminating foreign or "enemy" from "nonhostile tissue", then moving quickly to engage the target and destroy it. Part of the machinery necessary to perform these functions includes a wide variety of receptors found on the cell surface by which leukocytes sense their environment and react accordingly. One major category of surface receptors is the adhesion molecules. Various families of these receptors are found on leukocytes, and the ligands for these receptors can be found on other leukocytes, endothelial and epithelial cells, in the plasma, and in the extracellular matrix. Many cell surface adhesion molecules are carefully regulated, appearing or becoming "activated" only in response to specific, local inflammatory stimuli and being shed, recycled, or "inactivated" at set time points, again in response to both extracellular and intracellular environmental conditions. These various receptors and their regulation give leukocytes the flexibility and specificity required for the trafficking, migration, homing, antigen presentation, and target killing necessary for a fully functional immune system. One important family of such adhesion molecules is the β2 integrins.

The β2 integrin family (CD11/CD18, as defined by the β2 subunit or CD18)[1] is unique among the integrins for two reasons. First, the three α subunits in this family, CD11a(αL), CD11b(αM) and CD11c(αX), have each been found only in association with β2. Second, this family is found exclusively on leukocytes. Several of the characteristics of these three β2 heterodimers are summarized in Table 1.

I. β2 INTEGRIN DEFICIENCY

The importance of the β2 family of integrins was discovered as the result of unraveling an "experiment of nature".[2] A series of reports had appeared in the medical literature describing children with a pathogenetically uncharacterized immunodeficiency. These children typically experienced recurrent bouts of life-threatening bacterial infections, often resulting in death in infancy. *In vitro* testing demonstrated impaired neutrophil function. Both the Fc-receptor and CR1 (complement receptor type 1, the receptor for C3b, one of the degradation products of the third component of complement) functioned normally. Whole cell lysates of detergent solubilized neutrophils from these children,

0-8493-4711-4/94/$0.00+$.50

Table 1   β2 (CD11/CD18) Integrins

Cluster Designation	Molecular Mass	Amino Acids	Alternative Name	Tissue Distribution
CD11a/CD18	170–185 kDa (α subunit)	1063	LFA-1, TA-1, αL/β2	All leukocytes
	95–105 kDa (β subunit)	769		
CD11b/CD18	155–165 kDa (α subunit)	1136–1137[a]	CR3, Mol, Mac 1,	Phagocytic leukocytes (PMN, monocytes,
	95–105 kDa (β subunit)	769	OKM1 αM/β2	macrophages (↓), large granular lynmphocytes
CD11c/CD18	130–150 kDa (α subunit)	1144	p150,95, CR4 LeuM5 α2X/β2	Phagocytic leukocytes (PMN, moncytes, macrophages (↑), some B cells, come cytoxic T cells)
	95–105 kDa (β subunit)	769		

[a]See text.

normally. Whole cell lysates of detergent solubilized neutrophils from these children, when run on SDS-polyacrylamide gel electrophoresis, were missing a prominent band found in normal controls.[3] Monoclonal antibodies were raised against this band which revealed this protein to be the αM subunit. Fluorescence analysis of leukocytes from these children, surface labeled with monoclonal antibodies (MAb) to the αL, αM, and αX subunits, as well as the β2 subunit, revealed that these leukocytes were missing the three β2 heterodimers.[4,5]

The importance of this family of surface proteins was made even more apparent by parallel studies which demonstrated the ability of monoclonal antibodies directed against the various subunits to reproduce in normal cells the same functional defects found in the leukocytes from the affected children. Monoclonal antibodies blocked binding of iC3b (a further degradation product of C3b which binds to αMβ2, also called CR3), chemotaxis, phagocytosis, adhesion to endothelium, homotypic aggregation, and particle-induced degranulation and superoxide generation. Phagocytic cells from these children did, however, exhibit a normal oxidative burst in response to soluble stimuli, such as phorbol myristate acetate (Table 2).

Fluorescence analysis of leukocytes from these immunodeficient children demonstrated two staining patterns using MAbs to the β2 family that correlated well with their clinical presentation.[2] One group of cells showed approximately 10% of the normal level of expression of the three β2 surface heterodimers, and a second group demonstrated no detectable staining for these heterodimers. The first group exhibits a more benign clinical course, many of these patients surviving beyond childhood with aggressive antibiotic therapy. The second group of patients rarely survives without bone marrow transplant. It is interesting to note that leukocytes from the parents of these children demonstrate approximately 50% of normal leukocyte staining for the β2 heterodimers;[3] yet, these parents are clinically normal and their in vitro leukocyte functions are within the normal range.

Table 2 β2-Dependent Functions[a]

Leukocyte Function	β2 Integrin Involved
Myeloid (Phagocytic Leukocytes)	
1. Receptor iC3b	αMβ2, αXβ2
2. Phagocytosis	αMβ2
3. Spreading/chemotaxis/random migration	αLβ2, αMβ2, αXβ2
4. Homotypic aggregation	αLβ2, αMβ2
5. Adhesion to "activated"endothelium	αLβ2, αMβ2, αXβ2
6. "Activated" leukocyte adhesion to quiescent endothelium	αMβ2, αXβ2
7. Particle-induced degranulation and oxidative burst	αMβ2
8. Antibody-dependent cellular cytotoxicity	αLβ2, αMβ2, αXβ2
Lymphoid (T and B cell)	
1. Antigen-, alloantigen-induced proliferation (2° to contact with antigen presenting cell)	αLβ2
2. Homotypic aggregation	αLβ2
3. Adhesion to endothelium	αLβ2
4. B cell Ig production (2° T-helper cells)	αLβ2
5. CTL-mediated killing (cytotoxic T lymphocyte)	αLβ2
Large Granular Lymphocytes	
1. Natural killing	αLβ2, αMβ2, αXβ2

[a]Determined by inhibition studies using anti-β2 MAbs with normal cells.

The assumption was made that because all three heterodimers were missing, the defect responsible for this immunodeficiency must reside in the β2 subunit. Indeed, this was found to be the case based on biosynthesis studies using lymphocytes derived from children from both groups.[6-8] It was found that leukocytes from most patients synthesized a normal appearing beta subunit precursor based on SDS-polyacrylamide gel electrophoresis of immunoprecipitates, yet the beta subunits from the patients' cells were not able to associate with the alpha subunits and mature to the fully glycosylated cell surface form. Cells from a few patients with the complete deficiency were discovered to make either no detectable β2 precursor protein or an abnormal, degraded protein. Further studies looking at the messenger RNA from patients showed that β2 integrin or Leu-CAM (leukocyte adhesion molecule) deficiency or LAD was a heterogeneous disease. This was borne out by the cloning and sequencing of β2 cDNA from many of these patients. A variety of point mutations as well as deletions were found, most of which congregated into two distinct regions of the β2 molecule.[9,10] This will be discussed in greater detail further on. When COS cells were transfected with the various mutant β2 cDNAs, it was found that the resulting β2 proteins were unable to associate normally with the appropriate wild-type alpha subunits and demonstrated both decreased cell surface heterodimer and decreased β2 heterodimer-mediated function in similar fashion to the leukocytes from these patients.[10,11]

II. MOLECULAR ANALYSIS OF THE β2 INTEGRINS

Molecular cloning of the genes encoding all four β2 integrin proteins has been accomplished. The β2 subunit, CD18, is composed of 769 amino acids including a 22 amino acid leader sequence and the gene encoding for this protein is found on chromosome 21.[12,13] The β2 cytoplasmic tail has 46 amino acids containing a tyrosine and several serine and threonine residues which may serve as phosphorylation sites. This finding is important as will be discussed in greater detail later in this chapter. The extracytoplasmic domain of β2 shares great homology to other integrin β subunits. A cysteine-rich domain comprised of 4 tandem repeats of an 8-cysteine motif is highly conserved in all integrin β subunits. This region is predicted to be highly rigid and may provide the tertiary structure of the β subunit and consequently the ability of the alpha and beta subunits to align properly for surface expression and function. N terminal to the cysteine rich area is another highly conserved region of 250 amino acids. In β3 integrins, this region contains the binding site for the tripeptide RGD (arg-gly-asp)-containing ligands. Two ligands which bind to αMβ2, iC3b and fibrinogen, contain a noncontiguous RGD sequence (K/RXXGD). iC3b and fibrinogen binding to αMβ2 has not been shown, however, to be inhibited by RGD-containing peptides. Mutation of this sequence in iC3b does not hinder normal binding to αMβ2.[14] It is far more likely that these two ligands share close binding sites on the αM subunit (binding of fibrinogen to αMβ2 can be inhibited by iC3b). This line of reasoning is further supported by the fact that all three β2 integrins share the β2 subunit, but only αM and αL bind iC3b and fibrinogen, and anti-αM MAbs and not anti-αL MAbs inhibit iC3b binding. Moreover wild-type β2 transfected alone into COS cells does not bind iC3b-coated sheep erythrocytes, whereas COS cells similarly transfected with wild-type αM do bind iC3b, although not as well as the normally associated heterodimer (N. Dana, D. Fathallah, and M. A. Arnaout, unpublished data). This last set of results may be due to the necessity of heterodimer formation to achieve the optimal tertiary structure for ligand binding.

The three β2 integrin α subunits share approximately 40% identity to each other, with αM and αX being more homologous to each other than to αL.[15-20] This is reflected in their cell distribution, in that αL is found on all leukocytes, whereas αM and αX are limited to phagocytic cells. It is also reflected in analysis of their functions (as discussed below and in Table 2). The sizes of the three alpha subunits are 1063, 1136 to 1137, and 1144 amino acids for the αL, αM, and αX, respectively. The extra amino acid in some αM proteins is coded for at the beginning of an exon and may be the result of alternative splicing. The three alpha subunits share several features with the other integrin α subunits. The N-terminal extracytoplasmic end contains seven homologous tandem repeats, each approximately 60 amino acids long. In some integrin alpha subunits, including the β2 family, the three more C-terminal repeats contain a metal-binding consensus sequence. All known β2 integrin-mediated functions require the presence of divalent cations. A unique feature shared by αL, αM, and αX is an approximately 200-amino acid region, encoded for by 4 exons, which is found between the 2 and 3 repeats just described. This region, alternatively called either the "A" domain (because of its homology to the three A-domains of von Willebrand factor where this motif was first described), or the "I" domain (because it appears to be inserted into the protein), is thought to play an important role in function of the intact heterodimer. Regions homologous to the A-domain are found in only two β1 integrin alpha subunits, α1,[21] α2,[22] and in a β7 integrin, αE.[61] Other proteins that are involved in matrix or the complement and clotting cascades, but are not themselves integrins, also have A-domain homologous sequences; these include cartilage matrix protein (which binds collagen),[23] the complement proteins C2 and Factor B,[24,25] type VI, XII, and XIV collagen,[26-28] and von Willebrand factor (which binds to heparin, collagen, and platelet Ib).[29,30] Existing data map several MAbs to αM which inhibit heterodimer function and αM/β2 ligands to the αM I-domain. The three β2 integrin α subunits also share one other property with α1 and α2; all five α subunits lack an extracellular dibasic protease cleavage site found close to the membrane in all other integrin α subunits. This is due to a deletion of 20 to 30 amino acids present in the other integrin α subunits (with the exception of α4).[31] The three genes encoding αL, αM, and αX are clustered on chromosome 16, band p11 to p11.2, suggesting that they arose from one ancestral gene.[32,33]

The cytoplasmic tails of the three β2 integrin α subunit proteins do not share significant homology, and this may in part explain their different and distinct distribution in leukocytes, intracellular storage, and functions. There are multiple threonine and serine residues found in αL and αX. αM has one serine and one tyrosine residue. Some of these amino acids may serve as potential phosphorylation sites in the activation or deactivation of these heterodimeric receptors.[34-36]

In 1988, Nermut et al.[37] determined by electron microscopy of the human fibronection receptor that the tertiary structure of this integrin heterodimer assumes a configuration that is similar to a mushroom. That is, the two subunits associate extracellularly, probably near their N-terminal portion and form three-dimensional pockets which allow for the "fit" of divalent cations. Binding of divalent cations is postulated to confer the tertiary structure that allows for ligand binding. The C-terminal portion of the extracellular domains of the two subunits form the stalk which "tunnels" through the membrane and emerges in the cytoplasm of the cell. In the cytoplasm, association with various cytoskeletal proteins and members of the G-protein family could be postulated to participate in the cascade of signal "transmission" from the extracellular millieu (ligand binding), intracellularly to mediate cell function.[38] Indeed, several integrins have been shown to associate with cytosketelal proteins,[39,40] and our own laboratory and others have shown association of β2 to actin, α-actinin, vinculin, actin-binding protein-280, and talin.[41]

In the case of the β2 integrins, one could postulate that the A-domain of the α subunit must somehow be involved in the "mushroom head" where ligand binding is assumed to occur. This hypothesis is based on observations in this laboratory that show that some anti-αM MAbs that inhibit function bind to the A-domain.[100-102] Amino acid sequencing of β2 from patients deficient in β2 integrin surface expression revealed clustering of mutations in two highly conserved areas, suggesting that these two domains are of critical importance in intact heterodimer surface expression.[9] There is 1 region of 250 amino acids found in the more N-terminal part of the extracellular portion of the molecule which may be important in association with the α subunit to form the "head" of the heterodimer. The second region, also highly conserved, is the cysteine-rich region which is found in the more C-terminal region of the extracellular portion of β2. This domain with its many cysteine residues could provide the rigidity necessary for determining the β2 tertiary structure and thus may play an important role in providing the conformation needed for both optimal α/β association and ligand binding.

III. β2 INTEGRIN FUNCTION

The function of the β2 integrin family, like other integrins, is to mediate cell adhesion. Specific ligands, some expressed on other cell types, have been found for these leukocyte receptors; many are listed in Table 3. It is clear, however, especially for αM and αX, that not all the ligands have as yet been discovered. Several of the cell surface ligands for β2 integrins, including ICAM-1 (CD54) and ICAM-2, belong to a supergene family of immunoglobulin-like molecules (IgSF) (ICAM-3, a third ligand of αLβ2, is not a IgSF member). The various receptors involved in leukocyte trafficking, chemotaxis, and targeting will be briefly addressed later in this chapter. It has been found that many integrin molecules seem to have multiple ligand binding sites on their surface and that a ligand also may have multiple binding sites for different integrins. This allows the system tremendous flexibility in mediating adhesion.

αMβ2 provides an excellent model of functional diversity found in the β2 integrins. Epitope mapping of the α subunit, using a panel of functionally active MAbs to αM,

Table 3 Ligands to the β2-Integrin Receptors

β2 Integrin	Known Ligands
αLβ2	CD54(ICAM-1), ICAM-2, ICAM-3
αMβ2	iC3b, CD54, Factor X, Fibrinogen, Leishmania gp63, Histoplasma capsulatum (?polysaccharides), Lipopolysaccharides
αXβ2	iC3b, fibrinogen

revealed that this heterodimer has structurally and functionally distinct domains.[42] One group of antibodies inhibited iC3b-mediated functions but had no effect on so-called "adhesion-mediated" functions such as chemotaxis, aggregation, and adhesion to endothelial monolayers. Another set of MAbs had the reciprocal effects on function. Furthermore, it was found that the molecule exists in different forms or "activation states" as determined by new ligand and MAb binding to the heterodimer found only upon exposure to specific cytokines or cell activators such as PMA, ADP, or FMLP[43,44] or certain divalent cations.[45] The MAb 7E3, raised to the αII/β3 integrin on platelets, binds to αMβ2 only when the monocyte or neutrophil is first activated by ADP. Mn^{2+} alone can circumvent the need for receptor activation.[46] This is true also for the two ligands fibrinogen and Factor X. The binding site on the activated heterodimer for these two ligands must be spatially close to the binding site for iC3b, since iC3b can inhibit their binding to αMβ2.

αLβ2 plays an important role in lymphocyte adhesion to a variety of cells. *In vitro* functional studies demonstrate that MAbs to αL, if added early in culture, are capable of blocking the cell-cell adhesion step necessary for target cell killing.[47,48] Such MAbs also inhibit antigen-, mitogen-, or alloantigen-induced proliferation (all of which require contact of the T cell with an antigen presenting cell), T and B cell aggregation, T-helper activity for immunoglobulin production, and firm attachment to the endothelium. Leukocyte binding to endothelium is a sequential event with β2 integrins participating in the last, stabilizing step which allows transmigration through the endothelium.[49,50] Selective MAb binding to αLβ2 on only some T cells[51] confirms that this heterodimer undergoes conformational changes with "activation", similar to what has been found for αMβ2. This S6F1 MAb-defined epitope can be induced on a subpopulation of T cells (approximately 60% of $CD8^+$ T cells and 15% of $CD4^+$ T cells) upon activation in a mixed lymphocyte culture. One MAb (NKI-L16) "activates" αLβ2 function, acting either as a ligand or possibly stabilizing the "active form".[52,53]

Two lines of evidence suggest that β2 also is capable of activation in similar fashion to the α subunits. Stimulus-induced phosphorylation of the β2 cytoplasmic tail has been documented by several laboratories and is discussed in greater detail below. There is also MAb binding data to suggest conformational changes in β2 with activation. KIM127 is a MAb specific to β2, and its binding is only detectable at 37°C. MAb KIM185 is capable of inducing KIM127 binding at 4°C, suggesting that KIM127 does recognize an epitope on β2 which is exposed only under discrete conditions.[54] Such documented conformational changes undergone by members of the β2 family, which can be induced by environmental manipulation, strongly suggest that these adhesion receptors exist in both activated and nonactivated states.

IV. REGULATION OF β2 INTEGRIN FUNCTION: QUANTITATIVE AND QUALITATIVE CHANGES

Functional regulation of the β2 integrin family leading to increased cell adhesion is known to occur in at least two ways: an upregulation of the quantity of the receptor on the cell surface (predominantly seen with αMβ2, αXβ2) and activation of the receptor (as described above), which results in a conformational change of the receptor and increases ligand-receptor affinity. Evidence for both mechanisms exists.

The bulk of both αMβ2 and αXβ2 are found in intracellular pools in the unactivated neutrophil and monocyte. Several studies of αMβ2 have shown that this heterodimer is found on the surface of the peroxidase-negative specific granules.[55-59] Upon cell activation, cytoplasmic granules fuse with the cell membrane, and the αMβ2 found in these granules are brought to the cell surface membrane. Depending on the stimulus, the increase in cell membrane αMβ2 can be three- to tenfold the amount found in the resting cell and occurs within minutes. The same upregulation in surface αXβ2 also occurs, presumably by a similar mechanism. αLβ2 is not stored in intracellular pools, and therefore the very rapid increase in this surface receptor, seen with the other two β2 heterodimers, does not occur.[55]

The importance of a quantitative increase of αMβ2 on the capacity of the cell to modulate its adhesive function is derived from two lines of evidence. The first is decreased adhesive function of neutrophils from patients with specific granule deficiency as well as from normal neonates.[57,60] Neutrophils from both these sources cannot upregulate surface β2 integrin receptors, the result of which is insufficient neutrophil surface αMβ2 and αXβ2, resulting in a decreased ability of such patients to ward off bacterial infection. Another example is those patients with the partial form of β2 integrin deficiency. Our study of neutrophils from one patient with 15% of the normal quantity of β2 integrin on his leukocytes found that by use of a sensitive iC3b binding assay we were able to detect normal iC3b-mediated binding, suggesting that the αMβ2 receptor functioned normally, but that the quantity of surface β2 integrin found on the child's leukocytes was responsible for the leukocyte abnormalities measured *in vitro* and demonstrated clinically *in vivo*.[61]

There is overwhelming evidence suggesting that modulation of β2 integrin function also involves qualitative changes of these receptors. Increased β2 integrin-mediated adhesion can occur before or without change in the numbers of surface receptors. Such changes in "activation states" involve all three heterodimers and unlike αMβ2 and αXβ2, which are stored in intracellular pools, such changes are the only way of modulating αLβ2 within minutes, allowing leukocytes the capacity to react quickly to their environment. (Memory T cells do have increased surface αLβ2 on their surface, but this process is not rapid and most likely involves new protein synthesis.[62]) Qualitative changes of integrins may involve a number of different mechanisms resulting in changes in any part of the heterodimer: the cytoplasmic tail of one or both subunits, the extracellular domains of either subunit, or possibly the avidity of the α-β association and tertiary structure of the receptor. Any one of these changes would result in alterations of existing ligand-receptor affinities as well as new ligand or cation binding sites and could change the association of the receptor with other membrane receptors or structures as well as the heterodimer's association to cytoplasmic proteins such as cytoskeletal proteins or several of the many proteins involved in the signaling cascade.

Several observations point to at least two states of β2 integrin existence. The first has already been alluded to, the rapid changes in αLβ2-mediated function that cannot be accounted for by increased surface receptors. There is also a dissociation between the time in which increased αMβ2-mediated adhesion can be measured and the increased surface receptors are detected.[63] As already mentioned above, extracellular conformational changes are documented by the binding of new ligands or MAbs to the heterodimer on an activated cell, but not on the resting cell. The signal for conformational change may come from the extracellular environment by the binding of ligands to other membrane receptors, which then start a cascade of intracellular events, ultimately leading to a change in the β2 integrins. This may be the case with the S6F1 MAb binding site on αLβ2 which emerges during the mixed lymphocyte reaction.

Intracellular changes may also result in conformational changes, which allow for new ligand binding by the receptor. This may be the case with intracellular changes in ADP concentration, which result in the generation of the MAb 7E3 binding epitope on αMβ2 as well as the generation of fibrinogen and Factor X binding sites. Changes in the tertiary structure of the receptor would allow binding by a variety of different ligands and ultimately allow the cell to react to changes in the extracellular environment.

With the cloning and amino acid sequencing of all four subunits of the β2 integrins, it is known that several potential phosphorylation sites exist in all four cytoplasmic tails. Several studies have documented that the β2 integrin α subunit cytoplasmic tails are constitutively phosphorylated on their serine and threonine residues.[34–36] The β2 cytoplasmic domain, however, depends on an extrinsic stimulus for phosphorylation; this phosphorylation is transient and its time course is stimulus dependent. In monocytes, both PMA and FMLP induce phosphorylation of the β subunit tail within seconds, at serine, threonine, and tyrosine residues. Under these conditions no change in α subunit phosphorylation was detected. Whereas PMA directly activates protein kinase C (PKC), FMLP may work through the generation of diacylglycerol (DAG) which in turn activates, in part, PKC. Kinetic studies of β2 phosphorylation have shown PMA-induced phosphorylation to last for as long as 30 min, whereas FMLP-induced phosphorylation is much more transitory, lasting only minutes. The kinetics of β2 phosphorylation parallel the kinetics of the leukocyte adhesion observed *in vitro*. Taken together with the data derived from previous studies, in which decreased leukocyte adhesion was found in the presence of PKC inhibitors, these observations suggest that phosphorylation of β2 may play a role in β2-mediated functions such as adhesion, at least in response to PMA and FMLP. The role of α subunit phosphorylation and the response of β2 to other stimuli, most specifically the known ligands to these receptors, are still unknown. However, it is clear that the flexibility leukocytes need for chemotaxis, adhesion, transmigration through the endothelium, and targeting must require mechanisms that can be turned on and off rapidly in response to the extracellular environment. Phosphorylation/dephosphorylation of the cytoplasmic tails of these receptors may be one such system.

The importance of qualitative changes in β2 integrins is supported by some old and new observations. It has previously been shown that PMA induces the colocalization of αLβ2 and talin,[64] a cytoskeletal protein, and also induces surface aggregation of αMβ2 on the neutrophil surface.[65] The αMβ2-mediated adhesion observed secondarily to PMA stimulation can be inhibited by cytochalasin B.[66] These observations suggest an association of the β2 integrins with cytoskeletal proteins and that this association is crucial to the modulation of the β2 integrin-mediated functions. Recently, evidence from this laboratory has shown a direct association between the cytoplasmic tail of β2 and actin, α-actinin, and actin-binding protein-280.[41] Association of a select group of αMβ2 surface receptors with the cytoskeleton would offer one explanation for the observation that αMβ2 exists in two "forms" on the cell surface; one form is firmly anchored in the membrane and a second form is freely movable, allowing it to migrate to anti-αM MAbs attached to tissue culture plastic.[67] The ability of the integrins to exist in different forms (be it activation/deactivation and all that it implies in terms of receptor function and signaling) would allow for the rapidity of the flexibility found in leukocyte adhesion.

V. β2 INTEGRIN LIGANDS AND OTHER MEMBRANE ADHESION MOLECULES

MAb technology has uncovered many other leukocyte adhesion molecules. These antibodies made against either activated or non-activated endothelial cells and also against a variety of leukocytes and leukocyte cell lines were tested for their ability to inhibit adhesion. Using this approach, many other cell surface receptors were found, and the discovery of these receptors has allowed cell biologists to construct a working model of how these receptors function in concert to mediate adhesion, transmigration, and targeting.

Some other adhesion-mediating receptors on leukocytes are also integrins, such as the β1 family, also known as the VLA antigens.[68] These integrins are found on many cell types besides leukocytes, such as endothelial cells, fibroblasts, and platelets and play a vital role in the adhesion of cells to matrix, such as collagen, laminin, and fibronectin as well as to other cells. β1 integrins have diverse and vital roles early in life in the targeting of cells during morphogenesis and, later, in leukocyte homing. Other integrin families may also be found on some undifferentiated leukocyte cell lines, but their biological significance is not known.

A second gene family contributing several members important in leukocyte adhesion is the immunoglobulin-like (Ig) supergene family. ICAM-1 and -2 are two cell surface ligands for the β2 integrins. ICAM-1 (CD54) has five Ig-like domains; αLβ2 binds to the two most N-terminal Ig domains[69] whereas αMβ2 binds to the third Ig-like domain on the same molecule.[70] CD54 has a wide tissue distribution, but, significantly, is expressed in low amounts on resting endothelial cells, requires *de novo* synthesis, and maximally appears on the cell surface 12 to 24 h after stimulation by such inflammatory mediators as IL-1-β, TNF-α, and IFN-γ and is markedly diminished after 48 h.[71] CD54 also serves as the receptor for rhinovirus as well as one of two receptors for the infected erythrocyte stage of *P. falciparum*.[72,73] The malaria adhesins also seem to bind to the first Ig-like domain but to its "backside", which is 180° away from the αLβ2 binding site.[74] The rhinovirus competes for binding with αLβ2, suggesting that the binding sites on CD54 for αLβ2 and rhinovirus are either identical or at least in close proximity. ICAM-2 is a weak ligand for αLβ2, is constitutively expressed on endothelial cells, and cannot be upregulated during inflammation.[75,76] Of note is that soluble CD54 is found in the circulation in patients with various inflammatory conditions. It has been detected in such varied clinical states as heart transplants,[77] rheumatoid arthritis, Kawasaki's disease, and melanoma.[78,79]

ICAM-3 was originally found by adhesion studies of the SKW3 T cell lymphoma line to αLβ2.[80] It has a strong presence on all resting, nonactivated monocytic cells, much more than either CD54 or ICAM-2 and is not found on either resting or activated endothelial cells. Anti-ICAM-3 MAb strongly inhibit unactivated T cells from binding to αLβ2, suggesting a major role for ICAM-3 in αLβ2-mediated leukocyte binding to lymphocytes. The specific roles of ICAM-1, -2, and -3 in cell adhesion are further evidence of ligand-specific integrin-mediated binding to specific cell targets.

A second member of the Ig superfamily, VCAM-1, also plays an important role in leukocyte adhesion.[38] VCAM-1 binds to α4β1 (VLA-4, CD49d/CD29), a member of the β1 integrin family which is found on all leukocytes except neutrophils. VCAM-1 is not present on the quiescent endothelium and appears on the surface of cytokine-stimulated endothelial cells with similar kinetics to those of ICAM-1.[71]

A third family involved in leukocyte adhesion is referred to as the LEC-CAMs or selectins.[81] In this family, two members, E-selectin (ELAM-1) and P-selectin (GMP-140, CD62, PADGEM), are found on endothelial cells, and L-selectin (LAM-1, Mel-14, Leu8) is found on leukocytes. P-selectin is also stored in intracellular pools in platelets. Members of the selectin/LEC-CAM family share a Ca^{2+}-dependent lectin-like domain that makes up approximately the 200 amino acids in the N-terminal part of the molecule. The midregion of the proteins is homologous to epidermal growth factor, and the C terminus contains between two to nine cysteine-rich tandem repeats that are common to many of the complement regulatory proteins; hence; the name LEC-CAM.

P-selectin is found in the Weibel-Pallade bodies of endothelial cells and is translocated to the surface within seconds of endothelial stimulation by such mediators as thrombin, histamine, and leukotriene B_4 (LTB$_4$). *In vitro*, P-selectin is rapidly internalized, disappearing after 30 min.[81] One study has found that with endothelial activation by specific oxidants there is impaired P-selectin internalization, resulting in a more prolonged presence of P-selectin on the endothelial cell surface and suggesting that *in vivo*, in the presence of pathogens, P-selectin existence on the cell surface may follow different kinetics than those found *in vitro*.[82] Like the endothelial cell, the translocation of P-selectin from the platelet secretory granules to the surface occurs within seconds after stimulation; however, there is a sustained presence of P-selectin on the platelet surface.

E-selectin (ELAM-1) is not found on unactivated endothelial cells, but can be induced on the surface within 2 h by cytokines such as IL-1-β, LPS, and TNF-α. *In vitro*, its peak surface expression is found between 4 to 6 h after stimulation and disappears by 24 h (at a time when maximal cell surface ICAM-1 is found). The ligand for E-selectin is found on both neutrophils and monocytes and is the carbohydrate sialyl-Lewis^x (s-Le^x), a terminal carbohydrate which contains sialic acid and fucose and is found on glycolipids and glycoproteins in the leukocyte cell membrane.[83] P-selectin binds to both s-Le^x and Le^x.[83,84] Neutrophil activation is not necessary for adhesion to the E-selectin-positive endothelium. Studies using soluble E-selectin (with removal of the cytoplasmic and transmembrane portions) have found that E-selectin induces chemotaxis as well as the activation of β2 integrin molecules.[85] This is in distinction to P-selectin which has not been demonstrated to activate the β2 family.[86]

L-selectin (MEL-14, LAM-1, Leu-8, LECAM-1), unlike the other two members of the selectin family, is found on leukocytes and not endothelium. With leukocyte activation, the avidity of L-selectin to its ligand is rapidly increased (receptor activation?) and then L-selectin is rapidly cleaved near the transmembrane spanning region and shed from the neutrophil surface.[87,88] The ligands for L-selectin are found on activated endothelial cells and are thought to be variants of sLe^x. One candidate on activated endothelium may be E-selectin, where the sLe^x of the neutrophil L-selectin may function as one of the counterstructures for E-selectin.[89] The role of L-selectin on neutrophils appears quite different from that on lymphocytes. Unlike L-selectin on the neutrophil, lymphocyte L-selectin does not have sLe^x. In a select subset of lymphocytes, L-selectin appears to mediate homing of these cells to high endothelial cell venules of peripheral lymph nodes. How L-selectin can function as a homing receptor on these cells is an active subject of investigation. Certain addressins, unique to the peripheral lymph node, have been identified as members of the mucin family and one, sgp$_{50}$,[90] which serves as a ligand for L-selectin on lymphocytes can be specifically inhibited. Leukocyte adhesion to 24-h stimulated endothelium cannot be inhibited by a MAb specific for the sgp$_{50}$ addressin (MECA79), suggesting that other ligands to L-selectin exist on endothelial cells.[91]

VI. THE ADHESION CASCADE

The sequence of events that lead a leukocyte from the vascular space through the endothelium and to its final destination has been made clearer by recent experiments, including those which utilize time-lapse photography. These processes involve at least four groups of molecules: the selectins, the mucins, the integrins, and finally the Ig super-family.

It is believed that in most situations *in vivo* leukocytes do not initially engage the endothelium through their β2 integrin receptors, an association which results in strong adherence to and spreading on the vascular endothelial surface. Rather, local activation of the vascular endothelium results initially in the surface expression of selectins on both the endothelium and leukocytes. Within minutes following activation by such stimuli as thrombin, histamine, or LTB$_4$, P-selectin is released from the Weibel-Pallade bodies of endothelial cells, and L-selectin is found on the surface of leukocytes. The presence of P-selectin on the endothelial cell and L-selectin on the neutrophil surface is short lived. The disappearance of P-selectin on the endothelium is followed by the surface induction of another selectin, E-selectin, in response to a variety of cytokines responsible for mediating the inflammatory response. As mentioned, the ligands for the selectin family are carbohydrate moieties found on "flexible" protein cores. This flexibility allows for the capture of leukocytes, slowing them and bringing them in contact with the endothelium. This phenomenon has been termed "rolling".[92-95] The leukocytes are then "tethered" to the endothelium through association of the selectins with their "flexible" surface ligands.

Firm adhesion to the endothelium requires the activation of the integrins, primarily αLβ2 and α4β1 on lymphocytes, and αLβ2 and αMβ2 on phagocytic cells.[38] Activation and upregulation of the integrins increases their affinity to their ligands on the endothelial cells. Integrin activation is not necessarily dependent on the selectins, but some experiments show that soluble E-selectin can activate β2 integrins. The adhesion phenomenon is further amplified as some integrin ligands, such as CD54 and VCAM-1, members of the Ig supergene family, are also induced on the surface of endothelial cells by cytokine stimulation. The migration of leukocytes through the endothelium followed by chemotaxis to the target appears to be integrin-mediated as well, since anti-integrin MAbs can inhibit this leukocyte movement. It is of some interest to note that the presence of α4β1 on lymphocytes and its role in these cells' adhesion to endothelium may explain why patients suffering from β2 integrin deficiency present clinically with a phagocytic cell defect, but little functional consequence of β2 integrin deficiency on their T and B cells.[1] One can speculate that the α4β1 (and perhaps other as yet unidentified receptors) on their lymphocytes compensates for the missing or diminished αLβ2, whereas this redundancy is lacking phagocytic leukocytes.

VII. IMPLICATIONS FOR MODULATING IMMUNE RECEPTORS

Understanding the importance of the β2 integrins in the inflammatory response has led to efforts to manipulate these receptors *in vivo*. This has been done using murine MAbs directed to the alpha or beta chains in several animal models of neutrophil-induced tissue injury. In some models, such as tissue injury induced by ischemia-reperfusion secondary to vascular occlusion or thermal injury, reperfusion allows neutrophils to congregate and become activated by cytokines released locally in response to damaged tissue. The influx of neutrophils and their activation results in the release of a myriad of proteases and oxidants, eventually leading to further tissue injury.[96] In such animal models, anti-β2 integrin MAbs have been shown to be of benefit in decreasing tissue injury secondary to neutrophil influx.[97-99] In animal models of neutrophil-mediated acute glomerulonephritis, arthritis, and ulcerative colitis, these MAbs have also been demonstrated to be of benefit. Because of the success of these MAbs in animal studies, trials in humans using these or "humanized" anti-β2 integrin MAbs are in progress. Currently, an anti-CD54 MAb is being tested in patients receiving kidney allografts. Whether the favorable results found in phase 1 trials are related to the antibody's effects on rejection or inhibition of the host's reaction to the ischemia of the transplanted kidney is unclear.

The pivotal role of the β2 integrins in inflammation and other important biologic processes makes these receptors a focus for the development of novel approaches with potent and specific therapeutic agents for their clinical modulation. The involvement of integrins in such varied systems renders their further study and the elucidation of their ability to transmit information from the extracellular milieu into the cell, resulting in the activation of discrete cell functions critical to our understanding of both normal physiology and the pathophysiology of diverse disorders.

ACKNOWLEDGMENTS

We wish to thank Ms. Robin Parsons for secretarial help. The authors are supported by a Biomedical Science Award from the Arthritis Foundation and grants AI-21964 and AI-28465 from the National Institutes of Health and by a March of Dimes research grant.

REFERENCES

1. Arnaout, M. A., Structure and function of the leukocyte adhesion molecules CD11/CD18, *Blood* 75, 1037, 1990.
2. Arnaout, M. A., Leukocyte adhesion molecules deficiency: its structural basis, pathophysiology and implications for modulating the inflammatory response, *Immunol. Rev.*, 114, 145, 1990.
3. Arnaout, M. A., Pitt, J., Cohen, H. J., Melamed, J., Rosen, F. S., and Colten, H. R., Deficiency of a granulocyte-membrane glycoprotein (gp 150) in a boy with recurrent bacterial infections, *N. Engl. J. Med.*, 306, 693, 1982.
4. Dana, N., Todd, III, R. F., Pitt, J., Springer, T., and Arnaout, M. A., Deficiency of a surface membrane glycoprotein (Mol) in man, *J. Clin. Invest.*, 73, 153, 1984.
5. Dana, N. and Arnaout, M. A., Leukocyte adhesion molecular (CD11/CD18) deficiency, Kazatchine, M., Ed., in *Bailliere's Clinical Immunology and Allergy*, W.B. Saunders, New York, 1988.
6. Dana, N., Tennen, D., Clayton, L., Pierce, M., Lachmann, P., and Arnaout, M. A., Leukocytes from four patients with complete or partial Leu-CAM deficiency contain the common beta subunit precursor and beta subunit messenger RNA, *J. Clin. Invest.*, 79, 1010, 1987.
7. Kishimoto, T. K., Hollander, N., Roberts, T. M., Anderson, D. C., and Springer, T. A., Heterogenous mutations in the beta subunit common to the LFA-1, Mac-1, and p150,95 glycoproteins cause leukocyte adhesion deficiency, *Cell*, 50,1987.
8. Dimanche, M. T., LeDeist, F., Fischer, A., Arnaout, M. A., Griscelli, C., and Lisowska-Grospierre, B., LFA-1 beta-chain synthesis and degradation in patients with leukocytes adhesive protein deficiency, *Eur. J. Immunol.*, 17, 417, 1987.
9. Dana, N., and Arnaout, M. A., Leukocyte adhesion molecule deficiency, *Curr. Op. Pediatr.*, 2, 916, 1990.
10. Arnaout, M. A., Dana, N., Gupta, S. K., Tenen, D. G., and Fathallah, D. F., Point mutations impairing cell surface expression of the common β subunit (CD18) in a patient with Leu-CAM deficiency, *J. Clin. Invest.*, 85, 977, 1990.
11. Nelson, C., Rabb, H., and Arnaout, M. A., Genetic cause of leukocyte adhesion molecule deficiency: abnormal splicing and a misense mutation in a conserved region of CD18 impair cell surface expression of β2 integrins, *J. Biol. Chem.*, 267, 3351, 1992.

12. Law, S. K. A., Gagnon, J., Hildreth, J. E. K., Wells, C. E., Willis, A. C., and Wong, A. J., The primary structure of the beta subunit of the cell surface adhesion glycoproteins LFA-1, CR3 and p150,95 and its relationship to the fibronectin receptor, *EMBO J.*, 6, 915, 1987.

13. Kishimoto, T. K., O'Conner, K., Lee, A., Roberts, T. M., and Springer, T. A., Cloning of the beta subunit of the leukocyte adhesion proteins: homology to extracellular matrix receptor defines a novel supergene family, *Cell*, 48, 681, 1987.

14. Taniguchi-Sidle, A. and Isenman, D. E., Mutagenesis of the Arg-Gly-Asp triplet in human complement component C3 does not abolish binding of iC3b to the leukocyte integrin complement receptor type III (CR3, CD11b/CD18), *J. Bio. Chem.*, 267, 635, 1992.

15. Corbi, A. L., Miller, L. J., O'Connor, K., Larson, R. S., and Springer, T. A., cDNA cloning and complete primary structure of the subunit of a leukocyte adhesion glyco-protein, p150,95, *EMBO J.*, 6, 4023, 1987.

16. Arnaout, M. A., Gupta, S. K., Pierce, M. W., and Tenen, D. G., Amino acid sequence of the alpha subunit of human leukocyte adhesion receptor Mo1 (Complement receptor type 3), *J. Cell Biol.*, 106, 2153, 1988.

17. Corbi, A. L., Kishimoto, T. K., Miller, L. J., and Springer, T. A., The human leukocyte adhesion glycoprotein Mac-1 (complement receptor type 3, CD11b) α subunit, *J. Biol. Chem.*, 263, 12403, 1988.

18. Pytela, R., Amino acid sequence of murine Mac-lα chain reveals homology with the integrin family and an additional domain related to von Willebrand factor (VWF), *EMBO J.*, 7, 1371, 1988.

19. Hickstein, D. D., Hickey, M. J., Ozols, J., Baker, D. M., Back, A. L., and Roth, G. J., cDNA sequence for the αM subunit of the human neutrophil adherence receptor indicates homology to integrin α subunits, *Proc. Natl. Acad. Sci. U.S.A.*, 86, 257, 1989.

20. Larson, R. S., Corbi, A. L., Berman, L., and Springer, T., Primary structure of the leukocyte function-associated molecule–1 α subunit. An integrin with an embedded domain defining a protein superfamily, *J. Cell Biol.*, 108, 703, 1989.

21. Ignatius, M. J., Large, T. H., Houde, M., Tawil, J. W., Burton, A., Esch, F., Carbonetto, S., and Reichardt, L. F., Molecular cloning of the rat integrin α1-subunit: a receptor for laminin and collagen, *J. Cell Biol.*, 111, 709, 1990.

22. Takada, Y. and Hemler, M. E., The primary structure of the VLA-2/collagen receptor α2 subunit (plateletGPIa): homology to other integrins and the presence of a possible collagen-binding domain, *J. Cell Biol.*, 109, 397, 1989.

23. Argraves, W. S., Deak, F., Sparks, K. J., Kiss, I., and Goetinck, P. F., Structural features of cartilage matrix protein deduced from cDNA, *Proc. Natl. Acad. Sci. U.S.A.*, 84, 464, 1987.

24. Lambris, J. D., and Muller-Eberhard, H. J., Isolation and characterization of a 33,000-Dalton fragment of complement factor B with catalytic and C3b binding activity, *J. Biol. Chem.*, 259, 12685, 1984.

25. Shelton-Inoloes, B. B., Titani, K., and Sadler, J. E., cDNA sequences for human von Willebrand factor reveal five types of repeated domains and five possible protein sequence polymorphisms, *Biochemistry*, 25, 3164, 1986.

26. Bonaldo, P., Russo, V., Bucciotti, F., Bressan, G. M., and Colombatti, A., Alpha-1 chain of chick type VI collagen, *J. Biol. Chem.*, 264, 5575, 1989.

27. Wieslander, J., Langeveld, J., Butkowski, R., Jodlowski, M., Noelken, M., and Hudson, B. G., Physical and immunochemical studies of the globular domain of type IV collagen., *J. Biol. Chem.*, 260, 8564, 1985.

28. Yamagata, M., Yamada, K. M., Yamada, S. S., Shinomura, T., Tanaka, H., Nishida,

Y., Obara, M., and Kimata, K., The complete primary structure of type XII collagen shows a chimeric molecule with reiterated fibronectin type III motifs, von Willebrand factor A motifs, a domain homologous to a noncollagenous region of type IX collagen, and short collagenous domains with an Arg-Gly-Asp site, *J. Cell Biol.*, 115, 209, 1991.

29. Girma, J.-P., Meyer, D., Verweij, C. L., Pannekoek, H., and Sixma, J. J., Structure-function relationship of human von Willebrand factor, *Blood*, 70, 605, 1987.

30. Colombatti, A. and Bonaldo, P., The superfamily of proteins with von Willebrand factor type A-like domains: one theme common to components of extracellular matrix, hemostasis, cellular adhesion, and defense mechanisms, *Blood*, 77, 2305, 1991.

31. Hemler, M. E., VLA proteins in the integrin family: structures, functions, and their role on leukocytes, *Annu. Rev. Immunol.*, 8, 365, 1990.

32. Arnaout, M. A., Remold-O'Donnell, E., Pierce, M. W., Harris, P., and Tenen, D. G., Molecular cloning of the human and guinea pig alpha subunit of leukocyte adhesion glycoprotein, Mo1. chromosomal localization and homology to the alpha subunits of integrins, *Proc. Natl. Acad. Sci. U.S.A.*, 85, 2776, 1988.

33. Corbi, A. L., Larson, R. S., Kishimoto, T. K., Springer, T. A., and Morton, C. C., Chromosomal location of the genes encoding the leukocyte adhesion receptors LFA-1, Mac-1 and p150,95. Identification of a gene cluster involved in cell adhesion, *J. Exp. Med.*, 167, 1597, 1988.

34. Chatila, T., Geha, R. S., and Arnaout, M. A., Constitutive and stimulus-induced phosphorylation of CD11/CD18 leukocyte adhesion molecules, *J. Cell Biol.*, 109, 3435, 1989.

35. Buyon, J. P., Slade, S. G., Reibman, J., Abramson, S. B., Philips, M. R., Weismann, G., and Winchester, R., Constitutive and induced phosphorylation of the α- and β-chains of the CD11/CD18 leukocyte integrin family: relationship to adhesion-dependent functions, *J. Immunol.*, 144, 191, 1990.

36. Hibbs, M. L., Jakes, S., Stacker, S. A., Wallace, R. W., and Springer, T. A., The cytoplasmic domain of the integrin lymphocyte-function-associated antigen 1 β subunit: sites required for binding to intercellular adhesion molecule 1 and the phorbol ester-stimulated phosphorylation site, *J. Exp. Med.*, 174, 1227, 1991.

37. Nermut, M. V., Green, N. M., Eason, P., Yamada, S. S., and Yamada, K. M., Electron microscopy and structural model of human fibronectin receptor, *EMBO J.*, 7, 4093, 1988.

38. Hynes, R. O., Integrins: versatility, modulation, signaling and cell adhesion, *Cell*, 69, 11, 1992.

39. Horwitz, A., Duggan, K., Buck, C., Beckerle, M. C., and Burridge, K., Interactions of plasma membrane fibronectin receptor with talin—a transmembrane linkage, *Nature (London)*, 320, 531, 1986.

40. Tamkun, J. W., DeSimone, D. W., Fonda, D., Pateb, R. S., Buck, C., Horwitz, A. R., and Hynes, R. D., Structure of integrin, a glycoprotein involved in the transmembrane linkage between fibronectin and actin, *Cell*, 46, 271, 1986.

41. Sharma, C. P., Magil, S., and Arnaout, M. A., Microdomains in the cytoplasmic tail of CD18 involved in binding to cytoskeleton, *Clin Res.*, 40, 343, 1992.

42. Dana, N., Styrt, B., Griffin, G. D., Todd, III, R. F., Klempner, M. S., and Arnaout, M. A., Two functional domains in the phagocyte membrane glycoprotein Mo1 identified with monoclonal antibodies, *J. Immunol.*, 137, 3259, 1986.

43. Altieri, D. C., Bader, R., Mannucci, P. M., and Edgington, T. S., Oligospecificity of the cellular adhesion receptor MAC-1 encompasses an inducible recognition specificity for fibrinogen, *J. Cell Biol.*, 107, 1893, 1988.

44. Altieri, D. C. and Edgington, T. S., The saturable high affinity association of factor X to ADP-stimulated monocytes defines a novel function of the MAC-1 receptor, *J. Biol. Chem.*, 263, 7007, 1988.

45. Dransfield, I. and Hogg, N., Regulated expression of Mg^{2+} binding epitope on leukocyte integrin a subunits, *EMBO J.*, 8, 3759, 1989.

46. Altieri, D. C., Occupancy of CD11b/CD18 (Mac-1) divalent ion binding site(s) induces leukocyte adhesion, *J. Immunol.*, 147, 1891, 1991.

47. Hildreth, J. E. K. and August, J. T., The human lymphocyte function-associated (HLFA) antigen and a related function macrophage differentiation antigen (HMac-1): functional effects of subunit-specific monoclonal antibodies, *J. Immunol.*, 134, 3272, 1985.

48. Springer, T. A., Dustin, M. L., Kishimoto, T. K., and Marlin, S. D., The lymphocyte function-associated LFA-1, CD2 and LFA-3 molecules: cell adhesion receptors of the immune system, *Annu. Rev. Immunol.*, 5, 223, 1987.

49. Carlos, T. M. and Harlan, J. M., Membrane proteins involved in phagocyte adherence to endothelium, *Immunol. Rev.*, 114, 5, 1990.

50. Osborn, L., Leukocyte adhesion to endothelium in inflammation, *Cell*, 62, 3, 1990.

51. Morimoto, C., Rudd, C. E., Letvin, N. L., and Schlossman, S. F., A novel epitope of the LFA-1 antigen which can distinguish killer effector and suppressor cells in human CD8 cells, *Nature*, 330, 479, 1987.

52. Keizer, G. D., Visser, W., Vliem, M., and Figdor, C. G., A monoclonal antibody (NKI-L16) directed against a unique epitope of the α-chain of human leukocyte function-associated antigen 1 induces homotypic cell-cell interactions, *J. Immunol.*, 140, 1393, 1988.

53. van Kooyk, Y., Weder, P., Hogervorst, F., Verhoeven, A. J., van Seventer, G., te Velde, A. A., Borst, J., Keizer, G. D., and Figdor, C. G., Activation of LFA-1 through a Ca^{+2}-dependent epitope stimulates lymphocyte adhesion, *J. Cell Biol.*, 112, 345, 1991.

54. Robinson, M. K., Andrew, D., Rosen, H., Brown, D., Ortlepp, S., Stephens, P., Butcher, E. C., Antibody against the Leu-CAM β-chain (CD18) promotes both LFA-1 and CR3-dependent adhesion events, *J. Immunol.*, 148, 1080, 1992.

55. Arnaout, M. A., Spits, H., Terhorst, C., Pitt, J., and Todd, III, R. F., Deficiency of a leukocyte surface glycoprotein (LFA-1) in two patients with Mo1 deficiency: effects of cell activation on Mo1/LFA-1 surface expression in normal and deficient leukocytes, *J. Clin. Invest.*, 74, 1291, 1984.

56. Todd, III, R. F., Arnaout, M. A., Rosin, R. E., Crowley, C. A., Peters, W. A., Curnutte, J. T., and Babior, B. M., Subcellular localization of the subunit of Mo1 (Mo1 alpha; formerly gp110), a surface glycoprotein associated with neutrophil adhesion, *J. Clin. Invest.*, 74, 1280, 1984.

57. O'Shea, J. J., Brown, E. J., Seligman, B. E., Metcalf, J. A., Frank, M. M., and Gallin, J. I., Evidence of distinct intracellular pools of receptors for C3b and C3bi in human neutrophils, *J. Immunol.*, 134, 2580, 1985.

58. Bainton, D. F., Miller, L. J., Kishimoto, T. K., and Springer, T. A., Leukocyte adhesion receptors are stored in peroxidase-negative granules of human neutrophils, *J. Exp. Med.*, 166, 1641, 1987.

59. Fryer, D. R., Morganroth, M. L., Rogers, C. E., Arnaout, M. A., and Todd, III, R. F., Regulation of surface glycoproteins CD11/CD18 (Mo1, LFA-1, p150/95) by human mononuclear phagocytes, *J. Clin. Immunol. Immunopathol.*, 46, 272, 1988.

60. Anderson, D. C., Becker-Freeman, K. L., Heerdt, B., Hughes, B. J., Jack, R. M., and Smith, C. W., Abnormal stimulated adherence of neonatal granulocytes: impaired

induction of surface Mac-1 by chemotactic factors or secretagogues, *Blood*, 70, 740, 1987.

61. Shaw, S., Cepek, K. L., Murphy, E. A., Ressell, G. J., Brenner, M. B., and Parker, C. M., Molecular cloning of the human mucosal lymphocyte integrin αE subunit, *J. Biol. Chem.*, 269, 6016, 1994.

62. Sanders, M. E., Makgoba, M. W., Sharrow, S. O., Stephany, D., Springer, T. A., Young, H. O., and Shaw, S., Human memory T lymphocytes express increased levels of three cell adhesion molecules (LFA-3, CD2 and LFA-1) and three other molecules (UCHL1, CDw29, and Pgp-1) and have enhanced IFN-production, *J. Immunol.*, 140, 1401, 1988.

63. Buyon, J. P., Abramson, S. B., Philips, M. R., Slade, S. G., Ross, G. D., Weismann, G., and Winchester, R. J., Dissociation between increased expression of Gp165/95 and homotypic neutrophil aggregation, *J. Immunol.*, 140, 3156, 1988.

64. Burns, P., Kupfer, A., and Singer, S. J., Dynamic membrane-cytoskeletal interactions: specific association of integrin and talin arises in vivo after phorbol ester treatment of peripheral blood lymphocytes, *Proc. Natl. Acad. Sci. U.S.A.*, 85, 497, 1988.

65. Detmers, P. A., Wright, S. D., Olsen, E., Kimball, B., and Cohn, Z. A., Aggregation of complement receptors on human neutrophils in the absence of ligand, *J. Cell. Biol.*, 105, 1137, 1987.

66. Patarroyo, M., Jondal, M., Gordon, J., and Klein, E., Characterization of the phorbol ester 12,13-dibutyrate (P(Bu)2)-induced binding between human lymphocytes, *Cell. Immunol.*, 113, 278, 1983.

67. Graham, I. L., Gresham, H. D., and Brown, E. J., An immobile subset of plasma membrane CD11b/CD18 (Mac-1) is involved in phagocytosis of targets recognized by multiple receptors, *J. Immunol.*, 142, 2352, 1989.

68. Hemler, M. E., Huang, C., and Schwarz, L., The VLA protein family: characterization of five distinct cell surface heterodimers each with a common 130,000 molecular weight beta subunit, *J. Biol. Chem.*, 262, 7660, 1987.

69. Staunton, D. E., Dustin, M. L., Erickson, H. P., and Springer, T. A., The arrangement of the immunoglobulin-like domains of ICAM-1 and the binding sites for LFA-1 and rhinovirus, *Cell*, 61, 243, 1990.

70. Diamond, M. S., Staunton, D. E., Marlin, S. D., and Springer, T. A., Binding of the integrin Mac-1 (CD11b/CD18) to the third immunoglobulin-like domain of ICAM-1 (CD54) and its regulation by glycosylation, *Cell*, 65, 961, 1991.

71. Pober, J. S., and Cotran, R. S., The role of endothelial cells in inflammation, *Transplantation*, 50, 537, 1990.

72. Greve, J. M., Davis, G., Meyer, A. M., Forte, C. P., Yost, S. C., Marlor, C. W., Kamarck, M. E., and McClelland, A., The major human Rhinovirus receptor is ICAM-1, *Cell*, 56, 839, 1989.

73. Staunton, D. E., Merluzzi, V. J., Rothlein, R., Barton, R., Marlin, S. D., and Springer, T. A., A cell adhesion molecule, ICAM-1, is the major surface receptor for Rhino viruses, *Cell*, 56, 849, 1989.

74. Berendt, A. R., McDowell, A., Craig, A. G., Bates, P. A., Sternberg, M. J. E., Marsh, K., Newbold, C., and Hogg, N., The binding site on ICAM-1 for Plasmodium falciparum-infected erythrocytes overlaps, but is distinct from the LFA-1 binding site, *Cell*, 68, 71, 1992.

75. Staunton, D. E., Dustin, M. L., and Springer, T. A., Molecular characterization of ICAM-1 and ICAM-2: alternate ligands for LFA-1, *Fed. Proc.*, 3, a446, 1989.

76. Staunton, D. E., Dustin, M. L., and Springer, T. A., Functional cloning of ICAM-2, a cell adhesion ligand for LFA-1 homologous to ICAM-1, *Nature*, 339, 58, 1989.

77. Ballantyne, C. M., Mainolfi, R., Young, J. B., Windsor, N. T., Cocanougher, B., Lawrence, E. C., Anderson, D. C., and Rothlein, R., Prognostic levels of increased levels of circulating intracellular adhesion molecule-1 after heart transplant, *Clin. Res.*, 39, 285A, 1991.

78. Harning R., Mainolfi, E., Bystryn, J-C., Milagros, H., Merluzzi, V. J., and Rothlein, R., Serum levels of circulating intracellular adhesion molecule 1 in human malignant melanoma, *Cancer Res.*, 51, 5003, 1991.

79. Seth, R., Raymond, F. D., and Makgoba, M. W., Circulating ICAM-1 isoforms: diagnostic prospects for inflammatory and immune disorders, *Lancet*, 338, 83, 1991.

80. de Fougerolles, A. R. and Springer, T. A., Intercellular adhesion molecule 3, a third adhesion counter-receptor for lymphocyte function-associated molecule 1 on resting lymphocytes, *J. Exp. Med.*, 175, 185, 1991.

81. Zimmerman, G. A., Prescott, S. M., and McIntyre, T. M., Endothelial cell interactions with granulocytes: tethering and signaling molecules, *Immunol. Today*, 13, 93, 1992.

82. Patel, K. P., Patel, D., Zimmerman, G. A., Prescott, S. M., McEver, R. P., and McIntyre, T. M., Oxygen radicals induce human endothelial cells to express GMP-140 and bind neutrophils, *J. Cell Biol.*, 112, 749, 1991.

83. Polley, M. J., Phillips, M. L., Wayner, E., Nudelman, E., Singhal, A. K., Hakomori, S-E., and Paulson, J. C., CD62 and endothelial cell-leukocyte adhesion molecule 1 (ELAM-1) recognize the same carbohydrate ligand, sialyl-Lewis x, *Proc. Natl. Acad. Sci. U.S.A.*, 88, 6224, 1991.

84. Larsen, E., Palabrica, T., Sajer, S., Gilbert, G. E., Wagner, D. D., Furie, B. C., and Furie, B., PADGEM-dependent adhesion of platelets to monocytes and neutrophils is mediated by a lineage-specific carbohydrate, LNF III (CD15), *Cell*, 63, 467, 1990.

85. Lo, S. K., Lee, S., Ramos, R. A., Lobb, R., Rosa, M., Chi-Rosso, G., and Wright, S. D., Endothelial-leukocyte adhesion molecule 1 stimulates the adhesive activity of leukocyte integrin CR3 (CD11b/CD18, Mac-1, $\alpha_m\beta2$) on human neutrophils, *J. Exp. Med.*, 173, 1493, 1991.

86. Lorant, D. E., Patel, K. P., McIntyre, T. M., McEver, R. P., Prescott, S. M., and Zimmerman, G. A., Coexpression of GMP-140 and PAF by endothelium stimulated by histamine or thrombin: a juxtacrine system for adhesion and activation of neutrophils, *J. Cell Biol.*, 115, 223, 1991.

87. Kishimoto, T. K., Jutila, M. A., Berg, E. L., and Butcher, E. C., Neutrophil Mac-1 and Mel-14 adhesion proteins are inversely regulated by chemotactic factors, *Science (Washington, D.C.)*, 245, 1238, 1989.

88. Spertini, O., Kansas, G. S., Munro, J. M., Griffin, J. D., and Tedder, T. F., Regulation of leukocyte migration by activation of the leukocyte adhesion molecule-1 (LAM-1) selectin, *Nature*, 349, 691, 1991.

89. Picker, L. J., Warnock, R. A., Burns, A. R., Doerschuk, C. M., Berg, E. L., and Butcher, E. C., The neutrophil selectin LECAM-1 presents carbohydrate ligands to the vascular selectins ELAM-1 and GMP-140, *Cell*, 66, 921, 1991.

90. True, D. D., Singer, M. S., Lasky, L. A., and Rosen, S. D., Requirement for sialic acid on the endothelial ligand of a lymphocyte homing receptor, *J. Cell Biol.*, 111, 2757, 1990.

91. Imai, Y., Singer, M. S., Fennie, C., Lasky, L. A., and Rosen, S. D., Identification of a carbohydrate-based endothelial ligand for a lymphocyte homing receptor, *J. Cell Biol.*, 113, 1213, 1991.

92. Ley, K., Gaehttgens, P., Fennie, C., Singer, M. S., Laskey, L. A., and Rosen, S. D., Lectin-like cell adhesion molecule1 mediates leukocyte rolling in mesenteric venules *In vivo*, *Blood*, 77, 2553, 1991.

93. von Adrian, U. H., Chambers, J. D., McEvoy, L. M., Bargatze, R. F., Arfors, K-E., and Butcher, E. C., Two-step model of leukocyte-endothelial cell interaction in inflammation: distinct roles for LECAM-1 and the leukocyte β2 integrins *in vivo*, *Proc. Natl. Acad. Sci. U.S.A.*, 88, 7538, 1991.

94. Williams, A. F., Out of equilibrium, *Nature*, 352, 473, 1991.

95. Hogg, N., Roll, roll, roll your leukocyte gently down the vein . . . , *Immunol. Today*, 13, 113, 1992.

96. Malech, H. L. and Gallin, J. I., Neutrophils in human diseases, *N. Engl. J. Med.*, 317, 687, 1987.

97. Hernandez, L. A., Grisham, M. B., Twohig, B., Arfors, K.-E., Harlan, J. M., and Granger, D. N., Role of neutrophils in ischemia-reperfusion-induced microvascular injury, *Am. J. Physiol.*, 253, H699, 1987.

98. Simpson, P. J., Todd, III, R. F., Fantone, J. C., Mickelson, J. K., Griffin, J. D., and Lucchesi, B. R., Reduction of experimental myocardial reperfusion injury by a monoclonal antibody (anti-Mo1, anti-CD11b) that inhibits leukocyte adhesion, *J. Clin. Invest.*, 81, 624, 1988.

99. Vedder, N. B., Winn, R. K., Rice, C. L., Chi, E. Y., Arfors, K. E., and Harlan, J. M., A monoclonal antibody to the adherence-promoting leukocyte glycoprotein, CD18, reduced organ injury and improves survival from hemorrhagic shock and resuscitation in rabbits, *J. Clin. Invest.*, 81, 939, 1988.

100. Dana, N., Fathallah, D. F., and Arnaout, M. A., Mapping of mAb-defined epitopes in the leukocyte β2 integrin CD11b/CD18 to the unique A-domain, *Clin. Res.*, 38, 467, 1990.

101. Diamond, M. S., Garcia-Aguilar, J., Bickford, J. K., Corbi, A. L., and Springer, T. A., The I domain is a major recognition site on the leukocyte integrin Mac-1 (CD11b/CD18) for four distinct adhesion ligands, *J. Cell. Biol.*, 120, 1031, 1993.

102. Michista, M., Videm, V., and Arnaout, M. A., A novel divalent cation-binding site in the A domain of the B_2 integrin CR3 (CD11b/CD18) is essential for ligand binding, *Cell*, 72, 857, 1993.

Chapter 3

Platelet Membrane αIIbβ3 (Glycoprotein IIb-IIIa)

Joseph C. Loftus

CONTENTS

I. INTRODUCTION

The capacity of platelets to adhere to other platelets and to specific elements within the extracellular matrix is a control point for the maintenance of normal hemostasis. Within the vasculature, platelets circulate without apparent affinity for other platelets; however, in areas of vessel or tissue trauma, platelets will rapidly attach, spread, and aggregate to prevent excessive loss of blood. Failure of this remarkably fine tuned system is associated with a host of serious side effects including severe bleeding and thrombosis. The membrane glycoprotein αIIbβ3 (GP IIb-IIIa) complex plays a central role in the aggregation response by mediating the interaction of platelets with fibrinogen and other adhesive proteins in plasma including fibronectin and the von Willebrand factor. The original observations indicating the essential role αIIbβ3 plays in platelet aggregation were derived from studies of patients with the inherited bleeding disorder, Glanzmann's thrombasthenia. Platelets from affected individuals possess marked deficiencies in αIIbβ3 content and/or function and are characterized by a lack of adhesive protein binding and, subsequently, the absence of platelet aggregation. In addition to its role in hemostatic processes, studies of αIIbβ3 structure-function have had broad implications for cell adhesion in general with the realization of the existence of the integrin superfamily of functionally and structurally related adhesion receptors.[1-4] αIIbβ3 (GPIIb-IIIa in the platelet nomenclature) was a charter member of this adhesion receptor family, has figured prominently in its establishment, and has provided fundamental insights into the molecular mechanisms underlying cell adhesion. Indeed, αIIbβ3 was among the first of the integrins to be identified,[5] purified,[6] cloned, and sequenced,[7,8] and the first to be expressed in a completely recombinant form.[9]

II. αIIbβ3 (GLYCOPROTEIN IIb-IIIa) COMPLEX

αIIbβ3 is the major platelet membrane glycoprotein accounting for as much as 2% of the total platelet protein.[6,10] The two subunits exist as a noncovalently associated calcium-dependent heterodimer. Based on the results of the binding of radiolabeled monoclonal antibodies specific for the αIIbβ3 complex, normal platelets possess 40 to 50,000 copies of the receptor complex on the platelet surface.[11–14] In addition to the platelet surface membrane, αIIbβ3 has also been localized to the membranes of the surface-connected open cannalicular system and to the membranes of the platelet α granule.[15–17] Following agonist stimulation of platelets, this intracellular pool of αIIbβ3 complexes is recruited to the platelet surface where they participate in adhesive protein binding and platelet aggregation.[18]

αIIb and β3 noncovalently associate in a 1:1 stoichiometry to form a heterodimer with an apparent molecular weight of 265 kDa. The physical properties of the individual subunits and the receptor complex have been studied in detail.[6] β3 is a single chain polypeptide with an apparent molecular weight of 90 kDa on nonreduced SDS polyacrylamide gels. Following reduction of disulfide bonds, there is an apparent increase in the observed molecular weight to approximately 114 kDa, consistent with a high cysteine content. αIIb is the larger of the two subunits and exhibits a molecular weight of approximately 140 kDa on SDS polyacrylamide gels in the absence of reducing agents. αIIb is composed of two disulfide-linked subunits, an αIIb heavy chain with a molecular weight of 125 kDa, and an αIIb light chain of molecular weight 23 kDa, which are resolved on SDS polyacrylamide gels following reduction of disulfide bonds.[19]

Morphology of the αIIbβ3 complex has been investigated by electron microscopy utilizing rotary shadowing.[20] The results have provided a model of the receptor consisting of two domains. The first domain consists of an oblong globular head ~8 × 10 nm. The second domain consists of two rod-like tails ~14 to 17 nm long. At low detergent concentrations, the purified αIIbβ3 complexes aggregated into clusters with interactions between the tips of the tails. As is discussed later, the deduced sequence of both αIIb and β3 indicates the presence of a single transmembrane domain within each subunit. Therefore, this association of the purified receptors is likely mediated by hydrophobic interactions between the two transmembrane domains and suggests that each rod-like stalk represents a distinct portion of each subunit. The globular head domain would then be composed of portions of both αIIb and β3. That the globular head is composed of portions of both subunits is consistent with the results obtained in the peptide cross-linking studies (see below). Utilizing similar approaches, Nermut et al.[21] proposed a similar structural model of the fibronectin receptor. Studies on the tertiary structure of β3 utilizing proteolytic digestion and immunoblotting with specific monoclonal antibodies have contributed to the identification of the region of β3 which is contained within the globular head.[22,23] These studies have indicated that the amino terminal region of β3 is disulfide linked to the rest of the molecule in such a manner as to create a large disulfide loop of at least 325 amino acids. The importance of this loop in the formation of the ligand binding site is demonstrated by the fact that both functionally significant epitopes and the RGD peptide cross-linking site have been mapped to this loop (see below).

III. αIIbβ3 GENE STRUCTURE

A. β3 SUBUNIT (GLYCOPROTEIN IIIa)

The primary structure of β3 has been deduced from cDNA clones isolated from expression libraries constructed with mRNA from human umbilical vein endothelial cells[8] as well as from the human erythroleukemic (HEL) cell line.[24,25] These cDNAs contained a single open reading frame encoding 788 amino acid residues which, aside from sequence errors, was the same in both cell types. That the deduced sequences from HEL cells and endothelial cells are identical to authentic β3 present on platelets is supported by the observation that numerous sequences derived from proteolytic fragments of platelet

$\beta 3^{8,24,26}$ are contained within the deduced sequences. In addition, the sequence of fragments of platelet $\beta 3$ derived by amplification of platelet RNA utilizing the polymerase chain reaction matched exactly the cDNA derived sequence for these regions.[27,28] The deduced sequence of $\beta 3$ has the following characteristics. A 26-amino acid stretch beginning with methionine and consisting of predominantly hydrophobic amino acids that precedes the derived sequence of the amino terminus of platelet $\beta 3^{29}$ presumably represents a signal peptide and translation initiation site. This region is followed by an extracellular domain of 692 amino acids, a 29-residue transmembrane domain, and a cytoplasmic domain of 41 amino acids. The extracellular domain is remarkable for its high content of cysteine residues. There is a total of 56 cysteine residues in the mature protein. The majority of these residues, 31, are grouped within 4 homologous tandem repeats consisting of approximately 40 residues. Each repeat contains 8 cysteine residues except the first repeat which contains only 7. This motif of cysteine-rich repeats is a distinguishing feature of the integrin β subunits, and the location of individual cysteines is conserved amongst the known integrin β subunits.[30–37] The extracellular domain also contains six concensus sequences (N-X-S/T) for N-linked glycosylation. An additional site is found within the presumed cytoplasmic segment. Comparison of the sequence of $\beta 3$ with that of the other known integrin β subunits indicates an overall homology of 40 to 46%, except for $\beta 5$ with which it exhibits a homology of 56%. This is noteworthy since $\beta 5$, like $\beta 3$, forms a heterodimer with αv. However, the ligand recognition specificity of $\alpha v \beta 5$ is restricted to vitronectin, whereas $\alpha v \beta 3$ binds fibrinogen, fibronectin, and von Willebrand factor in addition to vitronectin.[38] Thus, those regions that exhibit differences are candidate regions for contributing to the ligand specificity.

The cytoplasmic domain of $\beta 3$ contains two tyrosine residues, one of which resides within a sequence similar to that which surrounds the site of tyrosine phosphorylation in the human epidermal growth factor receptor.[39] A tyrosine residue in a similar sequence located in the cytoplasmic domain of $\beta 1$ integrin[40] may be the site of tyrosine phosphorylation observed by Hirst el al.[41] in chicken cells transformed by Rous sarcoma virus. While phosphorylation of $\beta 3$ on tyrosine has not been described, it has been demonstrated that $\beta 3$ in intact platelets is phosphorylated, predominantly on threonine and to a lesser degree serine.[42,43] $\beta 3$ phosphorylation is increased following stimulation of platelets with agonists such as thrombin or phorbol 12-myristate 13-acetate (PMA). Prostacyclin, an inhibitor of platelet activation, does not induce phosphorylation of $\beta 3$. Thrombin-induced phosphorylation could be blocked by pretreatment of platelets with staurosporine, an inhibitor of protein kinase C. Further studies by Hillery et al.[44] determined the stoichiometry of $\beta 3$ phosphorylation in both resting and agonist-stimulated platelets. The stoichiometry of $\beta 3$ phosphorylation in resting platelets is low, increases minimally in activated platelets, and is unaffected by occupancy of the receptor by fibrinogen. These studies also suggested that the observed phosphorylation of $\beta 3$ in intact platelets is not directly mediated by protein kinase C. Thus, while the precise functional significance of $\beta 3$ phosphorylation remains to be determined, it is unlikely that it mediates conversion of $\alpha IIb\beta 3$ into an active receptor.

van Kuppeveldt et al.[45] identified an alternatively spliced form of $\beta 3$ from a placental cDNA library which contains a different cytoplasmic domain. Determination of the intron-exon structure of $\beta 3$ (see below) has demonstrated that this form arises from nonsplicing between exons 13 and 14. The first 20 amino acids of the cytoplasmic domain are identical to those of the previously identified form; however, the following 21 residues have been replaced with an alternative 13 amino acids that show little homology to the cytoplasmic domains of the other known integrin β subunits. Utilizing the technique of reverse transcriptase/polymerase chain reaction, this alternative form of $\beta 3$, referred to as $\beta 3'$, was found to be expressed at the RNA level in the human placenta as well as within the human MG63 osteosarcoma and HEL erythroleukemia cell lines. Northern blot analysis of mRNA from HEL cells and umbilical vein endothelial cells[25,46] has indicated that this variant form of β3 represents a minor form of β3, and its detection on the cell surface has not been described.

The organization of the gene for β3 has been determined. Utilizing somatic hybrid cell lines[47] as well as dual laser chromosome sorting[24] the gene for β3 has been localized to chromosome 17q21-23. The intron-exon structure of the gene was delineated from clones isolated from human genomic DNA libraries.[48] This study revealed that the gene encompasses approximately 46 kilobases and is divided into at least 15 exons. The ambiguity concerning the exact number of exons arises from the fact that the exon(s) containing the signal peptide and the 5′ untranslated regions have not been identified. A potential polyadenylation signal, which was not identified in any of the cDNA clones, was identified which would result in a full-length β3 transcript of 5.9 kb consistent with results of Northern blots. Examination of the intron-exon organization failed to identify any strict correlation between exons and potential functional domains of the mature protein. For example, the four cysteine-rich repeats are unevenly encoded within two exons. Nevertheless, it is noteworthy that the 63-amino acid residue fragment of β3 to which bound RGD peptides can be chemically cross-linked is contained entirely within exon 3 which encodes approximately 84 amino acids. As this region is amongst the most highly conserved regions of the integrin β subunits and has been implicated in the RGD recognition function, it will be interesting to observe whether the corresponding region within the other β subunits is also contained within a separate exon. Preliminary mapping and sequence analysis of β1 integrin has indicated that there is a good correspondence between the intron-exon junctions of the two genes[49] supporting a common evolutionary origin of the integrin β subunits. Further comparisons await the determination of the genomic structure of the other integrins.

B. αIIb SUBUNIT (GLYCOPROTEIN IIb)

Unlike β3 which is expressed on a number of different cell types, expression of αIIb is limited to cells of megakaryocyte lineage and to date has only been described as being expressed on the HEL cell line. The entire coding region of αIIb was deduced from cDNA clones derived from HEL cell mRNA.[50] Partial cDNA clones for αIIb were also isolated from a cDNA library prepared with mRNA from human megakaryocytes obtained from blood cell concentrates from patients with chronic myelogeneous leukemia.[51] The nucleotide sequence of the overlapping region from the two cell types was identical. The full length cDNA for αIIb contained a portion of the 5′ untranslated sequence, a short 3′ untranslated sequence, a polyadenylation signal, and a poly(A) tail. The single open reading frame encodes 1039 amino acids, which includes both the αIIb heavy and light chains. This is consistent with earlier studies which demonstrated that both subunits are derived from a αIIb precursor synthesized from a single mRNA sequence of approximately 3.4 kb.[46] The αIIb heavy chain consists of 859 amino acids and is preceded by a stretch of 30 hydrophobic amino acids with characteristics of a signal peptide. The αIIb light chain consists of 149 amino acids and contains a 26-amino acid residue transmembrane domain near its carboxyl terminus. The transmembrane domain is followed by a 20-residue cytoplasmic domain enriched in acidic residues. Five potential glycosylation sites are present in the extracellular domain, four located on the αIIb heavy chain and one on the αIIb light chain. The deduced sequence contains 18 cysteine residues with 15 located on the larger subunit and 3 located on the smaller chain. Calvette et al.[52] have established that each cysteine residue is disulfide bonded to its nearest neighbor in the amino acid sequence and therefore form relatively compact disulfide loops. This pattern would suggest that the disulfide bonds are more important in stabilizing local structure rather than the overall folding of the mature protein. αIIb heavy and light chains are joined by a single interchain disulfide bond $cys_{826}-cys_{880}$.

Alignment of the deduced sequence of αIIb with the sequences of related integrin α subunits, such as those for the fibronectin receptor and the vitronectin receptor, indicates that cysteines are located at homologous positions.[53] These results would suggest that the disulfide bonds in these subunits also utilize nearest neighbor pairing and that the overall protein folding may be similar.

Four stretches of twelve amino acids in the αIIb heavy chain possess homology to the helix-loop-helix structure, the "EF Hand",[54] which is characteristic of known Ca^{2+} binding proteins such as calmodulin and troponin C. However, these regions exhibit some divergence from the classic EF hand motif. Secondary structure predictions indicate that these αIIb potential Ca^{2+} binding sites lack the flanking α helices typically present in these structures. Moreover, the -Z coordination position in these αIIb sequences is occupied by a small hydrophobic residue in place of the glutamic acid which is invariant in the EF hand. This substitution is significant since the glutamate residue in the -Z position contributes both oxygen atoms in its carboxylate side chain to the coordination of the Ca^{2+} ion in a bidentate manner.[55] In this regard, these sequences more closely resemble that of the "lock-washer" Ca^{2+} binding site in the bacterial galactose-binding protein.[56] In this protein, the missing glutamate at the -Z position is contributed from elsewhere in the receptor chain. Thus, it is possible that the oxygenated residue required for the -Z position in these αIIb sequences may be supplied from another region of αIIb or perhaps from β3.

Similar to β3, alternative splicing of αIIb mRNA has been observed.[57] The alternative splice involves the splicing out or in of exon 27 which results in the removal of 34 amino acids from the extracellular domain of the αIIb light chain. This variant form of αIIb mRNA, which is not readily detectable on Northern blots, is present in the cultured HEL cell line as well as in platelets and megakaryocytes as determined by the polymerase chain reaction. Whether this alternative form of αIIb light chain is present on the platelet surface and its functional consequences has not been determined, heterogeneity, however, both in size and pI of isolated αIIb light chain has been described by two-dimensional gel electrophoresis.[58]

The organization of the αIIb gene has been characterized by Heidenreich et al.[59] The αIIb gene contains 30 exons and spans approximately 17 kb. The gene has been mapped to chromosome 17 by direct hybridization to chromosomes isolated by dual laser chromosome sorting.[46] As noted previously, the gene for β3 has also been mapped to the long arm of chromosome 17. Bray et al.[60] have demonstrated that the 2 genes are physically linked within a 260-kb fragment, with the αIIb gene located on the 3′ side of the gene for β3. This close physical linkage suggests expression of the two genes may be under the same regulatory mechanisms.

Similar to β3, there is no obvious correlation between exons and the proposed functional domains within the mature protein. Indeed, while the coding sequence for the first calcium binding domain is contained within a single exon, the coding sequences for the remaining three calcium binding domains is divided between two exons. A combination of primer extension and RNase protection analysis utilizing mRNA derived from HEL cells, megakaryocytes, and platelets has been used to map the transcription start site 32 nucleotides 5′ of the initiator methionine.[59,61] Analysis of the 5′ flanking sequence upstream of the transcription start site reveals the somewhat unusual feature that there are no canonical concensus sequences corresponding to the TATA or CAAT boxes at their characteristic locations.

IV. BIOSYNTHETIC PROCESSING AND SURFACE EXPRESSION

The biosynthetic pathway leading to surface expression of αIIbβ3 has been studied in cell-free translation systems with mRNA derived from the HEL cell line[62,63] and human megakaryocytes.[64] αIIb is initially synthesized as a single chain precursor polypeptide of molecular weight 130 kDa with high mannose oligosaccharides being added during translation. The proteolytic cleavage of pro-αIIb into heavy and light chains and the conversion of the high mannose oligosaccharides into complex type carbohydrates are independent events occurring in the same Golgi compartment.[65] The cleavage site of pro-αIIb was originally thought to occur following a ser-arg sequence located at position 871. This placement was based on the direct amino acid sequencing of the isolated αIIb light chain.[29] Cleavage at this point would be somewhat unusual as proteolytic cleavage of proproteins after pairs of basic amino acids is the most commonly used processing mechanism.[66] Two such dibasic sequences are located 16 amino acid residues upstream from the predicted cleavage point within the sequence lys $_{855}$-arg-asp-arg-arg $_{859}$. Immunological analysis utilizing antipeptide antibodies indicated that pro-αIIb is primarily cleaved within this sequence while, in a minor proportion of αIIb molecules, cleavage appeared to occur at both sites.[67] These results were supported by the comparison between the deduced sequence of human αIIb with rodent αIIb.[68] The rodent αIIb molecule, which also exists as a disulfide bonded heavy and light chain protein, shares an overall 78% identity with its human homologue. However, while the predicted ser-arg cleavage sequence is not present in rodent αIIb, the upstream Arg-Arg sequence is present. The cleavage site was definitively identified as occurring after arg$_{859}$ through the use of oligonucleotide-directed mutagenesis coupled with expression in heterologous cells.[69] Interestingly, this study also demonstrated that αIIb, which does not undergo endoproteolysis, was expressed on the cell surface, indicating that cleavage does not appear to be a prerequisite for complex formation and surface expression. Furthermore, heterodimers containing uncleaved αIIb together with β3 retained the capacity to bind fibrinogen in a solid phase assay.

β3 is not posttranslationally processed, but differential glycosylation of β3 has been observed depending upon whether it is complexed with αIIb or av. When associated with αIIb, one oligosaccharide chain is processed to the complex type, but not when associated with αv.[65]

The processing and surface expression of αIIbβ3 has also been examined through transfection of recombinant forms of αIIb and β3 into heterologous cell types.[9] Cotransfection of COS cells with cDNAs encoding both subunits resulted in the efficient processing and surface expression of αIIbβ3 (Figure 1). In contrast, when COS were transfected with cDNAs encoding individual subunits, no surface expression was detected although intracellular synthesis of each subunit was apparent as determined by immunoprecipitation of metabolically labeled cells. When cells were transfected with αIIb alone, αIIb did not undergo its characteristic post-translational cleavage, indicating that the presence of β3 and presumably complex formation is required for light-heavy chain cleavage (Figure 2). In addition to being unprocessed, this pro-αIIb appeared to be less stable as it disappeared more rapidly than when associated with β3. In subsequent studies of the expression of recombinant αIIbβ3,[70,71] transfection of selected human cell types with individual subunits resulted in surface expression as a result of heterodimer formation with endogenous β3 integrins. Thus, the correct processing, stability, and surface expression of αIIb requires β3. These findings provide a ready explanation for the coordinate lack of both αIIb and β3 from the platelets of thrombasthenic patients. Moreover, these results also provide an alternative explanation for the deficiency of both glycoproteins in patients who bear mutations affecting only one subunit.

V. ADHESIVE PROTEIN BINDING
A. LIGAND RECOGNITION SIGNALS
The binding of adhesive proteins to αIIbβ3 has been extensively reviewed.[72-74] A comprehensive review of these interactions is beyond the scope of this chapter; rather, what

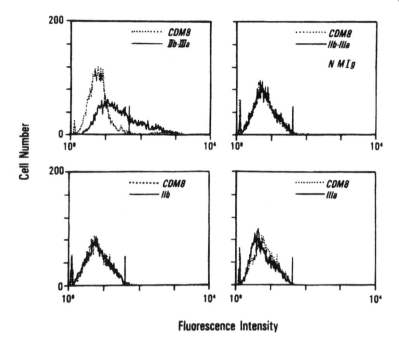

Figure 1 Flow cytometric analysis of COS cell transfectants. Full-length cDNA constructs encoding αIIb or β3 were transfected into COS cells individually or together by a DEAE-dextran procedure. Transfected cells were harvested after 48 h and stained with FITC-conjugated monoclonal antibodies specific for the individual receptor subunits or for the receptor complex. The FACS histograms obtained with specific transfectants (solid line) were superimposed upon the histograms obtained with wild-type CDM8 vector transfectants (dotted line). (From O'Toole, T., et al., *Blood*, 74, 14, 1988, with permission.)

follows is a brief review of the ligand recognition sequences. Two sets of peptides have been identified that define the recognition specificity of αIIbβ3 for adhesive proteins. The first set includes peptides that contain the tripeptide sequence arg-gly-asp (RGD). Originally identified as the sequence within fibronectin that mediates fibroblast adhesion,[75-77] subsequent studies demonstrated that RGD-containing peptides inhibited not only the binding of fibronectin to platelets[78] but also the binding of fibrinogen and von Willebrand factor.[79,80] Closely related to the RGD peptides are the "disintegrins", a family of low molecular weight proteins isolated from assorted snake venoms.[81-83] These highly homologous, disulfide-bonded proteins derive their inhibitory activity from the RGD sequence present within each protein; however, these proteins are 500 to 1000 times more potent than linear RGD-containing peptides in inhibiting fibrinogen binding and platelet aggregation. As reduction of the disulfide bonds results in a significant loss of activity, it is likely that the RGD sequence is conformationally constrained in these proteins in a manner that enhances their binding to αIIbβ3. Despite their increased potency, the therapeutic potential of these RGD-containing peptides may be limited due to their broad reactivity with other integrins. In contrast, a recently described venom isolate, barbourin, exhibits high specificity for αIIbβ3.[84] The structural feature responsible for this αIIbβ3 specificity is substitution of the RGD sequence by KGD. While linear KGD-containing peptides are poor αIIbβ3 inhibitors, these results indicate that the conformational presentation of the KGD sequence imparts potent αIIbβ3 antagonist activity. Indeed it has been previously demonstrated that the conformation of the RGD

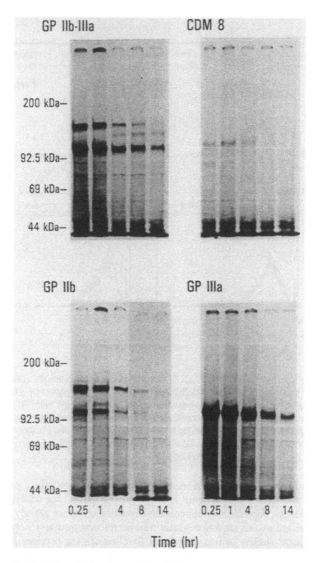

Figure 2 SDS-PAGE analysis of metabolically labeled COS cell transfectants. At 36 h after transfection, COS cells were metabolically labeled with ^{35}S-methionine for 10 min. At the indicated chase time points, cells were harvested and lysed in the presence of protease inhibitors. Cell extracts were immunoprecipitated with a polyclonal antiserum reactive with both subunits and analyzed by SDS-PAGE on a gradient gel under reducing conditions. The transfectant type is listed above each panel of the autoradiogram. (From O'Toole, T., et al., *Blood*, 74, 14, 1988, with permission.)

sequence within the peptide can alter the binding of these peptides to the fibronectin and vitronectin receptors.[85]

The second peptide set is the fibrinogen γ chain derivatives. Through proteolytic digestion fragments and synthetic peptide approaches it was determined that residues γ400 to 411 inhibited fibrinogen binding to platelets.[86,87] In addition to inhibition of fibrinogen binding, the γ chain peptide also inhibits the binding of fibronectin and von Willebrand factor to αIIbβ3 despite the fact that neither of these adhesive proteins contain homologous sequences. Furthermore, these two peptides inhibit the binding of each other. The basis for this cross-inhibition remains uncertain; however, the most likely explanation is that the two peptides bind to the same site or to a mutually exclusive site.[88]

B. THE ROLE OF DIVALENT CATIONS

The maintenance of the $\alpha IIb\beta 3$ complex is dependent upon Ca^{2+}.[89,90] This dependence on Ca^{2+} is illustrated by the fact that chelation of Ca^{2+} by EDTA or EGTA will dissociate the heterodimer into its constituent subunits.[91,92] This dissociation of the complex is accompanied by the loss of adhesive protein binding to the receptor and platelet aggregation. The Ca^{2+}-dependent structural transitions of the $\alpha IIb\beta 3$ complex have recently been investigated.[94] These investigators identified two Ca^{2+}-dependent transitions. The first transition, which occurred at about 60 μM Ca^{2+}, involved the dissociation of the complex into its subunits. This dissociation was dependent upon pH, temperature, and time. This dissociation could be prevented by Mg $^{2+}$; however, while Ca^{2+} was effective in promoting the reassociation of the subunits, Mg^{2+} could not induce the reassociation of the subunits. The second structural transition occurred at about 0.1 μM Ca^{2+} and involved the loss of the capacity of the dissociated subunits to reassociate into heterodimer complexes following the readdition of higher levels of Ca^{2+}. The failure to reassociate into a heterodimer appeared to be due to a Ca^{2+}-dependent change in $\beta 3$, resulting in the unfolding of the protein and the formation of high molecular weight aggregates. A tendency of αIIb to form aggregates during reassociation was also observed but at a lower Ca^{2+} concentration ($10^{-8}M$) than that observed for $\beta 3$. Therefore, although putative Ca^{2+} binding sequences have been identified in the sequence for αIIb, and direct Ca^{2+} binding to recombinant fragments of αIIb has been measured (see below), it appears that $\beta 3$ is also capable of binding Ca^{2+} and that this binding influences its structure.

As mentioned above, receptor structure and function is intimately associated with the capacity to interact with divalent cations; however, the precise number and location of these sites has remained largely undetermined. Early studies by Brass and Shattil[95,96] with intact platelets indicated that $\alpha IIb\beta 3$ contains 2 high affinity (kDa = 9 nM) and approximately 6 lower affinity (kDa = 400 nM) binding sites for Ca^{2+}. Recently, Rivas and Gonzalez-Rodriguez[97] have performed direct equilibrium dialysis measurements of the binding of Ca^{2+} to pure αIIb, $\beta 3$, and $\alpha IIb\beta 3$ in Triton X-100 solutions and to $\alpha IIb\beta 3$ reconstituted in liposomes. Their results indicate that both αIIb and $\beta 3$ in solution have low affinity (kDa = 200 to 300 μM) Ca^{2+} binding sites. The low affinity sites on αIIb were restricted to the αIIb heavy chain. In contrast, they determined that the $\alpha IIb\beta 3$ complex in solution has a single high affinity site (kDa = 80 nM) and 3 to 4 lower affinity sites (kDa = 40 μM). Reconstitution of $\alpha IIb\beta 3$ into liposomes decreased the kDa of the high affinity site (kDa = 9 nM) but did not affect the low affinity sites, indicating a stabilizing effect of the lipid bilayer on the structure of $\alpha IIb\beta 3$.

As determined from the deduced sequence of αIIb, there are four potential Ca^{2+} binding sites based on their homology to known Ca^{2+} binding regions of calmodulin. Gulino et al.[98] have verified the Ca^{2+} binding capacity of these domains through the use of recombinant technologies. A fragment of αIIb encoding residues 171 to 464 was expressed in a bacterial expression system resulting in the production of a 34-kDa protein that spans the 4 potential Ca^{2+} binding domains. The results of equilibrium dialysis experiments with the purified recombinant fragment demonstrate that the four binding sites can be occupied by Ca^{2+}. Ca^{2+} interacted with 2 sites of high affinity (with a kDa of 30 μM) and 2 sites of lower affinity (with a kDa of 100 μM). This interaction was competitively inhibited by Mg^{2+} and Mn^{2+}. The bacterially synthesized fragment of αIIb retained the capacity to interact with fibrinogen in a solid phase assay, and this interaction was maximal when all four sites were occupied. Occupancy of the high affinity sites alone was not sufficient to support fibrinogen binding consistent with the findings that optimal binding of fibrinogen[99-102] as well as RGD-containing peptides[103] requires millimolar concentrations of divalent cations. Taken together, these studies indicate that $\alpha IIb\beta 3$ receptor function is dependent on two distinct divalent cation requirements. The first requirement is in the μM range, is selective for Ca^{2+} and is related to maintaining the structural integrity of the heterodimer. The second requirement, fulfilled by mM concentrations of either Ca^{2+} or Mg^{2+}, is related to a low affinity cation binding site on the receptor being occupied for adhesive protein binding function.

VI. IDENTIFICATION OF FUNCTIONAL SITES WITHIN $\alpha IIb\beta 3$
A. THE USE OF MONOCLONAL ANTIBODIES

The identification of the domain(s) within $\alpha IIb\beta 3$ that mediate its interactions with fibrinogen as well as other adhesive proteins has utilized several approaches. The first of these approaches was to identify antibodies that would inhibit the binding of fibrinogen to platelets. Localization of the epitopes recognized by such antibodies would implicate these regions in the ligand recognition function. These regions would also then become candidates for mutagenesis studies utilizing recombinant forms of the receptor. The majority of the antibodies that have been identified that possess the ability to inhibit fibrinogen binding and platelet aggregation are operationally defined as "complex-specific" antibodies. This designation arises from the observation that while these antibodies react with the intact heterodimer, they fail to react with the individual subunits when the complex is dissociated and thus recognize a determinant formed by the association of the two subunits. This has precluded the localization of any of the epitopes recognized by these complex-specific antibodies. Therefore, it has not been determined whether the inhibitory activity of these antibodies results from a direct interaction with residues which comprise the ligand binding domain or is mediated indirectly through steric interactions. Nevertheless, two monoclonal antibodies have been identified which provide insight into the structural basis of ligand-receptor interaction. The first of these antibodies is PAC1, a monoclonal IgM antibody that binds to $\alpha IIb\beta 3$ only on activated platelets.[104] Further evidence that this antibody binds at or near the fibrinogen binding site on $\alpha IIb\beta 3$ is supported by the observations that fibrinogen and PAC1 compete with one another for binding to activated platelets and that the PAC1 binding is also inhibited by arg-gly-asp-containing peptides.[105,106] It was recently determined that the CDR3 of the PAC1 heavy chain contains a tripeptide sequence arg-tyr-asp and a synthetic peptide containing this sequence inhibited both the binding of PAC1 and fibrinogen to platelets.[107] A second monoclonal antibody, P37, completely inhibits ADP-induced platelet aggregation and recognizes an epitope which maps to a 23-kDa tryptic fragment of $\beta 3$ which contains the amino terminus[108] implicating this region in mediating fibrinogen binding.

B. THE USE OF SYNTHETIC PEPTIDES

Regions within $\beta 3$ have been further implicated in mediating the interaction with adhesive proteins through peptide cross-linking studies. Santoro and Lawing[109] demonstrated that the $\alpha IIb\beta 3$ complex on platelets was specifically labeled with photoactivatable derivatives of arg-gly-asp peptides and a dodecapeptide derived from the carboxyl terminus of the γ chain of fibrinogen. These authors reported that the RGDS probe labeled both αIIb and $\beta 3$, whereas the γ chain peptide probe specifically labeled only αIIb. D'Souza et al.[110] utilized a chemical cross-linking approach to demonstrate that an RGD-containing peptide could be cross-linked to both αIIb and $\beta 3$, but cross-linking to $\beta 3$ was preferential. In both cases, the cross-linking was saturable, activation dependent, and cation dependent. The residues within $\beta 3$ to which the specifically bound RGD peptides become cross-linked have been mapped[111] through a combination of proteolytic digestion and amino acid sequencing of peptide fragments to a discrete fragment of β3 defined as asp_{109}-glu_{171}. Alignment of this region of β3 with the corresponding regions of the other known integrin β subunits indicates that it is highly conserved. The identity within this region approaches 75% compared to an overall homology of approximately 45%, suggesting that the related sequences in the other β subunits may also be involved in their respective adhesive functions.

The results of the RGD peptide cross-linking studies together with the localization of the P37 epitope to the amino terminal portion of β3 support the importance of this region in mediating fibrinogen binding to the receptor. To further localize the fibrinogen binding site within the amino terminal portion of β3, Charo et al.[112] tested the capacity of a set of overlapping synthetic peptides spanning residues 1 to 288 of β3 to inhibit the binding of fibrinogen to purified αIIbβ3 in a solid phase microtiter assay. A peptide corresponding to $β3_{211-222}$ blocked the binding of fibrinogen to purified αIIbβ3. This peptide also blocked the binding of other adhesive proteins to αIIbβ3 including fibronectin, vitronectin, and von Willebrand factor and therefore may represent a portion of the adhesive protein binding domain of αIIbβ3. As was the case with the RGD cross-linking region, the $β3_{211-222}$ peptide sequence is also highly conserved among the integrin β subunits.

A similar cross-linking approach was used to localize the binding site on αIIbβ3 of the fibrinogen γ chain peptide.[113] The cross-linking site of the γ chain peptide was pinpointed to a 21-amino acid stretch of αIIb defined as ala_{294}-met_{314}. This location is particularly noteworthy as the identified region spans the second calcium binding repeat in αIIb. As mentioned earlier, these regions are highly conserved among the integrin α subunits, and the adhesive function of integrins is divalent cation dependent. An 11-residue peptide derived from this fragment of αIIb inhibits platelet aggregation and the binding of fibrinogen to platelets and to purified αIIbβ3 in a solid phase assay.[114] In addition, a direct interaction of fibrinogen with the peptide was demonstrated. These results further implicate this region in the ligand binding function of αIIbβ3.

β3 also functions as the β subunit of the vitronectin receptor. These two receptors bind many of the same adhesive protein ligands; however, it has been previously observed that a peptide derived from the fibrinogen γ chain will elute αIIbβ3, but not αvβ3, from an RGD affinity column,[115] suggesting differential peptide recognition by these two receptors. Recombinant β3 integrins expressed in heterologous cells also exhibited different peptide recognition specificities.[116] In particular, the fibrinogen $γ_{402-411}$ peptide was recognized better by αIIbβ3 than by αvβ3. In addition, endothelial cells which contain αvβ3 fail to adhere to fibrinogen γ chain peptides.[117,118] Since these receptors share the same β subunit, these experiments suggest that the ligand recognition specificities of these integrins is attributable to the α subunits. Inspection of the region of αv corresponding to $αIIb_{294-314}$ shows significant sequence substitutions. In particular, the replacement of αIIb Gly_{301} with an asp in the αv sequence appears significant since this Gly residue is invariant in the EF loop motif and contributes a β turn to that structure.[119] It is possible that this one change may alter the structure of this region in $α_v$ vs. αIIb; however, whether this difference is responsible for the different specificities will require additional studies. These questions could be addressed by swapping corresponding regions of αIIb and αv. The effect of these swaps on the ligand binding specificities of the chimeric receptors could then be directly compared following expression in heterologous cells.

C. THE USE OF NATURAL RECEPTOR VARIANTS

The ligand recognition sites of αIIbβ3 can be further characterized by the identification of mutations which disrupt ligand binding. Glanzmann's thrombasthenia is an autosomal recessive disorder characterized by prolonged bleeding time and lack of platelet aggregation.[5,120,121] Affected individuals exhibit either quantitative or qualitative defects in αIIbβ3.

The more prevalent type is characterized by a marked deficiency of αIIbβ3 on the platelet surface. Typically, the amounts of αIIb and β3 are coordinately decreased in these individuals. Indeed, the molecular basis of the defect in several of these patients has been elucidated. In each of these cases, a mutation affecting either αIIb or β3 resulted in the marked reduction of the αIIbβ3 complex on the cell surface.[122–124] This is not unexpected since expression studies[9,70,71] have demonstrated that the expression of αIIbβ3 requires both subunits. Thus, a mutation affecting the biosynthesis or processing of either of the two subunits would prevent the expression of the receptor on the platelet surface. In addition to those individuals possessing quantitative defects in the expression of αIIbβ3, functional variants possessing near normal levels of αIIbβ3 have been described[125–128] indicating an intrinsic dysfunction of αIIbβ3. Individuals with intrinsic dysfunction of αIIbβ3 would be predicted to have defects in either activation, ligand binding, or post-occupancy events. Thus, elucidation of the molecular defects in thrombasthenia due to dysfunctional αIIbβ3 is a powerful strategy to detect functionally significant mutations. The molecular basis for the defect in two of these variants has recently been determined and provides additional insight into the ligand binding domains of αIIbβ3. The first variant, designated CAM, is characterized by the inability of αIIbβ3 to recognize macromolecular[125] or synthetic peptide ligands.[129] In addition, αIIbβ3 from these patients exhibits abnormal reactivity with the monoclonal antibody PMI-1.[130,131] This antibody recognizes a defined epitope[132] on αIIb whose expression is modulated by divalent cations and inversely correlates with the capacity of the receptor to bind fibrinogen. At millimolar concentrations of Ca^{2+} or Mg^{2+}, which maximally support fibrinogen binding, the epitope is fully suppressed. Conversely, at divalent cation concentrations below 10 μM, which do not support fibrinogen binding, the epitope is fully expressed. These characteristics indicate that the mutation in the CAM receptor leads to defects in binding of both divalent cations and primary ligands. The structural basis of this defect was determined by amplification of mRNA isolated from platelets of affected individuals and found to be caused by a G-T mutation that resulted in the substitution of arg_{119} by Tyr in mature β3.[133] The sequence of a fragment of αIIb, ser_{226}-val_{454}, that contains the four divalent cation bind sites was indistinguishable from that present in normal individuals. That this single base change alone was sufficient to produce the CAM phenotype is illustrated by the results of expression studies with recombinant forms of the receptor following transfection into Chinese hamster ovary cells. The single base change alone did not affect surface expression of αIIbβ3; however, the mutant receptor failed to bind representative RGD peptide ligands as assessed by flow cytometry with the MAb anti-LIBS1[131] that recognizes an epitope on β3 that is dependent on occupancy of the receptor by the ligand (Figure 3). Furthermore, the mutant receptor had a divalent cation binding defect similar to that of the platelets of affected individuals. The mutation also did not impair complex formation of β3 with αv; however, this complex also failed to bind RGD ligands (Figure 3). Both mutant receptors also failed to bind to immobilized RGD peptides. These results indicate that this single amino acid substitution in β3 resulted in the loss of the RGD recognition function in both of the β3 integrins.

The affected residue, asp_{119}, is in close proximity to lys_{125}, the residue proposed to be cross-linked to the bound RGD peptide.[111] Alignment of this region of β3 with the corresponding regions of all the known β integrins shows that this asp residue is absolutely conserved (Figure 4). In addition, asp_{119} is located within a conserved cluster of oxygenated residues whose linear spacing approximates that of the residues in the calcium binding loop of EF hand proteins.[54,119] Significant sequence substitutions within this sequence suggest that the structure of this region will differ from the EF loop. Thus, further mutational analysis will be required to assign function to the other residues. Nevertheless, the proximity of the ligand binding site and this hypothesized cation

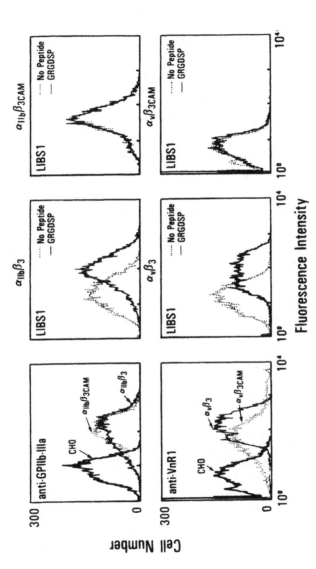

Figure 3 Flow cytometric analysis of the capacity of transfected cells to bind RGD ligand. FACS was performed on stable CHO cell lines containing the indicated transfected integrin. The GRGDSP peptide was at a final concentration of 1 mM and was added together with the anti-LIBS1 antibody (0.1 mM) in first incubation. Results are expressed as histograms of cell numbers (linear scale) on the ordinate softies vs. fluorescence intensity (log scale) on the abscissa. The reporting anntibodies are listed in the upper left of each panel. (From Loftus, J., et al., Science, 249, 915, 1990, with permission.)

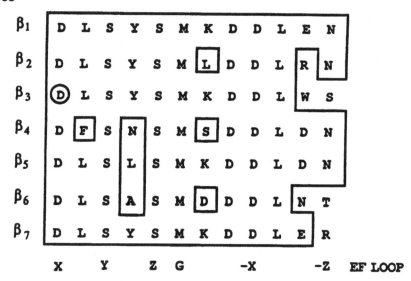

Figure 4 Alignment of the region of β3 containing the Cam mutation with the corresponding region from the other integrin β subunits. Asp₁₁₉, substituted by Tyr in Cam β3, is circled. Note the absolute conservation of this asp residue and the very high identity within this region. The consensus sequence for the EF Loop is shown on the bottom line. X, Y, Z, -X, and -Z denote the positions of oxygenated residues which coordinate the divalent cation. G is the invariant glycine which contributes a β turn to the loop.

binding site suggests the possibility of direct interactions between the bound cation and ligand which has been suggested for other integrins.[134,135] Moreover, the binding of ligands to αIIbβ3 modulates the binding of the monoclonal antibody PMI-1.[131] suggesting that ligand binding may perturb the interaction between bound divalent cation and the receptor. Given that, the function of all integrins is divalent cation dependent[99,102,103,136,137] and the conservation of the region surrounding asp₁₁₉ suggests that this may be a general feature of ligand binding to integrins and predicts that β subunit mutations affecting the asp residue corresponding to asp₁₁₉ in β3 would inhibit ligand binding function in other integrins. Preliminary support for this hypothesis has been provided by the observation that substitution of the corresponding residue in β1 integrin, asp₁₃₀, blocks the binding of α5β1 to the 110-kDa cell binding fragment of fibronectin as judged by ligand affinity chromatography.[138] Interestingly, this mutation also blocked the interaction of α5β1 with the bacterial protein invasin which occurs in a non-RGD dependent manner.[139] These results suggest that asp₁₃₀ is critical in the α5β1-ligand interaction regardless of whether the interaction is RGD recognition specific or not.

The identification of a point mutation in a second Glanzmann's variant has helped to further define a ligand binding domain in αIIbβ3 . Platelets from patient ET contain near normal levels of αIIbβ3 but fail to bind fibrinogen and aggregate following agonist stimulation.[128] The nature of the defect was defined by sequence analysis of amplified ET genomic DNA as a single G-A base change which encoded substitution of arg₂₁₄ by Gln in mature β3.[140] The affected residue is particularly noteworthy since it resides within the sequence of a synthetic peptide (β3₂₁₁₋₂₂₂) previously demonstrated to bind fibrinogen and inhibit fibrinogen binding to αIIbβ3.[112] Introduction of this point mutation into recombinant wild-type αIIbβ3 expressed in CHO cells reproduced the ET platelet deficits in the binding of fibrinogen and synthetic peptide ligands (Figure 5). Furthermore, the substitution of arg₂₁₄ by Gln in the synthetic peptide containing the β3₂₁₁₋₂₂₂ sequence

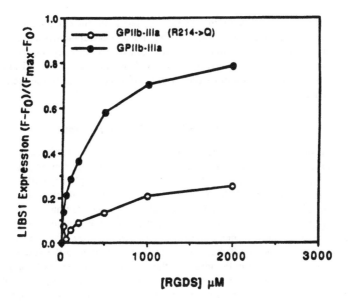

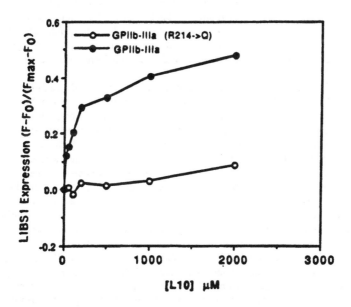

Figure 5 Flow cytometric analysis of the binding of peptides to recombinant αIIbβ3 and αIIbβ3 (arg$_{214}$→gln). The binding of ligand mimetic peptides RGDS or fibrinogen γ chain peptide L10 (LGGAKQAGDV) to recombinant receptors was analyzed with the monoclonal antibody LIBS1 which preferentially recognizes the occupied conformer of the receptor. Anti-LIBS1 binding was determined in the presence of the indicated dose of peptides. Results are depicted as $(F - F_0)/(F_{max} - F_0)$. F = mean fluoresence intensity, F_0 = mean fluoresence intensity in the absence of peptide, and F_{max} = mean fluoresence intensity of MAb 15 (anti-β3) binding to the wild type or mutant recombinant receptors in order to normalize to the total number of receptors per cell line. (From Bajt, M., et al., *J. Biol. Chem.*, 267, 3789, 1992, with permission.)

resulted in decreased ability of this peptide to block fibrinogen binding to purified αIIbβ3. The identification of this region by two independent approaches provides strong evidence for its role in ligand binding function. Since two other regions ($\alpha IIb_{294-314}$ and $\beta 3_{109-171}$) have also been implicated in ligand binding function by several independent approaches, it is likely that multiple sites within αIIbβ3 cooperate in the recognition of macromolecular ligands.

VII. AFFINITY MODULATION OF αIIbβ3

αIIbβ3 on resting circulating platelets will avidly bind small peptide ligands but exhibits very low affinity for soluble macromolecular ligands such as fibrinogen. Following stimulation with agonists such as thrombin, ADP, or epinephrine, αIIbβ3 will bind fibrinogen from solution with high affinity. The changes in platelets responsible for activation of αIIbβ3 is not fully understood. Intracellular signal transduction pathways involving G proteins, calcium, phospholipid metabolism, and kinases have been implicated in the acquisition of receptor competence.[141] Two theories have been proposed to explain the changes in αIIbβ3 that are required to convert it to the active state. The first theory is based upon the the idea that the microenvironment surrounding αIIbβ3 is altered following platelet activation facilitating access of large macromolecules to the ligand binding pocket. This theory presupposes that the receptor is functional on resting platelets but is unaccessible to macromolecular ligands because of their size. This model is supported by two experimental results. First, as already mentioned, small peptide ligands bind to the receptor on unstimulated platelets. Second, immunochemical analysis has demonstrated increased access of macromolecules following platelet activation.[142,143] The second theory is that platelet activation induces conformational changes in the αIIbβ3 complex. Conformational changes in the receptor have been detected immunochemically.[104,144-147]

Results of recent studies utilizing recombinant αIIbβ3 expressed on a different cellular background have helped to discriminate between these possibilities. Recombinant αIIbβ3 is expressed in an inactive state in heterologous cells and does not avidly bind soluble macromolecules.[9,71] However, it will mediate cell adhesion to immobilized fibrinogen[71] much as resting platelets adhere specifically to fibrinogen.[148,149] While recombinant αIIbβ3 cannot be activated with traditional platelet agonists, high affinity fibrinogen binding could be induced by certain anti-β3 monoclonal antibodies.[116] These activating antibodies are effective on both recombinant and native αIIbβ3 and exert their effect by acting directly on the receptor (Figure 6). These antibodies activated the receptor even in fixed cells which presumably lack physiologic signal transduction mechanisms. Moreover, these antibodies also activated the solubilized receptor. This would preclude a critical role for the membrane microenvironment. Similarly, Charo et al.[112] have observed that isolated, purified αIIbβ3 immobilized on microtiter wells will specifically and saturably bind fibrinogen. Thus, activation of αIIbβ3 is not an inherent property of platelets or their microenvironment but is an inherent property of the receptor itself.

What is the mechanism of the inside out transmembrane signal that results in the conformational change? O'Toole et al.[151] have recently shown that truncation of the cytoplasmic domain of αIIb, but not truncation of the cytoplasmic domain of β3, resulted in a receptor that was constitutively active with regard to the binding of soluble macromolecular ligands such as fibrinogen and the Mab PAC1. Replacement of the αIIb cytoplasmic domain with the α5 cytoplasmic domain failed to reverse the increased affinity. These results suggest that sequence(s) within the cytoplasmic domain of αIIb control the affinity state of αIIbβ3. One potential explanation is that these sequences may bind to an intracellular moiety to maintain the unactivated conformation. Modification of this moiety during activation could disrupt this interaction with αIIb, thereby inducing

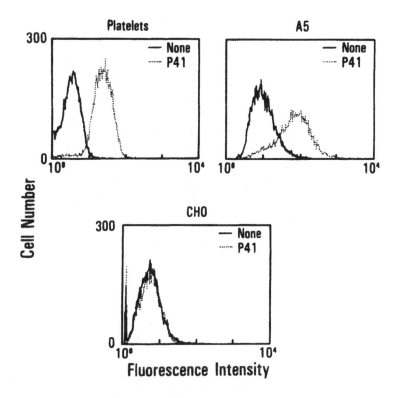

Figure 6 Activation of αIIbβ3 with a monoclonal antibody. The binding of FITC-conjugated monoclonal antibody PAC1 to platelets or stably transfected CHO cells expressing recombinant αIIbβ3 (A5 cells) was examined by flow cytometry. The solid line denotes binding of PAC1 in the absence of agonist, whereas the dotted line denotes binding in the presence of the anti-β3 monoclonal antibody P41. Both platelets and A5 cell transfectants bind PAC1 in the presence of P41. No binding is observed with wild-type CHO cells. (From O'Toole, T., et al., *Cell Reg.*, 1, 883, 1990, with permission.)

the high affinity state. The pathway leading to this modification may be cell type-specific since platelets, but not CHO cells, modulate αIIbβ3 in response to agonists. The role of the cytoplasmic domains in affinity modulation is further supported by the recent identification of a point mutation in the cytoplasmic domain of β3 in a variant type of Glanzmann's thrombasthenia.[151] Substitution of ser_{778} by Pro in the cytoplasmic tail of β3 is associated with the loss of fibrinogen binding following agonist stimulation. In contrast to the previously described variants which lack ligand recognition, αIIbβ3 from this patient retained the capacity to bind to an immobilized RGD peptide affinity column. Since phosphorylation of β3 on serine has been demonstrated, these results suggest that the cytoplasmic tail of β3 may be directly involved in activation of the receptor complex. Alternatively, the cytoplasmic domain of β3 may be involved indirectly via a structural motif disrupted by the significant Ser to Pro substitution.

Du et al.[152] recently demonstrated that the binding of ligand-mimetic peptides to αIIbβ3 provokes a change in the receptor associated with the acquisition of high affinity fibrinogen binding function. The structural specificity and dose response of the ligand-mimetic peptide activation of αIIbβ3 are similar to those for the inhibition of fibrinogen binding, suggesting that both events are a consequence of occupancy of the same ligand binding pocket. The activating effect of these peptides was observed with the receptor

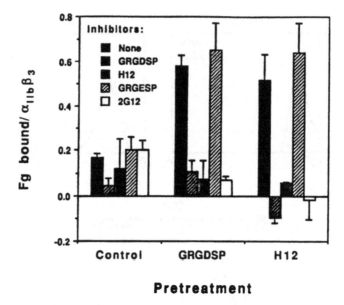

Pretreatment

Figure 7 Ligands activate purified αIIbβ3. αIIbβ3 purified from platelet membranes was incubated with buffer (control), with the peptide GRGDSP (1 m*M*), or with the peptide HHLGGAK-QAGDV (H12)(1 m*M*) and then was immobilized onto microtiter wells precoated with a monoclonal antibody against β3. ^{125}I-labeled fibrinogen (50 n*M*) binding was measured in the absence of inhibitor (none), in the presence of 0.5 m*M* concentrations of peptides GRGDSP, H12, or GRGESP, and in the presence of anti-αIIbβ3 monoclonal 2G12 (0.5 μ*M*). (From Du, X., et al., *Cell*, 65, 1, 1991, with permission.)

in fixed platelets as well as with the purified receptor (Figure 7). These results indicate that peptide activation of the receptor does not require any signal transduction event but rather appears to be due to a conformational change in the receptor itself. That occupancy of αIIbβ3 by ligand mimetic peptides can induce conformational changes in the receptor has been previously demonstrated both biophysically[153] and immunochemically.[131,154,155] Therefore, it is likely that the RGD sequences in ligands function both as part of the binding sites and as the trigger of secondary conformational changes leading to the expression of high affinity ligand binding.

VIII. CONCLUSIONS

In recent years, remarkable progress has been made toward an understanding of the molecular basis of platelet adhesive reactions. Many of these adhesive reactions are mediated by the αIIbβ3 complex, and, as such, it has served as a prototype adhesion receptor. The primary structure of these proteins has been determined and a detailed analysis of these sequences in conjunction with an analysis of genetic defects of the receptor has provided fundamental insights into structure-function relationships of the receptor. In particular, these studies have been instrumental in characterizing the ligand-receptor interaction. This understanding should provide new directions for the development of potential diagnostic and therapeutic agents for thrombotic disorders. As αIIbβ3 is a member of the integrin family of adhesion receptors, insight into αIIbβ3 structure and function has contributed to an elucidation of the molecular mechanism underlying cell adhesion in a much broader sense.

REFERENCES

1. Hynes, R. O., Integrins: a family of cell surface receptors, *Cell*, 48, 549, 1987.
2. Ruoslahti, E. and Pierschbacher, M. D., New perspectives in cell adhesion: RGD and integrins, *Science*, 491, 238, 1987.
3. Ginsberg, M. H., Loftus, J. C., and Plow, E. F., Cytoadhesins, integrins and platelets, *Thromb. Haemost.*, 59, 1, 1988.
4. Ruoslahti, E., Integrins, *J. Clin. Invest.*, 87, 1, 1991.
5. Nurden, A. T. and Caen, J. P., An abnormal glycoprotein pattern in three cases in Glanzmann's thrombasthenia, *Br. J. Haematol.*, 28, 253, 1974.
6. Jennings, L. K. and Phillips, D. R., Purification of glycoproteins IIb and IIIa from human platelet membranes and characterization of a calcium-dependent glycoprotein IIb-IIIa complex, *J. Biol. Chem.*, 257, 10458, 1982.
7. Poncz, M., Eisman, R., Heidenreich, Silver, S. M., Vilaire, G., Surrey, S., Schwartz, E., and Bennett, J. S., Structure of the platelet membrane glycoprotein IIb, *J. Biol. Chem.*, 262, 8476, 1987.
8. Fitzgerald, L. A., Steiner, B., Rall, S. C., Lo, S. S., and Phillips, D. R., Protein sequence of endothelial glycoprotein IIIa derived from a cDNA clone, *J. Biol. Chem.*, 262, 3939, 1987.
9. O'Toole, T. E., Loftus, J. C., Plow, E. F., Glass, A. A., Harper, J. R., and Ginsberg, M. H., Efficient surface expression of platelet GPIIb-IIIa requires both subunits, *Blood*, 74, 14, 1989.
10. Phillips, D. R., Charo, I. F., Parise, L. V., and Fitzgerald, L. A., The platelet membrane glycoprotein IIb-IIIa complex, *Blood*, 71, 831, 1988.
11. Bennett, J. S., Hoxie, J. A., Leitman, S. F., Vilaire, G., and Cines, D. B., Inhibition of fibrinogen binding to stimulated human platelets by a monoclonal antibody, *Proc. Natl. Acad. Sci. U.S.A.*, 80, 2417, 1983.
12. Coller, B. S., Peerschke, E. I., Scudder, L. E., and Sullivan, C. A., A murine monoclonal antibody that completely blocks the binding of fibrinogen to platelets produces a thrombasthenic-like state in normal platelets and binds to glycoproteins IIb and/or IIIa, *J. Clin. Invest.*, 72, 325, 1983.
13. McEver, R. P., Bennett, J. S., and Martin, M. N., Identification of two functionally distinct sites on human platelet membrane glycoprotein IIb-IIIa using monoclonal antibodies, *J. Biol. Chem.*, 258, 5369, 1983.
14. Pidard, D., Montgomery, R. R., Bennett, J. S., and Kunicki, T. J., Interaction of AP-2, a monoclonal antibody specific for the glycoprotein IIb-IIIa complex, with intact platelets, *J. Biol. Chem.*, 258, 12582, 1983.
15. Gogstad, G. O., Hagen, I., Korsmo, R., and Solum, N. O., Characterization of the proteins of isolated human platelet alpha-granules: evidence for a separate alpha-granule pool of the glycoproteins IIb and IIIa, *Biochim. Biophys. Acta*, 6709, 150, 1981.
16. Wencel-Drake, J. D., Plow, E. F., Kunicki, T. J., Woods, V. L., Keller, D. M., and Ginsberg, M. H., Localization of internal pools of membrane glycoproteins involved in platelet adhesive responses, *Am. J. Pathol.*, 124, 324, 1986.
17. Woods, V. L., Wolff, L. E., and Keller, D. M., Resting platelets contain a substantial centrally located pool of glycoprotein IIb-IIIa complex which may be accessible to some but not other extracellular proteins, *J. Biol. Chem.*, 261, 15242, 1986.
18. Niiya, K., Hodson, E., Bader, R., Byers-Ward, V., Koziol, J. A., Plow, E. F., and Ruggeri, Z. M., Increased surface expression of the membrane glycoprotein IIb/IIIa complex induced by platelet activation. Relationship to the binding of fibrinogen and platelet aggregation, *Blood*, 70, 475, 1987.

19. Phillips, D. R. and Agin, P. P., Platelet membrane glycoproteins: evidence for the presence of nonequivalent disulfide bonds using nonreduced-reduced two-dimensional gel electrophoresis, *J. Biol. Chem.*, 252, 2121, 1977.

20. Carrell, N. A., Fitzgerald, L. A., Steiner, B., Erickson, H. P., and Phillips, D. R., Structure of human platelet membrane glycoproteins IIb and IIIa as determined by electron microscopy, *J. Biol. Chem.*, 260, 1743, 1985.

21. Nermut, M. V., Green, N. M., Eason, P., Yamada, S. S., and Yamada, K. M., Electron microscopy and structural model of human fibronectin receptor, *EMBO J.*, 7, 4093, 1988.

22. Beer, J. and Coller, B. S., Evidence that platelet glycoprotein IIIa has a large disulfide bonded loop that is susceptible to proteolytic cleavage, *J. Biol. Chem.*, 264, 17564, 1989.

23. Kouns, W. C., Newman, P. J., Puckett, K. J., Miller, A. A., Wall, C. D., Fox, C. F., Seyer, J. M., and Jennings, L. K., Further characterization of the loop structure of platelet glycoprotein IIIa: partial mapping of functionally significant glycoprotein IIIa epitopes, *Blood*, 78, 3215, 1991.

24. Rosa, J. P., Bray, P. F., Gayet, O., Johnston, G. I., Cook, R. G., Jackson, K. W., Shuman, M. A., and McEver, R. P., Cloning of glycoprotein IIIa cDNA from human erythroleukemia cells and localization to chromosome 17, *Blood*, 72, 593, 1988.

25. Zimrin, A., Eisman, R., Vilaire, G., Schwartz, E., Bennett, J. S., and Poncz, M., The structure of platelet glycoprotein IIIa. A common subunit for two different membrane receptors, *J. Clin. Invest.*, 81, 1470, 1988.

26. Hiraiwa, A., Matsukage, A., Shiku, H., Takahashi, T., Naito, K., and Yamada, K., Purification and partial amino acid sequence of human platelet membrane glycoproteins IIb and IIIa, *Blood*, 69, 560, 1987.

27. Newman, P. J., Gorski, J., White, G. C., Gidwitz, S., Cretney, C. J., and Aster, R. H., Enzymatic amplification of platelet specific messenger RNA using the polymerase chain reaction, *J. Clin. Invest.*, 82, 739, 1988.

28. Newman, P. J., Derbes, R. S., and Aster, R. H., The human platelet alloantigens Pl[A1] and Pl[A2], are associated with a leucine[33]/proline[33] amino acid polymorphism in membrane glycoprotein IIIa, and are distinguishable by DNA typing, *J. Clin. Invest.*, 83, 1778, 1989.

29. Charo, I., Fitzgerald, L. A., Steiner, B., Rall, S. C., Bekeart, L. S., and Phillips, D. R., Platelet glycoproteins IIb and IIIa: evidence for a family of immunologically and structurally related glycoproteins in mammalian cells, *Proc. Natl. Acad. Sci. U.S.A.*, 83, 8351, 1986.

30. Argraves, W. S., Suzuki, S., Arai, H., Thompson, K., Pierschbacher, M. D., and Ruoslahti, E., Amino acid sequence of the human fibronectin receptor, *J. Cell Biol.*, 105, 1183, 1987.

31. Kishimoto, T. K., O'Connor, K., Lee, A., Roberts, T. M., and Springer, T. A., Cloning of the β subunit of the leukocyte adhesion proteins: homology to an extracellular matrix receptor defines a novel supergene family, *Cell*, 48, 681, 1987.

32. Suzuki, S. and Naitoh, Y., Amino acid sequence of a novel integrin β_4 subunit and primary expression of the mRNA in epithelial cells, *EMBO J.*, 9, 757, 1990.

33. McLean, J. W., Vestal, D., Cheresh, D. A., and Bodary, S. C., cDNA sequence of the human integrin β_5 subunit, *J. Biol. Chem.*, 265, 17126, 1990.

34. Ramaswamy, H., and Hemler, M., Cloning and primary structure and properties of a novel human integrin β subunit, *EMBO J.*, 9, 1561, 1990.

35. Hogervorst, F., Kuikman, I., von dem Borne, A. E. G. Kr., and Sonnenberg, A., Cloning and sequence analysis of beta-4 cDNA: an integrin subunit that contains a unique 118 kd cytoplasmic domain, *EMBO J.*, 9, 765, 1990.

36. Sheppard, D., Rozzo, C., Starr, L., Quaranta, V., Erle, D. J., and Pytela, R., Complete amino acid sequence of a novel integrin β subunit (β$_6$) identified in epithelial cells using the polymerase chain reaction, *J. Biol. Chem.*, 265, 11502, 1990.

37. Erle, D. J., Ruegg, C., Sheppard, D., and Pytela, R., Complete amino acid sequence of an integrin β subunit (β$_7$) identified in leukocytes, *J. Biol. Chem.*, 266, 11009, 1991.

38. Smith, J. W., Vestal, D., Irwin, S. V., Burke, T. A., and Cheresh, D. A., Purification and functional characterization of α$_v$β$_5$, an adhesion receptor for vitronectin, *J. Biol. Chem.*, 265, 11008, 1990.

39. Downard, J., Parker, P., and Waterfield, M. D., Autophosphorylation sites on the epidermal growth factor precursor, *Nature*, 311, 483, 1984.

40. Tankum, J. W., DeSimone, D. W., Fonda, D., Patel, R. S., Buck, C., Horwitz, A. F., and Hynes, R. O., Structure of integrin, a glycoprotein involved in the transmembrane linkage between fibronectin and actin, *Cell*, 46, 271, 1986.

41. Hirst, R., Horwitz, A., Buck, C., and Rohrschneider, L., Phosphorylation of the fibronectin receptor complex in cells transformed by oncogenes that encode tyrosine kinases, *Proc. Natl. Acad. Sci. U.S.A.*, 83, 6470, 1986.

42. Parise, L. V., Criss, A. B., Nannizzi, L., and Wardell, M. R., Glycoprotein IIIa is phosphorylated in intact human platelets, *Blood*, 75, 2363, 1990.

43. Freed, E., Gailit, J., van der Geer, P., Ruoslahti, E., and Hunter, T., A novel integrin α subunit is associated with the vitronectin receptor α subunit (α$_v$) in a human osteosarcoma cell line and is a substrate for protein kinase C, *EMBO J.*, 8, 2955, 1989.

44. Hillery, C. A., Smyth, S. S., and Parise, L. V., Phosphorylation of human platelet glycoprotein IIIa (GPIIIa). Dissociation from fibrinogen receptor activation and phosphorylation *in vitro*, *J. Biol. Chem.*, 266, 14663, 1991.

45. van Kuppevelt, T. H. M. S. M., Languino, L. R., Gailit, J. O., Suzuki, S., and Ruoslahti, E., An alternative cytoplasmic domain of integrin β$_3$ subunit, *Proc. Natl. Acad. Sci. U.S.A.*, 86, 5415, 1989.

46. Bray, P. F., Rosa, J. P., Johnston, G. I., Shiu, D. T., Cook, R. G., Lau, C., Kan, Y. W., McEver, R. P., and Shuman, M. A., Platelet glycoprotein IIb. Chromosomal localization and tissue expression, *J. Clin. Invest.*, 80, 1812, 1987.

47. Sosnoski, D. M., Emanuel, B. S., Hawkins, A. L., van Tuinen, P., Ledbetter, D. H., Nussbaum, R. L., Kaos, F.-T., Schwartz, E., Phillips, D. R., Bennett, J. S., Fitzgerald, L. A., and Poncz, M., Chromosomal localization of the genes for the vitronectin and fibronectin receptor α subunits and for platelet glycoproteins IIb and IIIa, *J. Clin. Invest.*, 81, 1993, 1988.

48. Zimrin, A. B., Gidwitz, S., Lord, S., Schwartz, E., Bennett, J. S., White, G. C., and Poncz, M., The genomic organization of platelet GPIIIa, *J. Biol. Chem.*, 265, 8590, 1990.

49. Lanza, F., Kieffer, N., Phillips, D. R., and Fitzgerald, L. A., Characterization of the human platelet glycoprotein IIIa gene. Comparison with the fibronectin receptor β subunit gene, *J. Biol. Chem.*, 265, 18098, 1990.

50. Poncz, M., Eisman, R., Heidenreich, Silver, S. M., Vilaire, G., Surrey, S., Schwartz, E., and Bennett, J. S., Structure of the platelet membrane glycoprotein, IIb, *J. Biol. Chem.*, 262, 8476, 1987.

51. Uzan, G., Frachet, P., Lajmanovich, A., Pradini, M. H., Denarier, E., Duperray, A., Loftus, J., Ginsberg, M., Plow, E., and Marguerie, G., cDNA clones for human platelet GPIIb corresponding to mRNA from megakaryocytes and HEL cells, *Eur. J. Biochem.*, 171, 87, 1988.

52. Calvete, J. J., Henchen, A., and Gonzalez-Rodriguez, J., Complete localization of the intrachain disulphide bonds and the N-glycoslation points in the α-subunit of human platelet glycoprotein IIb, *Biochem. J.*, 261, 561, 1989.

53. Fitzgerald, L. A., Poncz, M., Steiner, B., Rall, S. C., Bennett, J. S., and Phillips, D. R., Comparison of cDNA-derived protein sequences of the human fibronectin and vitronectin receptor α subunits and platelet glycoprotein IIb, *Biochemistry*, 26, 8158, 1987.

54. Tufty, R. M. and Kretsinger, R. H., Troponin and parvalbumin calcium binding regions predicted in myosin light chain and T4 lysozyme, *Science*, 187, 167, 1975.

55. Strynadka, N. C. J. and James, M. N. G., Crystal structures of the helix-loop-helix calcium-binding proteins, *Annu. Rev. Biochem.*, 58, 951, 1989.

56. Vyas, N. K., Vyas, M. N., and Quiocho, F. A., A novel calcium binding site in the galactose-binding protein of bacterial transport and chemotaxis, *Nature (London)*, 327, 635, 1987.

57. Bray, P. F., Leung, C. S.-I., and Shuman, M. A., Human platelets and megakaryocytes contain alternatively spliced glycoprotein IIb mRNAs, *J. Biol. Chem.*, 265, 9587, 1990.

58. Calvete, J. J., and Gonzalez-Rodriguez, J., Isolation and biochemical characterization of the α and β subunits of glycoprotein IIb of human platelet plasma membrane, *Biochem. J.*, 240, 155, 1986.

59. Heidenreich, R., Eisman, R., Surrey, S., Delgrosso, K., Bennett, J. S., Schwartz, E., and Poncz, M., Organization of the gene for platelet glycoprotein IIb, *Biochemistry*, 29, 1232, 1990.

60. Bray, P. F., Barsh, G., Rosa, J. P., Luo, X. Y., Magenis, E., and Shuman, M. A., Physical linkage of the genes for platelet membrane glycoproteins IIb and IIIa, *Proc. Natl. Acad. Sci. U.S.A.*, 85, 8683, 1988.

61. Prandini, M. H., Denarier, E., Frachet, P., Uzan, G., and Marguerie, G., Isolation of the human platelet glycoprotein IIb gene and characterization of the 5' flanking region, *Biochem. Biophys. Res. Commun.*, 156, 595, 1988.

62. Bray, P. F., Rosa, J. P., Lingappa, V., Kan, Y. W., McEver, R. P., and Shuman, M. A., Biogenesis of the platelet receptor for fibrinogen: evidence for separate precursors for glycoproteins GPIIb and IIIa, *Proc. Natl. Acad. Sci. U.S.A.*, 83, 1480, 1986.

63. Silver, S. M., McDonough, M., Vilaire, G., and Bennett, J. S., The in vitro synthesis of polypeptides for the platelet membrane glycoproteins IIb and IIIa, *Blood*, 69, 1031, 1987.

64. Duperray, A., Berthier, R., Chagnon, E., Ryckewart, J.-J., Ginsberg, M. H., Plow, E. F., and Marguerie, G., Biosynthesis and processing of platelet GPIIb-IIIa in human megakaryocytes, *J. Cell Biol.*, 104, 1665, 1987.

65. Troesch, A., Duperray, A., Polack, B., and Marguerie, G., Comparative study of the glycosylation of platelet glycoprotein GPIIb-IIIa and the vitronectin receptor, *Biochem. J.*, 268, 129, 1990.

66. Docherty, K. and Steiner, D. F., Post-translational proteolysis in polypeptide hormone biosynthesis, *Annu. Rev. Physiol.*, 44, 625, 1982.

67. Loftus, J. C., Plow, E. F., Jennings, L. K., and Ginsberg, M. H., Alternative proteolytic processing of platelet membrane glycoprotein IIb, *J. Biol. Chem.*, 263, 11025, 1988.

68. Poncz, M. and Newman, P. J., Analysis of rodent platelet GPIIb: evidence for evolutionarily conserved domains and alternative proteolytic processing, *Blood*, 75, 1282, 1990.

69. Kolodziej, M. A., Vilaire, G., Gonder, D., Poncz, M., and Bennett, J. S., Study of the endoproteolytic cleavage of glycoprotein IIb using oligonucleotide-mediated mutagenesis, *J. Biol. Chem.*, 266, 23499, 1991.

70. Bodary, S. C., Napier, M. A., and McLean, J. W., Expression of recombinant platelet glycoprotein IIB-IIIa results in a functional fibrinogen-binding complex, *J. Biol. Chem.*, 264, 18859, 1989.

71. Kieffer, N., Fitzgerald, L. A., Wolf, D., Cheresh, D. A., and Phillips, D. R., Adhesive properties of the β_3 integrins: comparison of GPIIb-IIIa and the vitronectin receptor individually expressed in human melanoma cells, *J. Cell Biol.*, 113, 451, 1991.
72. Plow, E. E., Ginsberg, M. H., and Marguerie, G. A., Expression and function of adhesive proteins on the platelet surface, in *Biochemistry of Platelets*, Phillips, D. R. and Shuman, M. A., Eds., Academic Press, Orlando, FL, 1986, 226.
73. Plow, E. F., Marguerie, G. A., and Ginsberg, M. H., Interaction of adhesive proteins with platelets: common features with distinct differences, in *Perspectives in Inflammation, Neoplasia, and Vascular Cell Biology*, Alan R. Liss, New York, 1987, 267.
74. Marguerie, G. A., Ginsberg, M. H., and Plow, E. F., The platelet fibrinogen receptor, in *Platelets in Biology and Pathology III*, MacIntyre, D. E. and Gordon, J. L., Eds., Elsevier Science Publishers, New York, 1987, 95.
75. Pierschbacher, M. D., and Ruoslahti, E., The cell attachment domain of fibronectin, *J. Biol. Chem.*, 257, 9593, 1982.
76. Pierschbacher, M. D., Hayman, E. G., and Ruoslahti, E., Synthetic peptide with cell attachment activity of fibronectin, *Proc. Natl. Acad. Sci. U.S.A.*, 80, 1224, 1983.
77. Pierschbacher, M. D. and Ruoslahti, E., Cell attachment activity of fibronectin can be duplicated by small synthetic fragments of the molecule, *Nature*, 309, 30, 1984.
78. Ginsberg, M. H., Pierschbacher, M. D., Ruoslahti, E., Marguerie, G. A., and Plow, E. F., Inhibition of fibronectin binding to platelets by proteolytic fragments and synthetic peptides which support fibroblast cell adhesion, *J. Biol. Chem.*, 260, 3931, 1985.
79. Haverstick, D. M., Cowan, J. F., Yamada, K. M., and Santoro, S. A., Inhibition of platelet adhesion to fibronectin, fibrinogen, and von Willebrand factor substrates by a synthetic tetrapeptide derived from the cell binding domain of fibronectin, *Blood*, 66, 946, 1985.
80. Plow, E. F., Pierschbacher, M. D., Ruoslahti, E., Marguerie, G. A., and Ginsberg, M. H., The effect of Arg-Gly-Asp containing peptides on fibrinogen and von Willebrand factor binding to platelets, *Proc. Natl. Acad. Sci. U.S.A.*, 82, 8057, 1985.
81. Huang, T.-F., Holt, J. C., Kirby, E. P., and Niewiarowski, S., Trigramin: primary structure and its inhibition of von Willebrand factor binding to glycoprotein IIb-IIIa complex on human platelets, *Biochemistry*, 28, 661, 1989.
82. Gan, Z.-R., Gould, R. J., Jacobs, J. W., Friedman, P. A., and Polokoff, M. A., Echistatin. A potent platelet aggregation inhibitor from the venom of the viper, *Echis Carinatus*, *J. Biol. Chem.*, 263, 19827, 1988.
83. Dennis, M. S., Henzel, W. J., Pitti, R. M., Lipari, M. T., Napier, M. A., Deisher, T. A., Bunting, S., and Lazarus, R. A., Platelet glycoprotein IIb-IIIa protein antagonists from snake venoms: evidence for a family of platelet-aggregation inhibitors, *Proc. Natl. Acad. Sci. U.S.A.*, 87, 2475, 1989.
84. Scarborough, R. M., Rose, J. W., Hsu, M. A., Phillips, D. R., Fried, V. A., Campbell, A. M., Nannizzi, L., and Charo, I. F., Barbourin. A GPIIb-IIIa specific integrin anatagonist from the venom of *Sistrus M. Barbouri*, *J. Biol. Chem.*, 266, 9359, 1991.
85. Pierschbacher. M. D. and Rouslahti, E., Influence of stereochemistry of the sequence Arg-Gly-Asp-Xaa on binding specificity in cell adhesion, *J. Biol. Chem.*, 262, 17294, 1987.
86. Kloczewiak, M., Timmons, S., and Hawiger, J., Localization of a site interacting with human platelet receptor on carboxy-terminal segment of human fibrinogen γ chain, *Biochem. Biophys. Res. Commun.*, 107, 181, 1982.
87. Kloczewiak, M., Timmons, S., Lukas, T. J., and Hawiger, J., Platelet receptor recognition site on human fibrinogen. Synthesis and structure function relationship of peptides corresponding to the carboxy-terminal segment of the γ chain, *Biochemistry*, 23, 1767, 1984.

88. Lam, S. C.-T., Plow, E. F., Smith, M. A., Andrieux, A., Ryckwaert, J.-J., Marguerie, G., and Ginsberg, M. H., Evidence that Arginyl-Glycyl-Aspartate peptides and fibrinogen γ chain peptides share a common binding site on platelets, *J. Biol. Chem.*, 262, 947, 1987.

89. Kunicki, T. J., Pidard, D., Rosa, J. P., and Nurden, A. T., The formation of Ca^{++} dependent complexes of platelet membrane glycoprotein IIb and IIIa in solution as determined by crossed/immunoelectrophoresis, *Blood*, 58, 268, 1981.

90. Fujimura, K. and Phillips, D. R., Calcium cation regulation of glycoprotein IIb-IIIa complex formation in platelet plasma membranes, *J. Biol. Chem.*, 258, 10247.

91. Fitzgerald, L. A. and Phillips, D. R., Calcium regulation of the platelet membrane glycoprotein IIb-IIIa complex, *J. Biol. Chem.*, 260, 11366, 1985.

92. Pidard, D., Didry, D., Kunicki, T. J., and Nurden, A. T., Temperature-dependent effects of EDTA on the membrane glycoprotein IIb-IIIa complex and platelet aggregability, *Blood*, 67, 604, 1986.

94. Steiner, B., Parise, L., Leung, B., and Phillips, D. R., Ca^{2+} dependent structural transitions of the platelet glycoprotein IIb-IIIa complex, *J. Biol. Chem.*, 266, 14986, 1991.

95. Brass, L. F. and Shattil, S. J., Changes in surface-bound and exchangeable calcium during platelet activation, *J. Biol. Chem.*, 257, 14000, 1982.

96. Brass, L. F. and Shattil, S. J., Identification and function of the high affinity binding sites for Ca^{2+} on the surface of platelets, *J. Clin. Invest.*, 73, 626, 1984.

97. Rivas, G. A. and Gonzalez-Rodriguez, J., Calcium binding to the platelet integrin GPIIb/IIIa and to its constituent glycoproteins. Effect of lipids and temperature, *Biochem. J.*, 276, 35, 1991.

98. Gulino, D., Boudignon, C., Zhang, L., Concord, E., Rabiet, M.-J., and Marguerie, G., Ca^{2+} binding properties of the platelet glycoprotein IIb ligand-interacting domain, *J. Biol. Chem.*, 267, 1001, 1992.

99. Bennett, J. S. and Vilaire, G., Exposure of platelet fibrinogen receptors by ADP and epinephrine, *J. Clin. Invest.*, 64, 1393, 1979.

100. Marguerie, G. A., Plow, E. F., and Edgington, T. S., Human platelets possess an inducible and saturable receptor specific for fibrinogen, *J. Biol. Chem.*, 254, 5357, 1979.

101. Plow, E. F. and Marguerie, G. A., Induction of the fibrinogen receptor on human platelets by epinephrine and the combination of epinephrine and ADP, *J. Biol. Chem.*, 255, 10971, 1980.

102. Phillips, D. R. and Baughan, A. K., Fibrinogen binding to human platelet plasma membranes identification of two steps requiring divalent cations, *J. Biol. Chem.*, 258, 10240, 1983.

103. Steiner, B., Cousot, D., Trzeciak, A., Gillessen, D., and Hadvary, P., Ca2+ dependent binding of a synthetic Arg-Gly-Asp (RGD) peptide to a single site on the purified platelet glycoprotein IIb-IIIa complex, *J. Biol. Chem.*, 264, 13102, 1989.

104. Shattil, S. J., Hoxie, J. A., Cunningham, M., and Brass, L. F., Changes in the platelet membrane glycoprotein IIb-IIIa complex during platelet activation, *J. Biol. Chem.*, 260, 11107, 1985.

105. Bennett, J. S., Shattil, S. J., Power, J. W., and Gartner, T. K., Interaction of fibrinogen with its platelet receptor. Differential effects of α and γ chain fibrinogen peptides on the glycoprotein IIb-IIIa complex, *J. Biol. Chem.*, 263, 12948, 1988.

106. Shattil, S. J., Motulsky, H. J., Insel, P. A., Flaherty, L., and Brass, L. F., Expression of fibrinogen receptors during activation and subsequent desensitization of human platelets by epinephrine, *Blood*, 68, 1224, 1986.

107. Taub, R., Gould, R. J., Garsky, V. M., Ciccarone, T. M., Hoxie, J., Friedman, P. A., and Shattil, S. J., A monoclonal antibody against the platelet fibrinogen receptor contains a sequence that mimics a receptor recognition domain in fibrinogen, *J. Biol. Chem.*, 264, 259, 1989.

108. Calvete, J. J., Rivas, G., Mauri, M., Alvarez, M., McGregor, J. L., Hew, C. L., and Gonzalez-Rodriguez, J., Tryptic fragment of human glycoprotein IIIa. Isolation and biochemical characterization of the 23 kDa N-terminal glycopeptide carrying the antigenic determinant for a monoclonal antibody (P37) which inhibits platelet aggregation, *Biochem. J.*, 260, 697, 1988.

109. Santoro, S. A. and Lawing, W. J., Competition for related but nonidentical binding sites on the glycoprotein IIb-IIIa complex by peptides derived from platelet adhesive proteins, *Cell*, 48, 867, 1987.

110. D'Souza, S. E., Ginsberg, M. H., Lam, S. C.-T., and Plow, E. F., Chemical cross-linking of Arginyl-Glycyl-Aspartic acid peptides to an adhesion receptor on platelets, *J. Biol. Chem.*, 263, 3943, 1988.

111. D'Souza, S. E., Ginsberg, M. H., Burke, T. T., Lam, C.-T., and Plow, E. F., Localization of an Arg-Gly-Asp recognition site within an integrin adhesion receptor, *Science*, 242, 91, 1988.

112. Charo, I. F., Nannizzi, L., Phillips, D. R., Hsu, M. A., and Scarborough, R. M., Inhibition of fibrinogen binding to GPIIb-IIIa by a GPIIIa peptide, *J. Biol. Chem.*, 266, 1415, 1991.

113. D'Souza, S. E., Ginsberg, M. H., Burke, T. A., and Plow, E. F., The ligand binding site of the platelet integrin receptor GPIIb-IIIa is proximal to the second calcium binding domain of its α subunit, *J. Biol. Chem.*, 265, 3440, 1990.

114. D'Souza, S. E., Ginsberg, M. H., Matsueda, G. R., and Plow, E. F., A discrete sequence in a platelet integrin is involved in ligand recognition, *Nature*, 350, 66, 1991.

115. Lam, S. C.-T., Plow, E. F., Cheresh, D. A., Frelinger, A. L., Ginsberg, M. H., and D'Souza, S. E., Isolation and characterization of a platelet membrane protein related to the vitronectin receptor, *J. Biol. Chem.*, 264, 3742, 1989.

116. O'Toole, T. E., Loftus, J. C., Du, X., Glass, A. A., Ruggeri, Z. M., Shattil, S. J., Plow, E. F., and Ginsberg, M. H., Affinity modulation of the $\alpha_{IIb}\beta_3$ integrin (platelet GPIIb-IIIa) is an intrinsic property of the receptor, *Cell Reg.*, 1, 883, 1990.

117. Cheresh, D. A., Berliner, S. A., Vincente, V., and Ruggeri, Z. M., Recognition of distinct adhesive sites on fibrinogen by related integrins on platelets and endothelial cells, *Cell*, 58, 945, 1989.

118. Smith, J. W., Ruggeri, Z. M., Kunicki, T. J., and Cheresh, D. A., Interactions of integrins $\alpha_v\beta_3$ and glycoprotein IIb-IIIa with fibrinogen. Differential peptide recognition accounts for distinct binding sites, *J. Biol. Chem.*, 265, 12267, 1990.

119. Kretsinger, R. H., Structure and evolution of calcium-modulated proteins, *CRC Crit. Rev. Biochem.*, 8, 119, 1980.

120. Phillips, D. R. and Agin, P. P., Platelet membrane defects in Glanzmann's thrombasthenia. Evidence for decreased amounts of two major glycoproteins, *J. Clin. Invest.*, 60, 535, 1977.

121. George, J. N., Caen, J. P., and Nurden, A. T., Glanzmann's thrombasthenia: the spectrum of clinical disease, *Blood*, 75, 1383, 1990.

122. Burk, C. D., Newman, P. J., Lyman, S., Gill, J., Coller, B. S., and Poncz, M. A., Deletion in the gene for glycoprotein IIb associated with Glanzmann's thrombasthenia, *J. Clin. Invest.*, 87, 270, 1991.

123. Bray, P. F. and Shuman, M. A., Identification of an abnormal gene for the GPIIIa subunit of the platelet fibrinogen receptor resulting in Glanzmann's thrombasthenia, *Blood*, 75, 881, 1990.

124. Newman, P. J., Seligsohn, U., Lyman, S., and Coller, B. S., The molecular genetic basis of Glanzmann's thrombasthenia in the Iraqi-Jewish and Arab populations in Israel, *Proc. Natl. Acad. Sci. U.S.A.*, 88, 3160, 1991.

125. Ginsberg, M. H., Lightsey, A., Kunicki, T. J., Kaufmann, A., Marguerie, G., and Plow, E. F., Divalent cation regulation of the surface orientation of platelet membrane glycoprotein GPIIb. Correlation with fibrinogen binding function and definition of a novel variant of Glanzmann's thrombasthenia, *J. Clin. Invest.*, 78, 1103, 1986.

126. Jung, S. M., Yoshida, N., Aoki, N., Tanoue, K., Yamazaki, H., and Moroi, M., Thrombasthenia with an abnormal platelet membrane glycoprotein IIb of different molecular weight, *Blood*, 71, 915, 1988.

127. Modderman, P. W., van Mourik, J. A., van Berkel, W., Cordell, J. L., Morel, M. C., Kaplan, C., Ouwehand, W. H., Huisman, J. G., and von dem Borne, A. E. G. Kr., Decreased stability and structural heterogeneity of the residual platelet glycoprotein IIb-IIIa complex in a variant of Glanzmann's thrombasthenia, *Br. J. Haematol.*, 73, 514, 1989.

128. Fournier, D., Kabral, A., Castaldi, P. A., and Berndt, M. C., A variant of Glazmann's thrombasthenia characterized by abnormal glycoprotein IIb-IIIa complex formation, *Thromb. Haemost.*, 62, 977, 1989.

129. Ginsberg, M. H., Frelinger, A. L., Lam, S. C.-T., Forsyth, J., McMillan, R., Plow, E. F., and Shattil, S. J., Analysis of platelet aggregation disorders based on flow cytometric analysis of membrane glycoprotein IIb-IIIa with conformation specific monoclonal antibodies, *Blood*, 76, 2017, 1990.

130. Shadle, P. J., Ginsberg, M. H., Plow, E. F., and Barondes, S. H., Platelet-collagen adhesion: inhibition by a monoclonal antibody that binds glycoprotein IIb, *J. Cell Biol.*, 99, 2056, 1984.

131. Frelinger, A. L., Lam, S. C.-T., Plow, E. F., Smith, M. A., Loftus, J. C., and Ginsberg, M. H., Occupancy of an adhesive glycoprotein receptor modulates expression of an antigenic site involved in cell adhesion, *J. Biol. Chem.*, 263, 12397, 1988.

132. Loftus, J. C., Plow, E. F., Frelinger, A. L., D'Souza, S. E., Dixon, D., Lacy, J., Sorge, J., and Ginsberg, M. H., Molecular cloning and chemical synthesis of a region of platelet GPIIb involved in adhesive function, *Proc. Natl. Acad. Sci. U.S.A.*, 840, 7114, 1987.

133. Loftus, J. C., O'Toole, T. E., Plow, E. F., Glass, A. A., Frelinger, A. L., and Ginsberg, M. H., A β_3 integrin mutation abolishes ligand binding and alters divalent cation dependent conformation, *Science*, 249, 915, 1990.

134. Lawler, J., Weinstein, R., and Hynes, R. O., Cell attachment to thrombospondin: the role of Arg-Gly-Asp, calcium and integrin receptors, *J. Cell Biol.*, 107, 2351, 1988.

135. Corbi, A. L., Miller, L. J., O'Conner, K., Larson, R. S., and Springer, T. A., cDNA cloning and complete primary structure of the alpha subunit of a leukocyte adhesion glycoprotein, p150,95, *EMBO J.*, 6, 4023, 1987.

136. Pierschbacher, M. D. and Ruoslahti, E., The cell attachment activity of fibronectin can be duplicated by small synthetic fragments of the molecule, *Nature*, 309, 30, 1984.

137. Dransfield, I. and Hogg, N., Regulated expression of Mg2+ binding epitope on leukocyte integrin alpha subunits, *EMBO J.*, 8, 3759, 1990.

138. Takada, Y., Ylanne, J., Mandelmann, D., Puzon, W., and Ginsberg, M. H., A point mutation of integrin β_1 subunit blocks binding of $\alpha_5\beta_1$ to fibronectin and invasin but not recruitment to adhesion plaques, *J. Cell Biol.*, submitted.

139. Isberg, R. R. and Leong, J. M., Multiple β_1 chain integrins are receptors for invasin, a protein that promotes bacterial penetration into mammalian cells, *Cell*, 60, 861, 1990.

140. Bajt, M. L., Ginsberg, M. H., Frelinger, A. L., Berndt, M. C., and Loftus, J. C., A spontaneous mutation of integrin $\alpha_{IIb}\beta_3$ (platelet GPIIb-IIIa) helps define a ligand binding site, *J. Biol. Chem.*, 267, 3789, 1992.

141. Shattil, S. J. and Brass, L. F., Induction of the fibrinogen receptor on human platelets by intracellular mediators, *J. Biol. Chem.*, 262, 992, 1987.

142. Coller, B. S., A new murine monoclonal antibody reports an activation-dependent change in the conformation and/or microenvironment of the platelet glycoprotein IIb-IIIa complex, *J. Clin. Invest.*, 76, 101, 1985.

143. Coller, B. S., Activation affects access to the platelet receptor for adhesive proteins, *J. Cell Biol.*, 103, 451, 1986.

144. Kouns, W. C., Wall, C. D., White, M. M., Fox, C. F., and Jennings, L. K., A conformation dependent epitope of human platelet glycoprotein IIIa, *J. Biol. Chem.*, 265, 20594, 1990.

145. Sims, P. J., Ginsberg, M. H., Plow, E. F., and Shattil, S. J., Effect of platelet activation on the conformation of the plasma membrane glycoprotein IIb-IIIa complex, *J. Biol. Chem.*, 266, 7345, 1991.

146. Andrieux, A., Rabiet, M. J., Chapel, A., Concord, E., and Marguerie, G., A highly conserved sequence of the Arg-Gly-Asp binding domain of the integrin β3 subunit is sensitive to stimulation, *J. Biol. Chem.*, 266, 14202, 1991.

147. Gulino, D., Ryckewaert, J. J., Andrieux, A., Rabiet, M. J., and Marguerie, G., Identification of a monoclonal antibody against platelet GPIIb that interacts with a calcium binding site and induces aggregation, *J. Biol. Chem.*, 265, 9575, 1990.

148. Coller, B. S., Interaction of normal, thrombasthenic, and Bernard-Soulier platelets with immobilized fibrinogen: defective platelet-fibrinogen interaction in thrombasthenia, *Blood*, 55, 169, 1980.

149. Lindon, J. N., McManama, G., Kushner, L., Merrill, E. W., and Salzman, E. N., Does the conformation of adsorbed fibrinogen dictate platelet interaction with artificial surfaces?, *Blood*, 68, 355, 1986.

150. O'Toole, T. E., Mandelmann, D., Forsyth, J., Shattil, S. J., Plow, E. F., and Ginsberg, M. H., Modification of the affinity of integrin $\alpha_{IIb}\beta_3$ (GPIIb-IIIa) by the cytoplasmic domain of GPIIb, *Science*, 254, 845, 1991.

151. Chen, Y., Diaffar, Pidard, D., Steiner, B., Kieffer, N., Caen, J. P., and Rosa, J.-P., The absence of activation of platelet GPIIb-IIIa, the integrin $\alpha_{IIb}\beta_3$, is associated with a point mutation of the cytoplasmic domain of GPIIIa (β_3) in a variant type of Glanzmann's thrombasthenia, *Blood*, 78, 279a, 1991.

152. Du, X., Plow, E. F., Frelinger, A. L., O'Toole, T. E., Loftus, J. C., and Ginsberg, M. H., Ligands activate integrin $\alpha_{IIb}\beta_3$ (platelet GPIIb-IIIa), *Cell*, 65, 1, 1991.

153. Parise, L. V., Helgerson, S. L., Steiner, B., Nannizzi, L., and Phillips, D. R., Synthetic peptides derived from fibrinogen and fibronectin change the conformation of purified platelet GPIIb-IIIa, *J. Biol. Chem.*, 262, 12597, 1987.

154. Frelinger, A. L., Cohen, I., Plow, E. F., Smith, M. A., Roberts, J., Lam, S. C.-T., and Ginsberg, M. H., Selective inhibition of integrin function by antibodies specific for ligand-occupied receptor conformers, *J. Biol. Chem.*, 104, 1655, 1990.

155. Frelinger, A. L., Du, X., Plow, E. F., and Ginsberg. M. H., Monoclonal antibodies to ligand-occupied conformers of integrin $\alpha_{IIb}\beta_3$ (glycoprotein IIb-IIIa) alter receptor affinity, specificity, and function, *J. Biol. Chem.*, 266, 17106, 1991.

Chapter 4

The αv Integrins

Candece L. Gladson and David A. Cheresh

CONTENTS

I. INTRODUCTION

The integrins are a large family of heterodimeric, divalent cation-dependent cell surface receptors that facilitate cell adhesion to extracellular matrix proteins and cell-cell adhesion.[1-3] They potentiate a number of other basic biologic events, such as cell migration, communication, proliferation, differentiation, and signaling from the extracellular environment to the intracellular compartment.[1-3] The integrin family is currently composed of eight beta (β) subunits and 13 alpha (α) subunits, whose heterodimeric pairing determines ligand specificity, in part.[1-3] Typically one β subunit pairs with multiple α subunits (see Figure 1).[1-3] These receptors, most notably the β1 integrin, are ubiquitous in distribution.[1-3] In addition, the integrins have been highly conserved throughout evolution; for example, the αv subunit is 83% identical in amino acid sequence between chick and man.[4]

The αv subunit is unique among the integrin α subunits in that it potentially associates with at least five structurally distinct β subunits (β3, β5, β1, β6, and β8).[1-16] Two additional β subunits that pair with the αv subunit have been identified (β100K and βS), and it remains to be clarified whether or not they are distinct from the β5 subunit.[4,17] The αv subunit and the β1 subunit are the most widely expressed integrin subunits *in vitro* and *in situ*. In the αv subfamily, αβ heterodimeric pairing determines ligand specificity, like the large integrin family. However, integrin αvβ3 is unique among the αv subfamily in that it is a promiscuous receptor.[1-7] The αv integrins play a role in diverse biologic processes, such as tumor cell invasion, metastasis proliferation,[1-3,18-24] retinal neurite outgrowth,[25,26] differentiation,[26-29] bone resorption,[30,31] and the immune response.[32-38] The entire integrin family, including the αv integrins, is depicted in Figure 1.

II. STRUCTURE OF THE αv INTEGRINS

Structurally the integrins are heterodimeric proteins with α and β subunits that span the lipid bilayer.[1-3] The α integrin subunits share an approximate 40% homology to each other and typically undergo posttranslational processing; for example, the αv subunit is

0-8493-4711-4/94/$0.00+$.50
© 1994 by CRC Press, Inc.

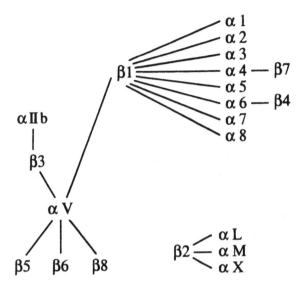

Figure 1 Integrin superfamily.

proteolytically cleaved into a heavy and light chain which become disulfide linked.[1-3,39,40] Like other integrin α subunits, the αv heavy chain contains four putative calcium binding sites likely involved in ligand binding.[4,39,40] This metal ion binding domain is composed of six homologous sequence elements whose consensus sequence $Dx^D/_NxDGxxD$ is homologous to the Ca^{2+} binding sites described in other proteins.[4,39,40] The β subunits share an overall 35 to 45% homology with each other and also undergo posttranslational processing events, such as glycosylation and intrasubunit disulfide bonding.[1-3,41] Typically, the β integrin subunits contain 56 cysteine residues, comprising three repeats each of which contain eight cysteine residues and one half repeat.[1-3,41] Both α and β subunits are composed of a large extracellular domain, a single transmembrane domain, and a short cytoplasmic tail.[1-3,39-41] Interestingly, three of the recently sequenced β subunits (β5, β6, and β8) which pair with the αv subunit contain an extra ten amino acids in the cytoplasmic tail domain, when compared to the β3 and β1 subunits.[10,11,14,16]

III. LIGAND RECOGNITION

All integrins appear to interact with their ligands in a cation-dependent fashion, and integrins are known to be differentially affected by divalent cations.[1-3,42] Integrin αvβ3 binds metal ions directly, and in the same concentration range necessary for ligand binding.[43] The presence of ligand is not necessary for αvβ3 metal ion interaction.[43] Both[54] Mn^{2+} and Ca^{2+} bind to αvβ3 at the same or mutually exclusive sites.[43] The coordination sphere of the bound metal ion directly affects ligand binding, since oxidative incorporation of Co(III) into the divalent metal binding sites inactivates αvβ3.[43]

It has been thought that integrin α subunits are primarily responsible for regulation of receptor binding by cations, since they contain metal binding sites.[1-3,39-44] However, the β subunit may also be an important determinant of cation-specific effects, since two integrins which share a common αv subunit and differ only in the β subunit (αvβ3 and αvβ1) significantly differ in their ability to bind RGD-containing peptides in the presence of Ca^{2+}.[45] In the presence of Mg^{2+}, Ca^{2+} inhibited the ability of integrin αvβ1 to bind RGD-containing peptides or its ligand fibronectin, while Ca^{2+} enhanced the binding of

Table 1 **The αv integrins**

β Subunit	Ligand(s)	Unique Features	References
β1	Fibronectin (vitronectin on a different cell type)	Ca²⁺ inhibits ligand binding	13 12
β3	Vitronectin	Promiscuous receptorᵃ involved in adhesion and migration	5,6,40,51
	Fibrinogen		6,51
	von Willebrand factor		6,51 51
	Thrombospondin		56,57
	Fibronectin		58,104
	Osteopontin		53
	Bone sialoprotein		53
	Thrombin		54
	Laminin Collagen types I and IV		59 60
β5	Vitronectin	Not found in focal contacts, involved in adhesion not migration	9–11
β6	Fibronectin	Potentially restricted to epithelial cells	14,15
β8	?		16

ᵃIt is unclear whether other αv integrins, in addition to the αvβ3 integrin, recognize more than one ligand on the same cell type. The αvβ1 integrin has been reported to recognize as ligands fibronectin and vitronectin on different cell types.

integrin αvβ3 to RGD-containing peptides or vitronectin.[45] This data along with that of other investigators[46] suggests Ca²⁺ may play a regulatory role in integrin ligand binding.

Other sequences within the extracellular domains of the integrin α and β subunits have been implicated in ligand binding, such as several short stretches of invariant amino acids identified in the amino-terminal half of the β integrin subunits.[1-3,42,47-50] Within this conserved region of the β3 subunit, photoaffinity cross-linking experiments of αvβ3 have shown amino acids 61 to 203 to be required for binding to RGD-containing peptides.[42,47] On the αv subunit, amino acids 139 to 3349 are required for RGD-containing peptide binding, and this amino acid sequence overlaps and precedes the amino-terminal side of the first of four calcium binding sites.[42,47] The amino acid sequences required for ligand binding in the other β subunits that pair with the αv subunit (β5, β1, β6, and β8) have not been reported. As noted above, there may be two additional β subunits that pair with αv (β100K and βS).[4,17,26]

In the αv subfamily, integrin αvβ3 is clearly a promiscuous receptor recognizing more than one ligand.[1-3,5-9,12,13] It is unclear whether the additional αv integrins are also promiscuous receptors. Integrin αvβ3 recognizes diverse ligands, such as vitronectin, fibrinogen, von Willebrand factor, fibronectin, thrombospondin, osteopontin, bone sialoprotein, thrombin, collagen types I and IV, and laminin, all of which contain an RGD peptide sequence (see Table 1).[1-3,5-7,51-60] Integrin αvβ1 recognizes fibronectin and potentially vitronectin, with ligand recognition apparently dependent upon the cell type.[12,13] In contrast, integrin αvβ5 recognizes vitronectin specifically.[8,9] Integrin αvβ100K binds vitronectin[26] and integrin αvβS recognizes the RGD peptide;[17] however, the identity or relationship of the β100K and βS integrin subunits to the β5 integrin subunit is unclear. Integrin αvβ6 recognizes at least fibronectin as a ligand,[15] and the ligand for integrin αvβ8 has not been reported. In general, αvβ3, αvβ1, αvβ5, and αvβ6 recognize their ligands in an RGD-dependent manner.[1-3,5-9,12,13,15] Integrin αvβ3 may recognize laminin in a non-RGD-dependent manner.[60] The biological significance of different integrins with similar or identical ligand specificities is not clear. Hypothetically, distinct integrins that recognize the same ligand may convey different signals to the cell, or be responsible for a distinct biologic response (such as adhesion vs. migration) or be of evolutionary necessity in the event of an integrin mutation. This is consistent with recent work demonstrating that αvβ3 is required for metastatic melanoma cell migration toward vitronectin, while αvβ5 fails to play a role in migration.[22]

Integrin αvβ3 binds to vitronectin and fibronectin in a two-step process which is initially RGD-dependent and dissociable, followed by a non-RGD-dependent stabilization between the receptor and ligand.[61] Non-RGD-containing sequences are required for molecular stabilization, since αvβ3 binding to a vitronectin-derived RGD-containing peptide (YPQVTRGDVFTMPED) failed to result in stabilization.[61] Mild glutaraldehyde fixation of purified αvβ3 also resulted in failure of molecular stabilization after binding to vitronectin,[61] suggesting receptor conformation is important in stabilization. Integrin αvβ3 has a 50-fold higher association rate constant for vitronectin as compared to that for the vitronectin-peptide, $9.0 \times 10^3 \text{ s}^{-1} \text{ M}^{-1}$ vs. $1.7 \times 10^2 \text{ s}^{-1} \text{ M}^{-1}$, respectively.[61] Receptor-ligand stabilization occurred after pretreatment of M21 melanoma cells with cytochalasin B (an inhibitor of actin polymerization), suggesting molecular stabilization is independent of the actin-cytoskeleton.[61] Taken together, these data suggest the mechanism of receptor-ligand stabilization involves conformationally dependent protein-protein interactions of the receptor and ligand. This concept is consistent with integrin receptor conformational changes postligand binding reported by other investigators.[25,62]

Integrin αvβ3 appears to be constitutively active. In contrast, the platelet-specific integrin αIIbβ3 (GPIIbIIIa) requires activation, and this function may reside in the cytoplasmic tail domain of the α subunit.[63] Integrin αIIbβ3 which binds fibrinogen after platelet activation appears to have increased ligand affinity in the absence of the α subunit cytoplasmic tail.[63] O'Toole et al.[63] have shown that a mutant molecule which lacks most of the α(IIb) subunit cytoplasmic tail domain is constitutively active. This increased ligand affinity was also detected in a chimeric integrin molecule constructed to contain the GPIIb α subunit extracellular and transmembrane domains, and the α5 subunit cytoplasmic tail.[63] These data suggest deletion of a specific sequence in the αIIb cytoplasmic tail domain is responsible for the increased affinity of the mutant αIIbβ3 integrin. In parallel with the above, Bodary et al.[64] demonstrated that cotransfection of cDNAs for the αIIb and β3 subunits into a human cell line produced a surface αIIbβ3 receptor that promoted adhesion of these cells to fibrinogen-coated substrates; however, the cotransfectants did not appear to bind soluble fibrinogen. This may reflect a need for activation of the αIIbβ3 receptor, as described on platelets.

For some integrins, such as the αLβ2 (LFA-1), the β subunit cytoplasmic tail domain appears to control ligand binding to intercellular adhesion molecules.[65] Hibbs et al.[65] demonstrated that truncation of the cytoplasmic domain of the LFA-1 β subunit eliminated binding to ICAM-1. However, all mutants with truncated αLβ2 α subunits, with the exception of one mutant, were capable of binding to ICAM-1.[65] Therefore, at least for the platelet αIIbβ2 and αLβ3 integrins, the cytoplasmic domain of the α or β subunit, respectively, helps to regulate ligand binding.

IV. *IN VITRO* FUNCTIONS OF THE αv INTEGRINS

In culture, the integrins provide a mechanism for cell adhesion and in general localize to focal contacts or cellular adhesion plaques when cells are adherent to the appropriate matrix protein.[66,67] Cell adhesion often results in cell spreading and involves cytoskeletal protein organization and alignment of stress fibers with focal contacts through integrin association with the actin-cytoskeleton.[1-3,67] For example, integrins αvβ3 and α5β1 localize to focal contacts when cells expressing these receptors are plated on vitronectin[6,7] and fibronectin,[1-3] respectively. The cytoplasmic tail of the β1 subunit has been implicated in focal contact formation.[1-3,68-73] Focal contact formation likely occurs, at least in part, through an interaction of the cytoplasmic domain of the integrin β subunit with cytoskeletal proteins such as talin, vinculin, and α-actinin,[68-73] resulting in integrin linkage of extracellular and intracellular structural elements. Integrin actin association leads to changes in cell shape, such as spreading.[1-3,67-72] Massia and Hubbell[74] have determined that 140 nm is necessary for focal contact and stress fiber formation of integrin αvβ3 on the appropriate ligand. Examples of integrins binding to the cytoskeleton include the chicken CSAT β integrin subunit, which *in vitro* binds talin[68] and an interaction between a β1 integrin cytoplasmic tail peptide and α-actinin.[70] In addition, α-actinin binds to intact heterodimeric αβ1 integrin and the platelet αIIbβ3 (GPIIbIIIa) integrin.[70]

Further evidence to support the involvement of the cytoplasmic tail in integrin localization to focal contacts is supplied by the work of Solowska et al.[75] These investigators constructed a mutant avian integrin β1 subunit lacking the cytoplasmic domain and transfected it into mouse cells.[75] The truncated exogenous β1 subunit formed heterodimers with endogenous α subunits that were exported to the cell surface and bound to fibronectin affinity columns.[75] However, immunofluorescent analysis of these cells adherent to fibronectin demonstrated little, if any, hybrid receptor in focal contacts (exogenous truncated β1 subunit in complex with an endogenous α subunit).[75] These data suggest the β1 cytoplasmic tail is required for localization to focal contacts.

The cytoplasmic tails of the β1 and β3 integrin subunits appear to be functionally interchangeable, based on a chimeric β1/β3 construct composed of the avian β1 subunit extracellular domains and the human β3 subunit transmembrane and cytoplasmic tail domains.[76] This chimeric β subunit in complex with one of three α subunits promoted adhesion and was targeted to focal contacts.[76] The β1/β3 chimeric subunit paired with α3, α5, and α6 subunits and possibly with the αv subunit, but not with the αIIb(GPIIb) subunit.[76] This report taken together with the study of Otey et al.[70] suggests that the cytoplasmic tail domain of the β3 integrin subunit, like the β1 subunit, may well bind or interact with the cytoskeletal protein α-actinin. In addition, the β3 subunit has an alternatively spliced cytoplasmic tail domain that may further enhance the complexity of events involving the cytoplasmic tail.[77]

In contrast to β1 and β3, integrin β5 on human melanoma and lung carcinoma cells adherent to vitronectin fails to localize to focal contacts, but forms a uniform/punctate staining pattern.[78] This suggests that integrin αvβ5 fails to associate with the actin-cytoskeleton.[78] This observation was further substantiated by studies of human lung and pancreatic carcinoma cells which both express integrin αvβ5 but fail to express integrin αvβ3.[22,78] These cells failed to spread when adherent to vitronectin, and integrin αvβ5 failed to localize to focal contacts.[78] This is further supported by the work of Leavesley et al.,[22] demonstrating that pancreatic carcinoma cells expressing integrin αvβ5 failed to change shape when adherent or migrate toward vitronectin; however, when transfected with β3 these cells formed focal contacts when adherent to vitronectin and migrated toward vitronectin.[22] The failure of αvβ5 to associate with the cytoskeleton may involve the β5 cytoplasmic tail. Both α and β integrin subunits typically contain a short cytoplasmic tail of <50 amino acids;[1-3] however, the β5 cytoplasmic tail has a 10-amino

ALIGNMENT OF CYTOPLASMIC TAILS FROM ß SUBUNITS PAIRING WITH αv

ß1 HDRREFAKFEKEKMNAKWDTGENPIYKSAVTTVVNPKYEGK
ß3 HDRKEFAKFEEERARAKWDTANNPLYKEATSTFTNITYRGT
ß5 HDRREFAKFQSERSRARYEMASNPLYRKPISTHTVDFTFNKFNKSYNGTVD
ß6 HDRKEVAKFEAERSKAKWQTGTNPLYRGSTSTFKNVTYKHREKQKVDLSTDC
ß8 WNSNKIKSSSDYRVSASKKDKLILQSVCTRAVTYRREKPEEIKMDISKLNAHETFRCCHF

Figure 2 Amino acids that appear to contribute to focal adhesions are underlined[79] (sequences: β1[105]; β3[41]; β5[10,11]; β6[14]; and β8[16]).

acid extension unlike that found on other known beta integrin subunits.[10,11] In addition, the cytoplasmic tail of the β5 subunit bears only a 10 to 20% homology to the β3 cytoplasmic tail, which is in contrast to the overall amino acid identity (56%) of the β5 and β3 subunits.[10,11] Therefore, the above data along with that of other investigators[68–73] supports the concept that the cytoplasmic tail may be involved in the organization of the cytoskeleton by integrins, and suggests postligand binding events cause integrins αvβ5 and αvβ3 to differentially distribute on the cell surface.

The amino acid sequences in the β1 cytoplasmic tail domain involved in cytoskeletal association have been identified by Reszka et al.[79] Using missense mutagenesis, the latter study demonstrated three clusters of amino acids (cyto-1, cyto-2, and cyto-3) primarily affect the localization of the β1 subunit to focal adhesions.[79] These sequences are highly conserved among integrin β subunits, excluding the β4 and β5 subunits.[79] The cyto-1 region contains four conserved amino acids and likely forms an alpha helix with all four amino acids on one side of the helix.[79] The cyto-2 region is comprised of the sequence NPXY and is thought to form a tight turn due to the proline residue.[79] The cyto-3 region which is similar to the cyto-2 region is comprised of the sequence, NPKY, and it also likely forms a tight turn; however, the cyto-3 region proline can be replaced without loss of focal contact localization.[79] The β5 integrin subunit which fails to localize to focal contacts contains the cyto-1, cyto-2, and cyto-3 regions; however, the cyto-3 region is shifted eight amino acid residues carboxy terminal from its location in other β subunits, suggesting the specific location of the cyto-3 region may be important for focal contact localization.[79] Both β1 and β3 integrins localize at the ends of actin stress fibers in focal adhesions, and the cytoplasmic domains of β1 and β3 integrins are 59% identical; however, in the cyto-1-3 regions the β1 and β3 cytoplasmic domains are 100% identical. The cytoplasmic tail sequences of the β1, β3, and β5 integrin subunits are depicted in Figure 2.

The αv subfamily may mediate cell signals *in vitro;* for example, integrin αvβ3 is involved in bone osteoclast calcium regulation.[53] RGD-containing peptides from bone proteins osteopontin and sialoprotein, as well as intact osteopontin, rapidly stimulate a Ca^{2+} efflux from osteoclasts by activating the plasma membrane Ca^{2+}-ATPase.[53] This Ca^{2+} efflux was inhibited by a monoclonal antibody directed to the ligand binding domain of integrin αvβ3.[53] Osteopontin is a known ligand for the osteoclast integrin αvβ3.[53] An inhibitor of calmodulin (calmodulin is a known regulator of the Ca^{+2}-ATPase) blocked the effect of osteopontin derived RGD-containing peptides, suggesting that ligand binding to osteoclast αvβ3 resulted in a calmodulin-dependent reduction in cytosolic Ca^{2+}.[53] Non-αv integrins have also been implicated in cell signaling, for example, the regulation of intracellular pH,[80–83] intracellular calcium,[84,85] and inositol turnover as well as tyrosine phosphorylation of intracellular proteins.[86]

In vitro, αv integrins appear to be involved in tumor cell migration or invasion. Leavesley et al.[22] recently showed that αvβ3 on human pancreatic carcinoma cells is required for tumor cell migration toward vitronectin.[22] The wild-type human pancreatic carcinoma cell line expressed the αvβ5 integrin but not integrin αvβ3, and the wild-type cells failed to migrate toward vitronectin or fibrinogen.[22] Transfecting the β3 cDNA into the wild-type cell line resulted in cell surface expression of the αvβ3 integrin and migration toward vitronectin and fibrinogen.[22] Previous studies have shown that melanoma, breast carcinoma, and fibrosarcoma cells migrate toward immobilized vitronectin, and this migration was partially inhibited by an antibody capable of blocking cell attachment to vitronectin.[87] This function of αvβ3 may be relevant *in situ*, since human melanoma cells derived from lymph node metastases utilize integrin αvβ3 to adhere to lymph node vitronectin.[23] This report, taken together with the known αvβ3 integrin expression on invasive melanoma cells[19] and the αvβ3 integrin recognition of the vitronectin RGD cell attachment domain,[6-8] suggests αvβ3 on melanoma cells is likely involved in melanoma cell migration toward vitronectin.

In contrast, Seftor et al.[24] have reported that binding of integrin αvβ3 on human melanoma cells by anti-αv or anti-αvβ3 antibodies, as well as the ligand vitronectin, resulted in increased tumor invasion through a Matrigel-coated membrane. The αv subunit appeared to be the subunit capable of mediating the putative invasion-enhancing signal, since an antibody directed to the β3 subunit failed to result in increased invasion.[24] The anti-αv antibody failed to block integrin αvβ3-dependent adhesion to vitronectin, which suggests that the increased invasiveness resulting from ligation of integrin αvβ3 can result from binding of another macromolecule to this integrin, independent of cell adhesion.[24] Antibody directed to integrin α5β1 (anti-fibronectin receptor) failed to stimulate invasion, suggesting the observed effect was specific for integrin αvβ3.[24] In addition, anti-αvβ3 antibodies incubated with these melanoma cells caused an increase in expresssion and secretion of type IV collagenase, while anti-α5β1 antibody failed to cause this effect.[24] These investigators have suggested that signal transduction through integrin αvβ3 could result in elevated expression of the collagenous IV metalloproteinase and increase invasive ability through Matrigel membranes.[24]

V. REGULATION OF THE αv INTEGRINS

Regulation of the integrin receptors is poorly understood. Phosphorylation of two integrin β subunits known to pair with αv has been reported, βS[17] and β1.[88,89] In an osteosarcoma cell line which expresses on its surface both integrins αvβS and αvβ3, the βS subunit and not the β3 subunit is phosphorylated when treated with the phorbol ester tumor promotor phorbol 12-myristate 13-acetate.[17] In addition, protein kinase C phosphorylates the βS subunit of the intact receptor *in vitro* at the same cytoplasmic domain serine.[17] Different amino acid sequences of the cytoplasmic domains of the β3 and βS integrin subunits could account for the phosphorylation of only one of these beta subunits; however, this remains to be determined. Hypothetically, based on the known cytoskeletal associations of some β subunits,[68,70,71] phosphorylation of the βS integrin subunit may serve to regulate the cytoskeletal associations of the αvβS receptor. Hillery et al.[90] have shown the β3 subunit in complex with αIIb(GPIIb) on resting platelets is phosphorylated. In this study, only a small fraction of the β3 (GPIIIa) subunit in the GPIIbIIIa complex was phosphorylated; however, this may be relevant since human platelet protein kinase C (PKC) phosphorylation of purified αIIbβ3 resulted in levels of β3 phosphorylation similar to that of intact platelets.[90] In addition, the incorporation of ^{32}P increased with platelet activation, suggesting the existence of a subpopulation of readily phosphorylated αIIbβ3 that may have distinct biologic properties.[90]

Upon oncogenic transformation, the β1 integrin on a chicken cell line can be phosphorylated on a cytoplasmic tail tyrosine residue.[89] To date, phosphorylation of the cytoskeletal tail of β subunits known to pair with αv has not resulted in altered ligand specificity or affinity, although phosphorylation of receptors in other families has been shown to regulate ligand binding.[17,88,89] This suggests phosphorylation of integrin β subunits may act to regulate biologic events such as cell migration or proliferation, rather than ligand binding. In contrast to the β integrin subunits, phosphorylation of the α6 integrin subunit helps to regulate macrophage adhesion to laminin.[91] Phosphorylation of αv has not been reported.

Protein kinase C (PKC)-dependent phosphorylation of integrins may help to regulate organization of the cytoskeleton and focal adhesion formation.[92] Woods and Couchman[92] have recently shown that PKC can promote focal adhesion formation of integrin β1 under conditions where focal adhesions do not normally form. Fibroblasts spread on fibronectin substrata in the presence of kinase inhibitors H7 and HA1004 (concentrations expected to inhibit PKC) demonstrated reduction in β1 focal adhesion formation.[92] In addition, inhibition of PKC, but not of cyclic AMP or cyclic GMP-dependent kinases, prevented stress fiber formation and induced dispersal of talin and vinculin.[92] In contrast, activation of PKC induced focal adhesion formation of integrin β1.[92]

Additional mechanisms by which αv integrins might be regulated include modulation by growth factors and cytokines.[93-96] Transforming growth factor-β1 (TGF-β1) incubated with transformed fibroblasts and osteogenic sarcoma cells has been reported to increase β3 mRNA resulting in increased αvβ3 expression on the cell surface.[93] In addition, TGF-β1 has been shown to enhance β1 integrin expression in WI-38 lung fibroblasts.[94] Bates et al.[95] demonstrated that TGF-β1, platelet-derived growth factor (PDGF), and tumor necrosis factor α (TNFα) failed to alter the ratio of four αv-containing integrins (αvβ3, αvβ1, and the αv subunit in association with two other beta subunits, likely β5 and β6) on the cell surface of embryonic fibroblasts.[95] In the same study, overnight culture in basic fibroblast growth factor (bFGF) decreased the level of expression of integrin αvβ1 and likely integrin αvβ5.[95] In contrast, the cytokine TNFα has been reported to downregulate the protein expression and mRNA levels of the β3 integrin subunit on confluent umbilical vein endothelial cells, while no effect of TNFα was observed on the αv integrin subunit.[96] The latter is consistent with the decreased ability of these endothelial cells to adhere to vitronectin while maintaining normal adherence to fibronectin.[96] Therefore, the possibility exists that growth factors, such as TGF-β1, PDGF, and bFGF, as well as cytokines, such as TNFα, may help to modulate αv integrin expression; however, their effect needs to be further studied on the same cell type, in both transformed and nontransformed cells. At present little is known about the mechanism(s) regulating the αv integrin repertoire on a single cell type. Conceivably, all β chains compete equally well for binding to αv. In such a case, an increase in the expression of one integrin may be the result of a reduction in one of the competing β subunits.

VI. CELL AND TISSUE EXPRESSION

The αvβ3 integrin has been reported on many cell types in culture, including osteosarcoma, melanoma, malignant astrocyte, endothelial, keratinocyte, chick retinal neuron, fibroblast, placenta, endometrial epithelial, monocytes, platelet, osteoclast, neuroblastoma, and pancreatic ductal carcinoma cells, and in some cases on these cells *in situ*, as well.[1-7,19-24,26-28,30,32,34,37,53] Messenger RNA and in some cases protein expression of the αv and β5 integrin subunits have been reported *in vitro* for erythroleukemia, lung carcinoma, embryonic kidney adenocarcinoma, osteosarcoma, cervical carcinoma, monocytic leukemia, hepatocarcinoma cells, and keratinocytes.[8-11,28,78] The αvβ5 integrin has not been reported *in situ*, possibly due to the difficulty in obtaining an antibody directed to the extracellular domains of the β5 subunit that does not cross-react with the highly homologous extracellular domains of the β3 integrin subunit. The αvβ1 integrin has been reported on fibroblasts, neuroblastoma, embryonic kidney, and embryonal carcinoma cells in culture, and αvβ1 protein expression has not been reported *in situ.*[12,13,27] Currently, antibodies directed toward the αvβ1 integrin complex are not available. Integrin αvβS has been identified on fibroblasts and MG-63 osteosarcoma cells in culture,[17] integrin αvβ6 has been identified on pancreatic carcinoma and endometrial cells,[14] and integrin αvβ8 has been identified on lung fibroblasts in culture.[16]

VII. EXAMPLES OF *IN SITU* FUNCTIONS
A. TRANSFORMATION AND TUMOR INVASION

The repertoire of integrins has been shown to be altered in transformed cells.[1-3,18-21,97-99] Of the αv integrins, αvβ3 has been most closely characterized in animal models of malignant tumors.[100-103] These animal studies have implicated integrin αvβ3 in melanoma cell proliferation.[100-103] For example, an antibody to the β3 integrin subunit has been reported to inhibit the growth of human melanoma cells in nude mice.[101] Similarly, RGD-containing peptides known to block the function of integrin αvβ3 have been shown to inhibit both melanoma tumor invasion *in vitro* and the development of experimental metastases in a murine melanoma model system.[100,102,103] RGD-containing peptides or their mimetics may prove to be clinically useful in blocking tumor invasion, since occupied integrin receptors on benign cells, such as in the vasculature, are probably highly stable[61] and the peptides would therefore be less likely to promote the reversal of the adhesive phenotype leading to an adverse clinical effect.

Consistent with the above results, Albelda et al.[19] have shown αvβ3 is not expressed on normal melanocytes, nevi, or horizontally spreading primary melanoma; however, vertically invasive primary melanoma and metastatic melanomas express integrin αvβ3. This finding is consistent with the recent report demonstrating the necessity of the αv integrin for melanoma cell tumorigenicity in the athymic nude mouse.[21] In the latter study, an αv negative variant melanoma cell line selected by fluorescent activated cell sorter analysis (FACS) was found to be minimally tumorigenic in athymic nude mice; however, transfection of the cDNA for the αv integrin subunit resulted in restoration of αvβ3 integrin expression on the cell surface and tumorigenicity similar to the wild-type melanoma cell line.[21] Integrin αvβ3 is not only important in melanoma cell proliferation, but also in melanoma cell metastasis to lymphnodes.[23] Nip et al.[23] have shown that metastatic melanoma cells utilize αvβ3 to attach to vitronectin within lymph nodes.

In the brain, we have shown integrin αvβ3 to be expressed on transformed glial cells *in situ* and *in vitro;* while, in contrast, it is not expressed on normal adult brain glia, suggesting it is a marker of the transformed phenotype.[20] In addition, a ligand for αvβ3, vitronectin, was also expressed in malignant glial-derived tumors and it colocalized with integrin αvβ3.[20] Vitronectin, like αvβ3, was not expressed in the normal adult brain.[20] Cultured glioblastoma cells (the most malignant glial-derived tumor cell) attached to the parenchyma of glioblastoma tumor cryostat sections where vitronectin was expressed, and this attachment was inhibited with antibodies directed toward integrin αvβ3 and vitronectin, as well as with RGD-containing peptides, confirming the availability of vitronectin to serve as an adhesive ligand in these tumors.[20] These data suggest integrin αvβ3 and its ligand vitronectin may play a role in malignant glial cell adhesion and potentially invasion of the normal brain.

B. DEVELOPMENT AND DIFFERENTIATION

In the chick retina, an αv integrin (αvβ100K) is involved in neurite outgrowth on the substrate vitronectin.[26] Retinal neuron attachment and neurite outgrowth is integrin αvβ100K-dependent in the late stages of chick retinal development, suggesting the receptor for vitronectin is developmentally regulated.[26] The expression of integrin αvβ100K corresponds to the localization of vitronectin in the optic stalk, optic nerve, inner limiting membrane (the basement membrane adjacent to the optic fiber layers), and surrounding neuroepithelial and ganglion cells. Like integrin αvβ100K, vitronectin is also expressed in the late stages of chick retinal development.[26] The β100K subunit may correspond to β5, βS, β6, or β8 subunits; however, it is clearly not the β1 or β3 integrin subunits.[26] Integrin αvβ3 has also been identified on retinal neurons; yet, it does not appear to be involved in retinal neurite attachment or process outgrowth. The αv subunit does not pair with the β1 subunit on chick retinal neurons, as described on several cell types *in vitro*.[26] Thus, vitronectin and its receptor may regulate cell adhesion, migration, and process outgrowth during chick retinal development.[26]

In contrast to the developing chick retina, expression of integrin αvβ1 was shown to correlate with neuronal differentiation of murine embryonal carcinoma cells stimulated with retinoic acid.[27] Neurite outgrowth was observed after *trans*-retinoic acid stimulation for 8 days and closely corresponded to an induction of the αvβ1 heterodimer (18-fold induction), and less so of the β3 subunit (3-fold induction).[27] Induction of the β1 subunit by retinoic acid occurred at the level of mRNA, with peak β1 mRNA induction at 6 days whereas the cell surface expression of β1 protein peaked at 8 days postexposure to retinoic acid.[27] The extent of induction of β1 protein expression at the cell surface was not entirely accounted for by a corresponding increase in β1 mRNA, suggesting retinoic acid may also affect the rate of processing and transport of integrins.[27] In the comparison untreated control cells, integrin αvβ1 was present in very small amounts on the cell surface.[27] Stimulation of a variant retinoic acid-resistant embryonal carcinoma cell line with retinoic acid failed to induce expression of αvβ1, suggesting expression of this integrin may be important in neuronal differentiation induced by retinoic acid.[27]

Surface expression of αv integrins shows changes in developing keratinocytes.[28] In culture these cells undergo terminal differentiation and fail to adhere to extracellular matrix proteins, consistent with the localization of integrins to only the basal cells by immunofluorescent staining.[28] In nondifferentiated keratinocytes in culture, the αv subunit pairs with β3, β1, and β5, while in terminally differentiated keratinocytes in culture, the αv subunit only pairs with β5.[28] This report suggests that at least in culture integrin αvβ3 is developmentally regulated on keratinocytes.

Another example of a developmental regulation of integrin αvβ3 is seen in the monthly hormonal cycle of the endometrial epithelium.[29] Expression of αvβ3 undergoes cycle-specific changes on endometrial epithelial cells.[29] By immunohistochemistry, these investigators demonstrated that expression of the αv subunit increased from day 1 of the monthly cycle, while the β3 subunit appeared abruptly on cycle day 20.[29] Infertility patients with a discordant luteal phase showed delayed epithelial β3 expression.[29] This report suggests that women who lack the β3 subunit may be infertile, and demonstrates a tight regulation of β3 expression in the endometrial cycle.[29] Expression of αvβ3 on endometrial epithelial cells may facilitate implantation of a fertilized egg. In the latter case, it is conceivable that antagonists of integrin αvβ3 may prove useful in a birth control model.

C. BONE RESORPTION

Osteoclasts are derived from bone marrow precursor cells and migrate to bone where they subsequently fuse to form osteoclasts.[30] A functional antigen, termed the osteoclast functional antigen (OFA), appears to be integrin αvβ3 and was identified on mature osteoclasts and their immediate precursors, but not on cells of the mononuclear phagocytic system.[30] *In situ* and *in vitro* studies have shown αvβ3 to be expressed by human, chick, and rat osteoclasts.[30] Integrin αvβ3 was not expressed on osteoblasts or other bone marrow cells *in situ* by immunohistochemistry,[30] although *in vitro* studies of osteosarcoma cell lines (which produce bone *in situ*) have demonstrated αvβ3 expression.[1-3] Blocking antibodies directed toward integrin αvβ3 as well as RGD-containing peptides have resulted in osteoclast retraction or inhibition of bone resorption, and inhibition of osteoclast motility.[30] Potential ligands for αvβ3 in bone include osteopontin, which contains an RGD peptide sequence.[28,30] Osteopontin is synthesized by osteoblasts and has been identified adjacent to bone immediately prior to osteoclast invasion, by *in situ* hybridization.[28,30] Taken together, these data suggest αvβ3 plays a role in osteoclast resorption of bone. Currently, several pharmaceutical firms are designing specific antagonists of the osteoclast αvβ3 to be used as potential inhibitors of osteoporosis.

D. IMMUNE RESPONSE

Macrophage recognition of apoptotic cells (both neutrophils and lymphocytes) is mediated by integrin αvβ3.[35] There appears to be a role for integrin αvβ3 in self-senescent-self-intercellular recognition leading to macrophage phagocytosis of cells undergoing apoptosis.[35] The macrophage αvβ3 recognizes aged neutrophils in a RGD-dependent and divalent cation-dependent manner.[35] The ligand on aged neutrophils that the macrophage αvβ3 recognizes is not known. Antibodies specific for the αv integrin subunit and the β3 integrin subunit specifically inhibit recognition of apoptotic or senescent neutrophils and lymphocytes by monocyte-derived macrophages.[35] The expression of integrin αvβ3 on monocyte-derived macrophages appears to be maturation related and apparently determines the capacity of these cells to recognize aged neutrophils and lymphocytes.[35] In contrast, freshly isolated peripheral blood monocytes fail to express αvβ3, but after 3 days in culture these monocytes begin to express integrin αvβ3 and the αv subunit paired with a β subunit denoted βx.[32] Integrin αvβx identified on cultured monocytes was not found on platelets, B lymphoblastoid, melanoma, endothelial, or erythroleukemia cell lines, suggesting the βx subunit may be unique to monocytic cells.[32] In addition, integrin αvβx could not be identified prior to nine days in culture, which is in contrast to integrin αvβ3.[32] The expression of αvβ3 and αvβx on monocytes may be differentially regulated.[32] Integrin αvβ3 is expressed on macrophages in a maturation-related process, and may represent a mechanism that limits tissue injury through clearance of neutrophils at an inflamed site.

In addition, a vitronectin receptor appears to be involved in the adhesion of complement to platelets or cells.[36] Vitronectin binds to the C5b-9 complement complex, and, when bound, vitronectin is recognized by a vitronectin receptor.[36] Skeletal muscle cell-derived myoblasts adhere to the C5b-9-vitronectin complex through a vitronectin receptor integrin, and this adhesion is inhibited by RGD-containing peptides, EDTA, and anti-vitronectin, as well as antivitronectin receptor antibodies.[36] Hypothetically, the vitronectin cell attachment site in the vitronectin-C5b-9 complement complex could form sites for platelet binding, as well as attachment sites for the regrowth of fibroblasts and other tissue.[36]

Integrin αvβ3 in the mouse may serve as an accessory molecule for T cell activation.[38] It is expressed on a variety of T cell lines and Con A-activated splenocytes, but not on resting T cells.[38] In addition, engagement of integrin αvβ3 by its ligand may be necessary, but not sufficient, for the induction of IL-4 production.[37]

VIII. SUMMARY

The αv integrins pair with seven potentially different β subunits to form receptors recognizing a number of different ligands. Integrin αvβ3 has been most closely studied and it has been shown to play a role in tumor invasion, proliferation and metastasis, bone resorption, differentiation, development, and the immune response. Future studies will likely continue to elicit *in vivo* functions for integrin αvβ3, define the biologic role of integrins αvβ5, αvβS, αvβ100K, αvβ1, αvβ6, αvβ8, and αvβx, as well as focus on the regulation of these receptors and their role in signal transduction.

REFERENCES

1. Hynes, R. O., Integrins: versatility, modulation and signaling in cell adhesion, *Cell*, 69, 11, 1992.
2. Ruoslahti, E., Integrins, *J. Clin. Invest.*, 87, 1, 1991.
3. Albelda, S. M. and Buck, C. L., Integrins and other cell adhesion molecules, *FASEB J.*, 4, 2868, 1990.
4. Bossy, B. and Reichardt, L. F., Chick integrin αv subunit molecular analysis reveals high conservation of structural domains and association with multiple beta subunits in embryo fibroblasts, *Biochemistry*, 29, 10191, 1990.
5. Pytela, R., Pierschbacher, M. D., and Ruoslahti, E., 125/115 kDa cell surface receptor specific for vitronectin interacts with the arginine-glycine-aspartic acid adhesion sequence derived from fibronectin, *Proc. Natl. Acad. Sci. U.S.A.*, 82, 5766, 1985.
6. Cheresh, D. A. and Spiro, R. C., Biosynthetic and functional properties of an arg-gly-asp-directed receptor involved in human melanoma cell attachment to vitronectin, fibrinogen and von Willebrand factor, *J. Biol. Chem.*, 262, 17703, 1987.
7. Cheresh, D. A., Structure, function and biological properties of integrin $\alpha_v\beta_3$ on human melanoma cells, *Cancer Metast. Rev.*, 10, 3, 1991.
8. Cheresh, D. A., Smith, J. W., Cooper, H. M., and Quaranta, V., A novel vitronectin receptor integrin (αvβx) is responsible for distinct adhesive properties of carcinoma cells, *Cell*, 57, 59, 1989.
9. Smith, J. W., Vestal, D. J., Irwin, S. V., Burke, T. A., and Cheresh, D. A., Purification and functional characterization of interin αvβ5. An adhesion receptor for vitronectin, *J. Biol. Chem.*, 265, 11008, 1990.
10. McLean, J. W., Vestal, D. J., Cheresh, D. A., and Bodary, S. C., cDNA sequence of the human integrin β5 subunit, *J. Biol. Chem.*, 265, 17126, 1990.
11. Ramaswamy, H. and Hemler, M. E., Cloning, primary structure and properties of a novel human integrin β subunit, *J. Eur. Mol. Biol. Organ.*, 9, 1561, 1990.
12. Bodary, S. C. and McLean, J. W., The integrin β1 subunit associates with the vitronectin receptor αv to form a novel vitronectin receptor in a human embryonic kidney cell line, *J. Biol. Chem.*, 265, 5938, 1990.
13. Vogel, B. E., Tarone, G., Giancotti, F. G., Gailit, J., and Ruoslahti, E., A novel fibronectin receptor with an unexpected subunit composition (αvβ1), *J. Biol. Chem.*, 265, 5934, 1990.
14. Sheppard, D., Rozzo, C., Starr, L., Quaranta, V., Erle, D. J., and Pytela, R., Complete amino acid sequence of a novel integrin β subunit (β6) identified in epithelial cells using the polymerase chain reaction, *J. Biol. Chem.*, 265, 11502, 1990.
15. Busk, M., Pytela, R., and Sheppard, D., Characterization of the integrin αvβ6 as a fibronectin binding protein, *J. Biol. Chem.*, 267, 5790, 1992.
16. Moyle, M., Napier, M. A., and McLean, J. W., Cloning and expression of a divergent integrin subunit β8, *J. Biol. Chem.*, 266, 19650, 1991.
17. Freed, E., Gailit, J., van der Geer, P., Ruoslahti, E., and Hunter, T., A novel integrin β subunit is associated with the vitronectin receptor α subunit (αv) in a human osteocarcinoma cell line and is a substrate for protein kinase C, *J. Eur. Mol. Biol. Organ.*, 8, 2955, 1989.
18. Ruoslahti, E. and Giancotti, F. G., Integrins and tumor cell dissemination, *Cancer Cells*, 1, 119, 1989.

19. Albeda, S. M., Mette, S. A., Elder, D. E., Stewart, R., Damjanovich, L., Herlyn, M., and Buck, C. A., Integrin distribution in malignant melanoma: association of the β3 subunit with tumor progression, *Cancer Res.*, 50, 6757, 1990.

20. Gladson, C. L. and Cheresh, D. A., Glioblastoma expression of vitronectin and the αvβ3 integrin: adhesion mechanism for transformed glial cells, *J. Clin. Invest.*, 88, 1924, 1991.

21. Felding-Habermann, B., Mueller, B. M., Romerdahl, C., and Cheresh, D. A., Requirement of integrin αv gene expression for human melanoma tumorigenicity, *J. Clin. Invest.*, 89, 2018, 1992.

22. Leavesley, D. I., Ferguson, G. D., Wayner, E. A., and Cheresh, D. A., Requirement of the integrin β3 subunit for carcinoma cell spreading or migration on vitronectin and fibronectin, *J. Cell Biol.*, 117, 1101, 1992.

23. Nip, J., Schibata, H., Loskutoff, D. J., Cheresh, D. A., and Brodt, P., Human melanoma cells derived from lymphnode metastases utilize integrin αvβ3 to adhere to lymphnode vitronectin, *J. Clin. Invest.*, in press.

24. Seftor, R. E., Seftor, E. A., Gehlsen, K. R., Stetler-Stevenson, W. G., Brown, P. D., Ruoslahti, E., and Hendrix, M. J., Role of the alpha v beta 3 integrin in human melanoma cell invasion, *Proc. Natl. Acad. Sci. U.S.A.*, 89, 1557, 1992.

25. Neugebauer, K. M., and Reichardt, L. F., Cell-surface regulation of β1-integrin activity on developing retinal neurons, *Nature*, 350, 68, 1991.

26. Neugebauer, K. M., Emmett, C. J., Venstrom, K. A., and Reichardt, L. F., Vitronectin and thrombospondin promote retinal neurite outgrowth: developmental regulation and role of integrins, *Neuron*, 6, 345, 1991.

27. Dedhar, S., Robertson, K., and Gray, V., Induction of expression of the αvβ1 and αvβ3 integrin heterodimers during retinoic acid-induced neuronal differentiation of murine embryonal carcinoma cells, *J. Biol. Chem.*, 266, 21846, 1991.

28. Adams, J. C. and Watt, F. M., Expression of β1, β3, β4 and β5 integrins by human epidermal keratinocytes and non-differentiating keratinocytes, *J. Cell Biol.*, 115, 829, 1991.

29. Lessey, B. A., Damgnovich, L., Coutifaris, C., Castelbaum, A., Abelda, S. M., and Buck, C. A., Integrin adhesion molecules in the human endometrium. Correlation with the normal and abnormal menstrual cycle, *J. Clin. Invest.*, 90, 188, 1992.

30. Davies, J., Warwick, J., Totty, N., Philp, R., Helfrich, M., and Horton, M., The osteoclast functional antigen, implication in the regulation of bone resorption, is biochemically related to the vitronectin receptor, *J. Cell Biol.*, 109, 1817, 1989.

31. Horton, M. A., Taylor, M. L., Arnett, T. R., and Helfrich, M. H., Arg-Gly-Asp (RGD) peptides and the anti-vitronectin receptor antibody 23C6 inhibit dentine resorption and cell spreading by osteoclasts, *Exp. Cell Res.*, 195, 368, 1991.

32. Krissansen, G. W., Elliot, M. J., Lucas, C. M., Stomski, F. C., Berndt, M. C., Cheresh, D. A., Lopez, A. F., and Burns, G. F., Identification of a novel integrin β subunit expressed on cultured monocytes (macrophages): evidence that one α subunit can associate with multiple β subunits, *J. Biol. Chem.*, 265, 823, 1991.

33. Klingermann, H.-G. and Dedhar, S., Distribution of integrins on human peripheral blood mononuclear cells, *Blood*, 74, 1348, 1989.

34. Maxfield, S. R., Moulder, K., Koning, F., Elbe, A., Stingl, G., Coligan, J. E., Shevach, E. M., and Yokoyama, W. M., Murine T cells express a cell surface receptor for multiple extracellular matrix proteins, *J. Exp. Med.*, 169, 2173, 1989.

35. Savill, J., Dransfield, J., Hogg, N., and Haslett, C., Vitronectin receptor-mediated phagocytosis of cells undergoing apoptosis, *Nature*, 343, 170, 1990.

36. Biesecker, G., The complement SC5b-9 complex mediates cell adhesion through a vitronectin receptor, *J. Immunol.*, 145, 209, 1990.

37. Roberts, K., Yokoyama, W. M., Kehn, P. J., and Shevach, E. M., The vitronectin receptor serves as an accessory molecule for the activation of a subset of gamma/delta T cells, *J. Exp. Med.*, 173, 231, 1991.

38. Moulder, K., Roberts, K., Shevach, E. M., and Coligan, J. E., The mouse vitronectin receptor is a T cell activation antigen, *J. Exp. Med.*, 173, 343, 1991.

39. Suzuki, S., Argraves, W. S., Pytela, R., Arai, H., Krusius, T., Pierschbacher, M. D., and Ruoslahti, E., cDNA and amino acid sequences of the cell adhesion protein receptor recognizing vitronectin reveal a transmembrane domain and homologies with other adhesion protein receptors, *Proc. Natl. Acad. Sci. U.S.A.*, 83, 8614, 1986.

40. Suzuki, A., Argraves, W. S., Arai, H., Languino, L. R., Pierschbacher, M. D., and Ruoslahti, E., Amino acid sequence of the vitronectin receptor α subunit and comparative expression of adhesion receptor mRNAs, *J. Biol. Chem.*, 262, 14080, 1987.

41. Fitzgerald, L. A., Steiner, B., Rall, S. C., Jr., Lo, S., and Phillips, D. R., Protein sequence of endothelial glycoprotein IIIa derived from a cDNA clone. Identity with platelet glycoprotein IIIa and similarity to integrin, *J. Biol. Chem.*, 262, 3936, 1987.

42. Smith, J. W. and Cheresh, D. A., Integrin (αvβ3) ligand interaction: identification of a divalent cation-dependent, heterodimeric RGD binding site on the vitronectin receptor, *J. Biol. Chem.*, 265, 2168, 1990.

43. Smith, J. W. and Cheresh, D. A., Labeling of integrin alpha v beta 3 with ^{58}Co(III). Evidence of metal ion coordination sphere involvement in ligand binding, *J. Biol. Chem.*, 266, 11429, 1991.

44. Gailit, J. and Ruoslahti, E., Regulation of the fibronectin receptor affinity by divalent cations, *J. Biol. Chem.*, 263, 12927, 1988.

45. Kirchhofer, D., Grzesiak, J., and Pierschbacher, M. D., Calcium as a potential physiological regulator of integrin-mediated cell adhesion, *J. Biol. Chem.*, 266, 4471, 1991.

46. Loftus, J. C., O'Toole, T. E., Plow, E. F., Glass, A., Frelinger, A. L., III, and Ginsberg, M. H., A β3 integrin mutation abolished ligand binding and alters divalent cation dependent conformation, *Science*, 249, 915, 1990.

47. Smith, J. W. and Cheresh, D. A., The Arg-Gly-Asp binding domain of the vitronectin receptor, *J. Biol. Chem.*, 4263, 18726, 1988.

48. D'Souza, S. E., Ginsberg, M. H., Lam, S., and Plow, E., Chemical crosslinking of arginyl-glycyl-aspartic acid peptides on adhesion receptors on platelet, *Science*, 242, 91, 1988.

49. D'Souza, S. E., Ginsberg, M. H., and Plow, E. F., Arginyl-glycyl-aspartic acid (RGD): a cell adhesion motif, *TIBS*, 16, 246, 1991.

50. Yamada, K. M., Adhesive recognition sequences, *J. Biol. Chem.*, 266, 12809, 1991.

51. Cheresh, D. A., Human endothelial cells synthesize and express an arg-gly-asp directed receptor involved in attachment to fibrinogen and von Willebrand factor, *Proc. Natl. Acad. Sci. U.S.A.*, 84, 6471, 1987.

52. Cheresh, D. A., Berliner, S. A., Vicente, V., and Ruggeri, Z. M., Recognition of distinct adhesive sites on fibrinogen by related integrins on platelet and endothelial cells, *Cell*, 58, 945, 1989.

53. Miyauchi, A., Alvarez, J., Greenfield, E. M., Teti, A., Grano, M., Colucci, S., Zambonin-Zallone, A., Ross, F. P., Teitelbaum, S. L., Cheresh, D., and Hruska, K. A., Recognition of osteopontin and related peptides by an $\alpha_v\beta_3$ integrin stimulates immediate cell signals in osteoclasts, *J. Biol. Chem.*, 266, 20369, 1991.

54. Bar-Shavit, R., Sabbah, V., Lampugnani, M. G., Marchisio, P. C., Fenton, J. W., Vlodavsky, I., and Dejana, E., An Arg-Gly-Asp sequence within thrombin promotes endothelial cell adhesion, *J. Cell Biol.*, 112, 335, 1991.

55. Kramer, R. H., Cheng, Y.-F., and Clyman, R., Human microvascular endothelial cells

use β1 and β3 integrin receptor complexes to attach to laminin, *J. Cell Biol.*, 111, 1233, 1990.

56. Lawler, J., Weistein, R., and Hynes, R. O., Cell attachment to thrombospondin: the role of RGD and integrin receptors, *J. Cell Biol.*, 107, 2351, 1988.

57. Sun, X., Skorstengaard, K., and Mosher, D. F., Disulfides modulate RGD-inhibitable cell adhesive activity of thrombospondin, *J. Cell Biol.*, 118, 693, 1992.

58. Cheng, Y.-F., Clyman, R. I., Enenstein, J., Waleh, N., Pytela, R., and Kramer, R. H., The integrin complex αvβ3 participates in the adhesion of microvascular endothelial cells to fibronectin, *Exp. Cell Res.*, 194, 69, 1991.

59. Kramer, R. H., Cheng, Y.-F., and Clyman, R., Human microvascular endothelial cells use β1 and β3 integrin receptor complexes to attach to laminin, *J. Cell Biol.*, 111, 1233, 1990.

60. Clyman, R. I., Mauray, F., and Kramer, R. H., β1 and β3 integrins have different roles in the adhesion and migration of vascular smooth muscle cells on extracellular matrix, *Exp. Cell Res.*, 200, 272, 1992.

61. Orlando, R. A. and Cheresh, D. A., Arg-gly-asp-binding leading to molecular stabilization between integrin $\alpha_v\beta_3$ and its ligand, *J. Biol. Chem.*, 266, 19543, 1991.

62. Frelinger, A. L., III, Lam, S. C.-T., Plow, E. F., Smith, M. A., Loftus, J. C., and Ginsberg, M. H., Occupancy of an adhesive glycoprotein receptor modulates expression of an antigenic site involved in cell adhesion, *J. Biol. Chem.*, 263, 12397, 1988.

63. O'Toole, T. E., Mandelman, D., Forsyth, I., Shattil, S. J., Plow, E. F., and Ginsberg, M. H., Modulation of the affinity of integrin alpha IIb beta 3 (GPIIb-IIIa) by the cytoplasmic domain of alpha IIb, *Science*, 254, 845, 1991.

64. Bodary, S. C., Napier, M. A., and McLean, J. W., Expression of recombinant platelet glycoprotein IIbIIIa results in a functional fibrinogen binding complex, *J. Biol. Chem.*, 264, 18859, 1989.

65. Hibbs, M. L., Stacker, S. A., and Springer, T. A., Regulation of adhesion of ICAM-1 by the cytoplasmic domain of LFA-1 integrin beta subunit, *Science*, 251, 1611, 1991.

66. Singer, I. I., Scott, S., Kawka, D. W., Kazazis, D. M., Gailit, J., and Ruoslahti, E., Cell surface distribution of fibronectin and vitronectin receptors depends on substrate composition and extracellular matrix accumulation, *J. Cell Biol.*, 106, 2171, 1988.

67. Burridge, K., Fath, K., Kelly, T., Nuckolls, G., and Turner, C., Focal adhesions: transmembrane junctions between the extracellular matrix and the cytoskeleton, *Annu. Rev. Cell Biol.*, 4, 487, 1988.

68. Horwitz, A., Duggan, K., Buck, C., Beckerle, M. C., and Burridge, K., Interaction of plasma membrane fibronectin receptor with talin, a transmembrane linkage, *Nature*, *(London)*, 320, 531, 1986.

69. Nuckolls, G. H., Turner, C. E., and Turner, K., Functional studies of the domains of talin, *J. Cell Biol.*, 110, 1635, 1990.

70. Otey, C. A., Pavalko, F. M., and Burridge, K., An interaction between α-actinin and the β1 integrin subunit *in vitro*, *J. Cell Biol.*, 11, 721, 1990.

71. Argraves, W. S., Dickerson, K., Burgess, W. H., and Ruoslahti, E., Fibulin, a novel protein that interacts with the fibronectin receptor beta subunit cytoplasmic domain, *Cell*, 58, 623, 1989.

72. Hayashi, Y., Haimovich, B., Reszka, A., Boettiger, D., and Horwitz, A., Expression and function of chicken integrin β1 subunit and its cytoplasmic domain in mouse NIH 3T3 cells, *J. Cell Biol.*, 110, 175, 1990.

73. Quaranta, V. and Jones, J. C. R., The internal affairs of an integrin, *Trends Cell Biol.*, 1, 2, 1991.

74. Massia, S. P. and Hubbell, J. A., An RGD spacing of 440 nm is sufficient for integrin αvβ3-mediated fibroblast spreading and 140 nm for focal contact and stress fiber formation, *J. Cell. Biol.*, 114, 1089, 1991.

75. Solowska, J., Guan, J.-L., Marcantonio, E. E., Trevithick, J. E., Buck, C. A., and Hynes, R. O., Expresssion of normal and mutant avian integrin subunits in rodent cells, *J. Cell Biol.*, 109, 853, 1989.

76. Solowska, J., Edelman, J. M., Abelda, S. M., and Buck, C. A., Cytoplasmic and transmembrane domains of integrin β1 and β3 subunits are functionally interchangeable, *J. Cell Biol.*, 114, 1079, 1991.

77. van Kuppevelt, T. H. M. S. M., Languino, L. R., Gailit, J. O., Suzuki, S., and Ruoslahti, E., An alternative cytoplasmic domain of the integrin β3 subunit, *Proc. Natl. Acad. Sci. U.S.A.*, 86, 5415, 1989.

78. Wayner, E. A., Orlando, R. A., and Cheresh, D. A., Integrins $\alpha_v\beta_3$ and $\alpha_v\beta_5$ contribute to cell attachment to vitronectin but differentially distribute on the cell surface, *J. Cell Biol.*, 113, 919, 1991.

79. Reszka, A. A., Hayashi, Y., and Horwitz, A. F., Identification of amino acid sequences in the integrin β1 cytoplasmic domain implicated in cytoskeletal association, *J. Cell Biol.*, 117, 1321, 1992.

80. Schwartz, M. A., Cragoe, E. J., and Lechene, C., Effect of cell spreading on cytoplasmic pH in normal and transformed fibroblasts, *Proc. Natl. Acad. Sci. U.S.A.*, 86, 4525, 1989.

81. Schwartz, M. A., Cragoe, E. J., and Lechene, C., pH regulation in spread and round cells, *J. Biol. Chem.*, 265, 1327, 1990.

82. Schwartz, M. A., Ingber, D. E., Lawrence, M., Springer, T. A., and Lechene, C., Multiple integrins share the ability to induce elevation of intracellular pH, *Exp. Cell Res.*, 195, 533, 1991.

83. Ingber, D. E., Prusty, D., Frangioni, J. J., Cragoe, E. J., Lechene, C. P., and Schwartz, M. A., Control of intracellular pH and growth by fibronectin in capillary endothelial cells, *J. Cell Biol.*, 110, 1803, 1990.

84. Ng-Sikorski, J., Andersson, R., Patarroyo, M., and Andersson, T., Calcium signaling capacity of the CDIIb/CD18 integrin on human neutrophils, *Exp. Cell Res.*, 195, 504, 1991.

85. Richter, J., Ng, S. J., Olsson, I., and Andersson, T., Tumor necrosis factor-induced degranulation in adherent human neutrophils is dependent on CDIIb/CD18-integrin-triggered oscillations of cytosolic free Ca^{2+}, *Proc. Natl. Acad. Sci. U.S.A.*, 87, 9472, 1990.

86. Conforti, G., Zanetti, A., Pasquali-Ronchetti, I., Quaglino, D., Neyroz, P., and Dejanna, E., Modulation of vitronectin receptor binding by membrane lipid composition, *J. Biol. Chem.*, 265, 4011, 1990.

87. Basara, M. L., McCarthy, J. B., Barnes, D. W., and Furcht, L. T., Stimulation of haptotaxis and migration of tumor cells by serum spreading factor, *Cancer Res.*, 45, 2487, 1985.

88. Haimovich, B., Ancskievich, B. J., and Boettiger, D., Cellular partitioning of β1 integrin and their phosphorylated forms is altered after transformation by Rous sarcoma virus or treatment with cytochalasin D, *Cell Regul.*, 2, 271, 1991.

89. Hirst, R., Horwitz, A. F., Buck, C., and Rohrschneider, L., Phosphorylation of the fibronectin receptor complex in cells transformed by oncogenes that encode tyrosine kinases, *Proc. Natl. Acad. Sci. U.S.A.*, 83, 6470, 1986.

90. Hillery, C. A., Smyth, S. S., and Parise, L. V., Phosphorylation of human platelet glycoprotein IIIa (GPIIIa). Dissociation from fibrinogen receptor activation and phosphorylation of GPIIIa in vitro, *J. Biol. Chem.*, 266, 14663, 1991.

91. Shaw, L. M., Messier, J. M., and Mercurio, A. M., The activation dependent adhesion of macrophages to laminin involves cytoskeletal anchoring and phophorylation of the $\alpha v \beta 1$ integrin, *J. Cell Biol.*, 110, 2167, 1990.

92. Woods, A. and Couchman, J. R., Protein kinase C involvement in focal adhesion formation, *J. Cell Sci.*, 101, 277, 1992.

93. Ignotz, R. A., Heino, J., and Massague, J., Regulation of cell adhesion receptors by transforming growth factor-β, *J. Biol. Chem.*, 264, 389, 1989.

94. Heino, J., Ignotz, R. A., Hemler, M. E., Crouse, C., and Massague, J., Regulation of cell adhesion receptors by transforming growth factor β, *J. Biol. Chem.*, 264, 380, 1989.

95. Bates, R. C., Rankin, L. M., Lucas, C. M., Scott, J. L., Krissansen, G. W., and Burns, G. F., Individual embryonic fibroblasts express multiple β chains in association with the αv integrin subunit, *J. Biol. Chem.*, 266, 18593, 1991.

96. Defilippi, P., Truffa, G., Stefanuto, G., Altruda, F., Silengo, L., and Tarone, G., Tumor necrosis factor alpha and interferon gamma modulate the expression of vitronectin receptor (integrin $\beta 3$) in human endothelial cells, *J. Biol. Chem.*, 266, 7638, 1991.

97. Plantefaber, L. C. and Hynes, R. O., Changes in integrin receptors on oncogenetically transformed cells, *Cell*, 56, 281, 1989.

98. Giancotti, F. G. and Ruoslahti, E., Elevated levels of the alpha 5 beta 1 fibronectin receptor suppress the transformed phenotype of Chinese hamster ovary cells, *Cell*, 60, 849, 1990.

99. Chan, B. M., Matsuura, N., Takada, Y., Zetler, B. R., and Hemler, M. E., *In vitro* and *in vivo* consequences of VLA-2 expression on rhabdomyosarcoma cells, *Science*, 251, 1600, 1991.

100. Gehlson, K. R., Argraves, W. S., Pierschbacher, M. D., and Ruoslahti, E., Inhibition of *in vitro* tumor invasion by arg-gly-asp-containing peptides, *J. Cell Biol.*, 106, 25, 1988.

101. Boukerche, H., Brethier-Vergnes, O., Bailly, M., Dore, J. F., Leung, L. K., and McGregor, J. L., A monoclonal antibody (LYP18) directed against the blood platelet glycoprotein IIb/IIIa complex inhibits human melanoma growth *in vivo*, *Blood*, 74, 909, 1989.

102. Humphries, M. J., Olden, K., and Yamada, K. M., A synthetic peptide from fibronectin inhibits experimental metastasis of murine melanoma cells, *Science*, 233, 467, 1986.

103. Humphries, M. J., Yamada, K. M., and Olden, K., Investigation of the biological effects of the anti-cell adhesion synthetic peptides that inhibit experimental metastasis of B16-F10 murine melanoma cells, *J. Clin. Invest.*, 81, 782, 1988.

104. Charo, I. F., Nannizzi, L, Smith, J. W., and Cheresh, D. A., The vitronectin receptor $\alpha v \beta 3$ binds fibronectin and acts in concert with $\alpha 5 \beta 1$ in promoting cellular attachment and spreading on fibronectin, *J. Cell Biol.*, 111, 2795, 1990.

105. Tamkun, J. W., DeSimone, D. W., Fonda, D., Patel, R. S., Buck, C., Horwitz, A. F., and Hynes, R. O., Structure of integrin, a glycoprotein involved in the transmembrane linkage between fibronectin and actin, *Cell*, 46, 271, 1986.

Chapter 5

Signaling to and from T Cell Integrins

Yoji Shimizu

CONTENTS

I. INTRODUCTION

In order to be able to mount an effective response against a foreign challenge, the immune system orchestrates precise and regulated interactions between different cell types. Physical interactions must therefore occur between cells, such as between T and B lymphocytes or between T lymphocytes and monocytes. These interactions are mediated by a multitude of adhesion molecules.[1-6] These cell-cell interactions also provide important intracellular signals to both interacting cells via the engagement of multiple receptors and ligands (or counter-receptors). Furthermore, the directed migration or homing of lymphocytes to specific anatomic sites is also mediated by multiple adhesion receptors interacting with ligands on endothelial cells and in the extracellular matrix (ECM). Consequently, adhesion molecules play a vital role in all aspects of lymphocyte function, from the local cell-to-cell contacts that result in the response of a specific T cell to foreign antigen to the more regional interactions that result in specific patterns of migration *in vivo*.

As with other cell types, members of the integrin family of adhesion molecules play a vital role in lymphocyte adhesion.[1,3,4,6] However, unlike many other cell types, the differentiation and function of lymphocytes requires perpetual movement among and between various anatomic sites. Thus, the analysis of the role of integrins on lymphocytes has resulted in important insights into aspects of the regulation of integrin function and the potential role that these integrins play in signal transduction. This chapter will assess

Table 1   **T Cell Integrins**

Name	Alt. Name	Expression Pattern	Refs.
α1β1	VLA-1	*In vitro* activated T cells, specialized T cell subsets *in vivo*	19,21
α2β1	VLA-2	*In vitro* activated T cells, specialized T cell subsets *in vivo*	6,24
α3β1	VLA-3	Low levels on resting T cells	6,20
α4β1	VLA-4	Resting peripheral T cells	20,28
α5β1	VLA-5	Resting peripheral T cells	6,20
α6β1	VLA-6	Resting peripheral T cells	6,20
α4β7	LPAM-1, α4βp	Mouse lymphocytes, human T cell subset, implicated as Peyer's patch homing receptor in the mouse	25–27
αEβ7	HML-1	Mucosal T cells	27,54,56
αLβ2	LFA-1	All T cells	7,8
αMβ2	Mac-1	Subset of CD8+ T cells	15
αXβ2	p150/95	Some cytotoxic T cell clones	16
αvβ3	VNR	Mouse T cell clones	51
α?β1		Activated peripheral T cells, thymic epithelial cells	57
gp140/95		Mouse cytotoxic T cell clones	58

our current state of knowledge regarding the role that integrins play in both responding to intracellular signals generated in lymphocytes and in generating signals when integrins are engaged by a relevant ligand. With a particular focus on integrins expressed on human T cells, three questions of fundamental importance to our understanding of integrin function will guide this review:

1. How is T cell integrin function regulated?
2. Do T cell integrins transduce signals?
3. What is the *in vivo* significance of signaling to and from T cell integrins?

II. EXPRESSION OF INTEGRINS ON T CELLS

Integrins are a large and increasingly complex group of cell surface proteins, each consisting of an α chain noncovalently associated with a β chain to form an αβ heterodimeric receptor.[2] There are multiple α chains and β chains. However, two important aspects of integrin αβ association have emerged: (1) each β chain associates with only a subset of the known α chains, and (2) some α chains have been found to associate with more than one β chain. The simplest integrin classification scheme revolves around the three most well-defined β chains, designated β1, β2, and β3. However, the identification of additional novel β chains (and also novel α chains) has recently served to make the integrin family considerably more complex. Studies of integrin expression on T cells have demonstrated that: (1) multiple integrins are expressed on T cells (Table 1); and (2) the specific integrin expressed is dependent on both the activation state and differentiation state of the T cell.

Table 2   **Ligands for T cell integrins**

Integrin	Cell Surface Ligands	Extracellular Matrix Ligands	Other Ligands	Refs.
α1β1		Collagen, laminin		213,214
α2β1	??	Collagen, laminin		29,30,215
α3β1	??	Collagen, laminin, fibronectin		183,215
α4β1	VCAM-1	Fibronectin, thrombospondin	Invasin	28,31,32,47
α5β1		Fibronectin, thrombospondin		20,47
α6β1		Laminin		20
α4β7	MadCAM-1, VCAM-1	Fibronectin		25,43–45
αEβ7				
αLβ2	ICAM-1, ICAM-2, ICAM-3			9–12
αMβ2	ICAM-1	Fibrinogen	C3bi, LPS	17,216
αXβ2	Endothelial ligand?	Fibrinogen	C3bi	18,217,218
αVβ3		Vitronectin, fibronectin, fibrinogen		51
α?β1				
gp140/95		Vitronectin, fibronectin, fibrinogen		58

A. β2 INTEGRINS

Members of the β2 integrin subfamily consist of the αLβ2 (LFA-1, also CD11a/CD18), αMβ2 (Mac-1 also CD11b/CD18), and αXβ2 (p150/95, also CD11c/CD18) integrins (Table 1). αLβ2 is expressed on essentially all leukocytes and was initially shown to be a molecule important in cytotoxic T cell recognition of target cells.[7,8] Consequently, of all the integrins expressed on T cells, αLβ2 has been the most extensively studied and has served as a paradigm for the analysis of other integrins expressed on lymphocytes. αLβ2 mediates lymphocyte adhesion by binding to three different ligands (Table 2), designated ICAM-1, ICAM-2, and ICAM-3.[9-14]

Both αMβ2 and αXβ2 are found on neutrophils, monocytes, and natural killer cells. However, αMβ2 is also expressed on a subset of CD8+ T cells[15] and αXβ2 expression has also been reported on some cytotoxic T cell clones.[16] αMβ2 has also recently been shown to bind ICAM-1,[17] and studies of αXβ2 function on activated human B cells have demonstrated αXβ2-dependent adhesion to the ECM protein fibrinogen (FB).[18]

B. β1 INTEGRINS

Members of the (VLA proteins) β1 integrin subfamily consist of the β1 chain (also designated CD29) associated with at least six different α chains to form integrins designated α1β1 to α6β1 (Table 1).[6] Initial studies of T cells identified α1β1 as an antigen that appeared on T cells that had been activated for several weeks *in vitro*.[19] The "VLA" nomenclature signifies this historical designation of α1β1 (VLA-1) as a "very late antigen" on activated T cells. However, subsequent analysis showed that the β chain in α1β1 is identical to the β chain in the classical fibronectin (FN) receptor,

$\alpha5\beta1$. Thus, it is important to note that: (1) $\beta1$ integrins are expressed on a multitude of other cell types besides lymphocytes, and (2) $\beta1$ integrins are expressed on resting as well as activated T cells.

While $\alpha1\beta1$ and $\alpha2\beta1$ are not expressed on resting peripheral human T cells, these cells do express low but detectable levels of $\alpha3\beta1$, and readily detectable levels of $\alpha4\beta1$, $\alpha5\beta1$, and $\alpha6\beta1$.[6,20] Activation of peripheral T cells *in vitro* results in distinct changes in $\beta1$ integrin expression; $\alpha1\beta1$ and $\alpha2\beta1$ expression is induced and $\alpha5\beta1$ expression increases, but $\alpha6\beta1$ expression decreases after activation.[6]

Several examples of $\alpha1\beta1$ and/or $\alpha2\beta1$ expression *in vivo* have been reported. Expression of $\alpha1\beta1$ has been detected in T cells found in the lower respiratory tract.[21] Furthermore, $\alpha1\beta1$ and/or $\alpha2\beta1$ expression has been reported on various T cells derived from patients with various diseases. For example, $\alpha1\beta1+$ T cells have been reported in the synovial fluid of rheumatoid arthritis patients[22,23] and $\alpha2\beta1$ has been reported to be expressed on a subset of CD8+ T cells isolated from the peripheral blood of AIDS patients.[24]

Of the well-defined $\beta1$ integrins ($\alpha1\beta1$ to $\alpha6\beta1$), the $\alpha4\beta1$ integrin has been of considerable interest because $\alpha4\beta1$ expression is restricted to cells of hematopoietic origin in the adult. This is in contrast to $\alpha1\beta1$, $\alpha2\beta1$, $\alpha3\beta1$, $\alpha5\beta1$, and $\alpha6\beta1$, all of which are also expressed on nonlymphoid cells. Further interest in $\alpha4\beta1$ has come from studies in the mouse, where the $\alpha4$ chain has been shown to associate with a novel β chain designated βp in the mouse[25] and found to be homologous to the recently described $\beta7$ chain in the human.[26,27] In the mouse, both $\alpha4\beta1$ and $\alpha4\beta p$ have been implicated in mediating the specific adhesion of lymphocytes to endothelium derived from Peyer's patches, suggesting a role for these integrins in lymphocyte homing to the gut.[25]

Most of the ligands for the $\beta1$ integrins have been shown to be components of the ECM, such as collagen, FN, and laminin (LN) (Table 2). It is important to stress three points regarding recognition of ligands by $\beta1$ integrins. First, one $\beta1$ integrin can recognize multiple ligands. Second, one ligand can have multiple integrin receptors. For example, T cells utilize both $\alpha4\beta1$ and $\alpha5\beta1$ to bind to FN.[20,28] Third, studies of $\beta1$ integrins suggest that there may be some cell specificity in the spectrum of ligands that are recognized by any $\beta1$ integrin.[29-31] Thus, some of the ligand specificities for $\beta1$ integrins that have been demonstrated with the use of nonlymphoid cells have yet to be confirmed for lymphocytes.

The $\alpha4\beta1$ integrin also distinguishes itself from the other VLA integrins with regard to ligand binding, since it was the first $\beta1$ integrin to be shown to bind to a cell surface as well as an ECM ligand. Thus, $\alpha4\beta1$ binds not only to FN but also to the inducible endothelial cell surface molecule VCAM-1.[32] *In vitro* studies demonstrate that T cell adhesion to activated endothelial cells is mediated in part by T cell $\alpha4\beta1$ binding to VCAM-1 on endothelium.[32,33] These *in vitro* studies have recently been extended to *in vivo* studies in the rat, where $\alpha4$ subunit-specific monoclonal antibodies (MAb) have been shown to inhibit lymphocyte migration to inflammatory sites.[34,35] Furthermore, the $\alpha4\beta1$/VCAM-1 interaction has been implicated in B cell development in germinal centers,[36] and recent studies demonstrate a role for the $\alpha4\beta1$/VCAM-1 interaction in myotube formation during mouse development.[37] Additional cell surface ligands in addition to VCAM-1 have been predicted, since: (1) $\alpha4\beta1$ MAbs induce $\alpha4\beta1$-dependent aggregation of cells that do not express VCAM-1;[38,39] (2) $\alpha4\beta1$ is involved in B cell binding to bone marrow-derived adherent cells that may not express VCAM-1;[40] (3) $\alpha4\beta1$-dependent cell adhesion to stimulated endothelial cells has been reported to be only partially blocked by VCAM-1-specific MAbs;[41] and (4) $\alpha4$ subunit MAbs can inhibit T cell-mediated cytotoxicity of VCAM-1-negative target cells.[42] The ligand on specialized endothelium derived from Peyer's patches that mediates lymphocyte adhesion via the α4βp integrin has recently been identified as MadCAM-1;[43] several reports also suggest that α4βp, like α4β1, can also bind to FN and VCAM-1.[44,45]

Several novel β1 integrin ligands have also recently been identified. Thrombospondin is an ECM protein that is expressed at high concentrations in damaged and inflamed tissue.[46] A variety of molecules have been implicated in cell adhesion to thrombospondin, including heparan sulfate proteoglycans, the CD36 antigen, and the α2β1, αvβ3, and αIIbβ3 integrins. Analysis of peripheral T cell adhesion has also recently demonstrated that T cell binding to thrombospondin is mediated by at least three different receptors, two of which are the α4β1 and α5β1 integrins.[47]

Perhaps the most intriguing recent aspect of identification of β1 integrin ligands is the realization that bacteria and viruses often utilize the normal adhesive functions of integrins and other adhesion molecules in order to attach to and penetrate host cells. The outer membrane protein invasin expressed on *Yersinia pseudotuberculosis* has been shown to be critical to bacterial attachment and penetration of human cells; invasin mediates bacterial attachment to human cells by binding to various β1 integrins, including α3β1, α4β1, α5β1, and α6β1.[48] Human T cell binding to invasin has recently been shown to be mediated predominantly by α4β1 and occurs in an activation-independent manner.[31] The human cell receptor for echovirus-1 has been identified as the α2β1 integrin[49] and β1 integrins have also been implicated in the attachment of *Trypanosoma cruzi* to macrophages.[50]

C. β3 INTEGRINS

Although the β3 integrins have been most extensively studied on platelets, there are several reports of β3 integrin expression on T cells. The αvβ3 integrin (vitronectin receptor) has been shown to be expressed on certain T cell clones in the mouse and also on primary T cells after activation for 1 to 3 weeks with mitogen or alloantigen.[51] These αvβ3+ T cells bind to the ECM proteins FN, FB, and vitronectin (VN) via the tripeptide sequence arg-gly-asp (RGD), which has been shown to be a critical integrin recognition sequence for several ECM ligands.[52] There has also been one report of β3 expression in human peripheral blood lymphocytes.[53]

D. OTHER INTEGRINS

Other less well-defined members of the integrin family of adhesion molecules have also recently been reported to be expressed on various types of T cells (Table 1). In addition to associating with the α4 chain, the β7 chain has also been reported to associate in the mouse with a novel α chain designated αM290[27] or αE.[54] The αEβ7 integrin is expressed primarily on mucosal T cells, particularly intraepithelial lymphocytes.[27] The αE chain appears to be homologous to the antigen recognized by the HML-1 MAb in the human, which also binds to mucosal lymphocytes.[55,56] The specific expression of this integrin on mucosal T cells suggests that αEβ7, like α4β7, may also be involved in lymphocyte homing. However, the exact function of αEβ7, and the ligand(s) it recognizes, remains to be elucidated.

In the human, a novel integrin composed of the β1 chain associated with an undefined α chain has been reported to be expressed on thymic epithelial cells and to be involved in thymocyte/thymic epithelial cell interactions.[57] This novel integrin has also been found to be expressed on peripheral blood lymphocytes cultured *in vitro* with interleukin-2 for 3 to 4 weeks. Thus, this integrin has a pattern of expression on human lymphocytes similar to α1β1 and α2β1. The ligands for this integrin remain undefined. In the mouse, an integrin designated gp 140/95 was found to be expressed on murine cytotoxic T cell clones and to mediate adhesion to FN, FB, and VN.[58]

E. DIFFERENTIAL INTEGRIN EXPRESSION ON HUMAN T CELLS

As discussed above, activation of resting T cells *in vitro* can result in dramatic changes in the expression of individual integrins, particularly members of the β1 integrin subfamily. Furthermore, the identification of specific lymphocyte populations that express integrins not normally found on resting peripheral T cells, such as α1β1, α2β1, and αEβ7, suggests a particularly important role for these adhesion molecules in mediating the specific functions or homing properties of these cells. For example, it has been postulated that expression of α1β1 on T cells in specific anatomic sites is a reflection of the chronic activation of these cells and results in retention of these cells at this site of chronic activation, presumably via α1β1 binding to a ligand such as collagen.[21] Differential integrin expression may also play a role in T cell differentiation in the thymus, since studies in the mouse demonstrate that while unfractioned thymocytes bind poorly to FN, the CD4-CD8- "double negative" thymocyte fraction binds quite strongly in a α4β1-dependent manner.[59] This differential adhesion correlates with differences in α4β1 expression in various thymocyte subsets.[59] A recent report demonstrating lack of α4β1 expression on B cells resident in tissue[60] contrasts sharply with the abundant expression of α4β1 on peripheral B cells,[6] and suggests a potential role for changes in α4β1 expression in the retention of B cells in tissue.

Differential expression of integrins, along with other adhesion molecules, is also a phenotypic hallmark of two distinct subsets of peripheral T cells that are thought to represent distinct stages of peripheral T cell differentiation. The β1-specific MAb 4B4 was one of the first markers of CD4+ T cells that are often designated naive and memory T cells.[61,62] The subset expressing lower levels of β1 is thought to represent naive T cells, cells which have been exported from the thymus but have not been activated in the periphery. The reciprocal subset is thought to represent memory T cells, cells that have been activated in the periphery and have reverted back to a resting state.[62-64] Isoforms of the CD45 molecule are now routinely used as the best marker to discriminate these two subsets. Phenotypic analysis of naive and memory T cells demonstrates greater expression of many adhesion molecules on memory T cells, including one- to twofold higher levels of αLβ2, and three- to fourfold higher levels of α3β1, α4β1, α5β1, and α6β1.[20,65,66] Greater integrin expression on memory T cells correlates with: (1) greater memory T cell adhesion to integrin ligands such as FN, LN, ICAM-1, and VCAM-1;(2) enhanced functional responsiveness of memory T cells; and (3) the preferential localization of memory T cells in inflammatory sites.[20,33,65,67-71]

III. SIGNALING TO T CELL INTEGRINS

Considerable recent interest has focused on the role that lymphocyte activation events play in regulating the functional activity of integrin molecules. Several early studies were critical to our current understanding of the role of activation in regulating integrin activity on T cells. First, a role for αLβ2 in mediating the interaction between a cytotoxic T cell clone and a target cell had been established through initial antibody blocking studies of T cell-mediated cytotoxicity.[7] However, it was subsequently demonstrated that significant adhesion occurred even when the target cell did not express the relevant antigen recognized by the T cell clone; this "antigen-independent" adhesion was shown to be mediated by αLβ2 and the CD2 adhesion molecule.[72,73] These studies helped to establish the importance of integrins such as αLβ2 in T cell adhesion. However, the use of T cell clones was rather fortuitous in hindsight, since later studies showed that resting peripheral T cells bind poorly to target cells.[74] Second, early analysis of the aggregation or homotypic adhesion of EBV-transformed B cell lines demonstrated that treatment of these cell lines with the phorbol ester PMA could increase the size of these aggregates.[75,76] Furthermore, this aggregation could be inhibited by αL subunit-specific

MAbs and was not due to increased expression of αLβ2, providing the first evidence for a role for activation in regulating the activity of αLβ2.[76] Third, PMA treatment was also shown to dramatically increase T cell adhesion to FN;[77] although this initial report did not characterize the receptors involved, subsequent studies demonstrated that the effect of PMA on T cell adhesion to FN involved similar upregulation of integrin activity that occurs in PMA-induced aggregation of B cell lines.

The seminal observation establishing a role for T cell activation events in upregulating T cell integrin function was made by Dustin and Springer using a purified form of the αLβ2 ligand ICAM-1.[74] These studies showed that resting T cells, which express high levels of αLβ2, bind poorly to ICAM-1. However, strong T cell adhesion to ICAM-1 occurred within minutes after the T cells were activated with PMA. This increase in adhesion occurs without a change in the level of expression of αLβ2. More significantly, similar upregulation of αLβ2 activity occurs when the antigen-specific CD3/T cell receptor (CD3/TCR) complex is engaged by MAb cross-linking using a CD3-specific MAb. Similar results were obtained when assessing T cell adhesion to ICAM-1+ cell lines, demonstrating the relevance of the observations using purified ICAM-1 to an actual cell-cell interaction. The dependence of αLβ2-mediated adhesion on T cell activation is in contrast to the other well-defined T cell adhesion pathway, the CD2/LFA-3 interaction, which has been shown to be constitutively active on resting T cells.[78]

Subsequent studies extended these observations on αLβ2 upregulation in three important ways. First, αLβ2 functional activity was also shown to be upregulated by pairs of MAbs specific for the CD2 molecule.[79] These results were consistent with earlier studies clearly demonstrating that such pairs of CD2-specific MAbs are capable of activating T cells.[80] Second, in addition to upregulating αLβ2 activity, PMA, CD3 cross-linking, and mitogenic pairs of CD2 MAbs were subsequently shown to also upregulate the functional activity of α4β1, α5β1, and α6β1 on human CD4+ T cells[20,65] as assessed by T cell adhesion to FN (via α4β1 and α5β1) and LN (via α6β1). These studies demonstrated that these activation signals coordinately upregulate multiple integrins expressed on the T cell surface. Furthermore, the use of purified ligands or differential antibody blocking has demonstrated that α4β1 adhesion to VCAM-1 and αLβ2 adhesion to ICAM-2 are also activation dependent.[33,81,82] Third, the greater expression of integrins on human memory T cells is associated with increased memory T cell adhesion to integrin ligands such as ICAM-1, FN, and LN.[20,65] The greater adhesion of memory T cells to integrin ligands compared to naive T cells was observed with both resting cells and cells activated by PMA, CD3 cross-linking, or CD2 MAbs.

Further analysis of the role of activation in upregulating integrin activity has clearly revealed two other important aspects of signaling to integrins: (1) multiple activation signals increase T cell integrin activity, and (2) activation also plays a role in upregulating integrin activity on other lymphoid cells (Table 3). Studies using purified integrin ligands have shown that multiple activation signals besides PMA, CD3-, and CD2-mediated activation can also upregulate T cell integrin activity. Treatment of T cells with the Ca^{2+} ionophore A23187 has been shown to result in integrin upregulation, implicating changes in intracellular Ca^{2+} in this cellular event.[82] Antibody cross-linking of the T cell-specific accessory molecules CD7 and CD28 has also been shown to upregulate the activity of αLβ2 and β1 integrins on CD4+ T cells,[82] consistent with other studies demonstrating a role for CD7 and CD28 in T cell signal transduction.[83–85] In contrast to integrin expression, CD7 expression is greater on naive CD4+ T cells than memory T cells. Thus, MAb cross-linking of CD7, in contrast to PMA, CD3, and CD2 activation, results in comparable adhesion of both naive and memory T cells to integrin ligands.[82] Studies of the CD31 molecule expressed on endothelial cells have suggested a potential role for CD31 in mediating cell-cell adhesion;[86] CD31 has also recently been found to be expressed on human T cells, and engagement of CD31 by MAb also upregulates αLβ2

Table 3 Molecules regulating integrin function on lymphoid cells

Regulatory Molecule	Cell Type	Integrin Receptor	Comments	Refs.
CD3/TCR	T	αLβ2,α4β1,α5β1,α6β1	Induction of adhesion is rapid but transient	20,74,79
CD2	T	αLβ2,α4β1,α5β1,α6β1	Induction of adhesion is rapid but prolonged, requires mitogenic pair of CD2 MAbs for upregulation	20,79
CD7	T	αLβ2,α4β1,α5β1,α6β1	Greater expression of CD7 on naive vs. memory CD4+ T cells	82
CD28	T	αLβ2,α4β1,α5β1,α6β1		82
CD31	T	αLβ2,α4β1,α5β1,α6β1	Preferential expression on CD8+ naive T cells, more efficient inducer of β1 integrins, particularly α4β1	87
α4β1	T,B	α4β1	α4β1 MAb induces aggregation inhibitable by other α4β1 MAbs	38,39
CD43	T	CD18	CD43 MAb induces aggregation inhibitable by β2 chain MAb but not αL chain MAb	89
CD44	T,B,MO	αLβ2	CD44 MAb induces cell aggregation	90,91
Surface Ig	B	αLβ2		97
HLA class II	B,MO	αLβ2	Induces sustained adhesion in monocytes, can also induce αLβ2 independent aggregation in B cells, upregulation can occur with either class II MAb or purified CD4	95,98,107
CD14	MO		Can induce αLβ2-dependent and -independent cell aggregation	96
CD19	B	αLβ2	Can induce αLβ2-dependent and -independent cell aggregation	98,99
CD20	B	αLβ2	Can induce αLβ2-dependent and -independent cell aggregation	98
CD39	B	αLβ2	Can induce αLβ2-dependent and -independent cell aggregation	98,100
CD40	B	αLβ2	Can induce αLβ2-dependent and -independent cell aggregation	98,101
CD4	T	αLβ2, α4β1, α5β1, α6β1	Inhibits activation-dependent integrin upregulation	102–104

Note: Abbreviations: T, T cell; B, B cell; MO, Monocyte.

and β1 integrin function.[87] Three aspects of CD31 upregulation deserve comment. First, CD31 is expressed differentially on T cell subsets, with the greatest expression on CD8+ naive T cells. Second, unlike CD3-, CD7-, and CD28-mediated activation, all of which require additional cross-linking in order to see integrin upregulation, no additional cross-linking is necessary in order to observe integrin upregulation using CD31 MAb. Third, CD31-mediated activation increases β1 integrin function on CD31+ T cells more efficiently than αLβ2 function.

Recent evidence suggests that cytokines may also be involved in rapidly upregulating T cell integrin functional activity. Treatment of human T cells with the cytokine MIP-1β, which is a member of the intercrine/chemokine family that includes IL-8 and RANTES, has recently been shown to upregulate α4β1-dependent adhesion to purified VCAM-1.[88] Several aspects of MIP-1β-induced upregulation of α4β1 activity are novel: (1) MIP-1β preferentially enhances α4β1 activity on CD8+ T cells, with a minimal effect on CD4+ T cells; (2) upregulation of α4β1 activity by MIP-1β can occur when the MIP-1β is "captured" by proteoglycans, including purified CD44 antigen; and (3) MIP-1β expression can be detected on endothelium in lymphoid tissue.

Integrin-dependent aggregation of human lymphocytes and lymphoid cell lines has also been a useful system for identifying additional cell surface structures that upregulate integrin activity. The addition of MAbs specific for the CD43[89] or CD44[90,91] molecules has been shown to induce the αLβ2-dependent aggregation of human lymphocytes. Furthermore, treatment of lymphocytes with a unique αL-specific MAb, L16, has also been shown to induce αLβ2-dependent cell aggregation.[92]

The α4β1 integrin also appears to play a role in homotypic adhesion events. Certain α4-specific MAbs are capable of inducing the aggregation of human lymphocytes:[38,39] other α4 MAbs can inhibit the aggregation, suggesting that the inducing α4 MAbs, like the αL-specific MAb L16, are capable of upregulating the functional activity of the integrin to which they bind. Extensive epitope analysis using a panel of α4-specific MAbs revealed three distinct epitopes involved in four functions of α4β1: binding to FN, binding to VCAM-1, triggering of aggregation, and blocking of the induced aggregation.[93] The existence of additional cell surface ligands for α4β1 is suggested since: (1) the cells that are induced to aggregate by α4 MAbs do not express VCAM-1, and (2) one epitope on α4β1 was found to be involved in blocking aggregation but not in triggering aggregation or blocking adhesion to FN or VCAM-1.[93] Both a α5-specific MAb and a β1-specific MAb have been shown to induce aggregation of the Jurkat T cell line but not a B cell line that expresses similar levels of α5β1.[94] However, since this aggregation phenomenon was not inhibited by β2 MAb or a α4 MAb, it is presently not known whether the α5β1-induced aggregation is integrin-mediated.

Activation of other lymphoid cells in addition to T cells has also been shown to result in upregulation of αLβ2 activity. Antibody engagement of HLA class II antigens[95] or the CD14 molecule[96] on human monocytes has been shown to result in αLβ2 upregulation. For B cells, MAbs against a number of B cell-specific cell surface antigens have been shown to result in homotypic aggregation. These include surface immunoglobulin,[97] CD19,[98,99] CD20,[98] CD39,[98,100] CD40,[98,101] and HLA class II.[98] These studies illustrate several important points. First, the engagement of multiple cell surface antigens on B cells by MAb can induce cell aggregation. Second, the differential aggregation responses of peripheral B cells and various B cell lines suggest that the phenomenon is dependent on the differentiation state of the B cell. Third, MAb blocking studies with αLβ2 and ICAM-1 MAbs suggest that certain signals can simultaneously induce αLβ2-dependent and αLβ2-independent aggregation.

Engagement of certain cell surface molecules may also have a negative effect on integrin activity. CD3-induced upregulation of αLβ2 activity and activation dependent binding of rat T cells to FN and LN has been shown to be inhibited by MAbs specific for the CD4 antigen on T cells, suggesting that the interaction of CD4 with its ligand, HLA class II, may be involved in transmission of a negative signal that downregulates αLβ2 function.[102–104]

It is important to note that a major limitation of many of these studies is that the activation stimulus has been delivered to cells by MAb, with the fundamental assumption that MAb engagement is a reasonable mimic for interaction of these antigens with a natural ligand. Notable exceptions are (1) three studies showing that activation of T cell clones or hybridomas with specific antigen results in upregulation of β1 integrin activity,[105] αLβ2 activity,[106] or adhesion to FN and LN,[104] consistent with the initial reports using CD3-specific MAbs;[20,74,79] and (2) one study demonstrating that interaction of a CD4/Ig fusion protein can induce αLβ2-dependent B cell aggregation by engaging MHC class II molecules.[107] While ligands for many of the molecules implicated in upregulating integrin activity have yet to be identified, ligands are known for others, such as CD2 (the LFA-3 molecule)[108] and CD28 (the B7 molecule).[85,109] It will be important to determine whether engagement of such molecules by a ligand can similarly increase integrin function.

It is currently unknown whether upregulation of T cell integrin activity by these various modes of activation occurs by similar or distinct intracellular signaling mechanisms. There is some evidence of differences in the various activation signals that upregulate integrin activity on T cells. First, the kinetics of integrin upregulation differs depending on the stimulus delivered. PMA and CD2 activation result in a rapid increase in adhesion that peaks at 10 min and remains strong over a time period of 30 to 60 min.[65,74,79] In contrast, CD3 activation also results in a peak in adhesion at 10 min, but the adhesion then rapidly decays, returning to baseline levels within 30 to 60 min.[74] Second, while CD31 activation upregulates both β1 and β2 integrins, β1 integrin activity, particularly α4β1, appears to be more efficiently upregulated.[87] Similarly, while macrophages bind constitutively to FN, α6β1-mediated adhesion to LN requires activation with PMA or interferon-γ.[110–112] Studies in the rat suggest differential utilization of αLβ2 ligands depending on the stimulus delivered; while both PMA and CD3 resulted in increased lymphocyte binding to rat endothelium, CD3 induced a αLβ2-dependent ICAM-1-dependent pathway while PMA induced a αLβ2-dependent ICAM-1-independent pathway.[113] Third, treatment of T cells with the protein kinase C inhibitor staurosporin was found to have a negligible effect on integrin upregulation mediated by the Ca^{2+} ionophore A23187, while it inhibited adhesion induced by PMA, CD3, CD7, and CD28 activation.[82]

It is important to emphasize that while activation is obviously a critical factor regulating integrin functional activity, recent studies also suggest important ligand-dependent differences in integrin activity. While resting T cells bind poorly to integrin ligands such as ICAM-1, VCAM-1, FN, and LN,[20,74] resting T cells bind strongly to the α4β1 bacterial ligand invasin.[31] Furthermore, activation of the T cell has a minimal effect on T cell adhesion to invasin. Thus, of all of the integrin ligands analyzed to date, invasin is unique in being able to promote activation-independent adhesion of human T cells. Differential adhesion to FN and VCAM-1 has also been used to define inactive, partially active, and fully active states of the α4β1 integrin,[114] again emphasizing the importance of the ligand in regulating integrin activity.

IV. POTENTIAL MECHANISMS OF INTEGRIN UPREGULATION ON T CELLS

The exact mechanism by which signaling to T cell integrins upregulates integrin activity remains undefined. Activation likely results in a qualitative change of the preexisting integrin receptors on a resting T cell, since integrin upregulation by activation occurs within minutes without a change in the level of integrin expression. Recent studies have focused on phosphorylation, the cytoskeleton, and integrin conformation as being particularly important in our understanding of this adhesive event. The majority of these studies have focused on αLβ2, with the assumption that similar mechanisms also upregulate β1 integrin activity.

A. ROLE OF PHOSPHORYLATION

Two lines of evidence have suggested that phosphorylation of integrin molecules may play a role in upregulating αLβ2 activity: the rapidity with which activation results in integrin upregulation, and the finding that the β2 chain is phosphorylated in response to PMA treatment.[115-117] PMA treatment has also been reported to phosphorylate the α6 chain in macrophages but not the β1 chain, which correlates with increased α6β1-mediated adhesion of PMA-activated macrophages to LN.[112] While the PMA-induced phosphorylation of β2 in T cells also correlates with increased αLβ2 activity, recent molecular studies have shown that phosphorylation of the β chain is not necessary to observe PMA-induced upregulation of αLβ2 activity.[118] The serine residue at position 756 has been shown to be the major phosphorylated site in the β2 chain; when this residue and other potential phosphorylated sites in the β2 chain were substituted, the resulting αLβ2 molecules expressed in αLβ2 deficient B lymphoblastoid cells still exhibited increased adhesion in response to PMA treatment.[118] These studies are consistent with earlier work demonstrating that αLβ2 expressed in COS cell transfectants is constitutively active,[119] even though PMA results in a substantial increase in β2 phosphorylation in these cells.[118] These results suggest that phosphorylation of the β2 subunit at these sites is not responsible for the PMA-induced upregulation of integrin activity. Nevertheless, it is important to realize that phosphorylation of other cytoplasmic proteins, such as cytoskeletal elements, may play a critical role in regulating integrin activity (see below). In addition, these studies have clearly shown that the cytoplasmic domain of β2 is required for PMA-induced upregulation of αLβ2 activity, particularly two regions in the carboxy-terminal third of the cytoplasmic domain.[118,120] Tyrosine phosphorylation may also be critical in regulating integrin function, since it has been reported that treatment of rat T cells with the tyrosine kinase inhibitor genistein inhibits the activation-dependent binding of these cells to FN and LN.[104] It will be important to continue to investigate the role of phosphorylation in regulating integrin function on T cells, particularly after activation via a more physiological signal such as CD3-cross-linking.

B. ROLE OF THE CYTOSKELETON

Several lines of evidence suggest an involvement of the cytoskeleton in regulating integrin activity on T cells. First, treatment of T cells with the cytoskeletal disrupting agent cytochalasin B has been shown to dramatically inhibit both αLβ2 and β1 integrin upregulation in response to PMA, CD2, CD3, CD7, or CD28.[76,82] Second, cocapping studies suggest an association between αLβ2 and the cytoskeletal protein talin after treatment with phorbol ester.[121] Third, in vitro studies have demonstrated the association of β1 with both talin and another cytoskeletal protein, α-actinin.[122,123] Fourth, the α6β1 integrin has been shown to be anchored to the cytoskeleton after PMA treatment of macrophages, which increases α6β1-mediated adhesion to LN.[112] These studies have led to models whereby integrin activity might be regulated by altered redistribution of integrin molecules to sites of adhesive interaction,[124-127] perhaps by activation-dependent association of integrins with the cytoskeleton. However, one study investigating the interaction of a cytotoxic T cell clone with an ICAM-1+ B cell line failed to see αLβ2 redistribution upon T cell activation.[78] Furthermore, no direct in vivo biochemical evidence for such integrin-cytoskeleton interactions in T cells currently exists. The definition of specific sites on the β2 chain that are important for αLβ2 function[118] provides important avenues for exploring this issue further.

C. ROLE OF INTEGRIN CONFORMATION

The regulation of integrin function on T cells may also involve changes in the conformational state of the receptor. A role for integrin conformation in regulating integrin activity is suggested by the identification of several MAbs that recognize unique epitopes on integrin α chains. One αL-specific MAb, designated L16, has been shown to be unique among αL MAbs in its ability to induce αLβ2-dependent homotypic adhesion.[92] The Ca^{2+}-dependent epitope recognized by L16 is not present on resting T cells but can be induced by activation with either PMA or CD3.[119] However, there is not a complete correlation between L16 expression and αLβ2 activity, since T cells activated with CD3 for 30 min, and thus no longer binding to ICAM-1, continue to express the L16 epitope.[119] In addition, T cell clones that express the L16 epitope do not show constitutive LFA-1-dependent aggregation.[128] Based on these results, a model has been proposed where three different states of LFA-1 can be expressed on T cells: low activity LFA-1 that is L16-negative, low activity LFA-1 that is L16-positive, and a fully active form of LFA-1 that is also L16-positive.[125,126]

Another MAb, designated MAb 24, recognizes a cation-dependent epitope present on all three α chains in the β2 integrin subfamily.[129,130] Analysis using this MAb has shown a correlation between expression of the MAb 24 epitope and αLβ2 functional activity.[130] In addition studies with MAb 24 indicate an important role for divalent cations in the regulation of αLβ2 activity. Both Mn^{2+} and Mg^{2+} were found to result in expression of the MAb 24 epitope and αLβ2-dependent adhesion of T cells to ICAM-1, although induction by Mg^{2+} required the removal of extracellular Ca^{2+}. In addition, Ca^{2+} was found to specifically inhibit MAb 24 binding and αLβ2 activity induced by Mn^{2+} or Mg^{2+}.[130] It has been proposed that the binding of specific divalent cations by αLβ2 may serve to regulate the functional activity of the molecule. For example, resting T cells may express αLβ2 bound tightly to Ca^{2+}, serving to keep the molecule in a low activity state. However, T cell activation may serve to cause a conformational change in αLβ2 that allows it to displace Ca^{2+} and bind Mg^{2+}, resulting in an increase in αLβ2 activity.[127,130]

Analysis of cation-dependent binding of novel integrin-specific MAbs has paralleled studies in many systems demonstrating the effects of modifying divalent cation conditions on integrin-mediated cell adhesion. The addition of Mn^{2+} in particular has been shown to increase the functional activity of numerous integrins, including α5β1,[131-133] αIIbβ3,[134] αLβ2,[130] α3β1,[135] α4β1,[114] and α6β1.[136] Ca^{2+} has also been shown to modulate integrin activity, although its effects appear to be dependent somewhat on the specific integrin receptor/ligand interaction being studied.[114,135,137] The ability of divalent cations to dramatically modulate integrin activity is consistent with the model that conformational changes in the integrin receptor can in fact modulate adhesion activity. Although the physiological significance of divalent cation modification of integrin activity remains to be defined, changes in Ca^{2+} and Mg^{2+} concentration at a site of tissue injury have been proposed to play a role in α2β1-mediated adhesion of platelets to collagen.[137] Selective upregulation of specific integrin receptor/ligand interactions may also be a critical function for divalent cations *in vivo*.

These studies with αLβ2 specific MAbs suggest that conformational changes in αLβ2 can occur under various conditions, with the implication that such changes may also occur when T cells are appropriately activated to upregulate αLβ2 function. These studies have been extended to other integrins as well, as unique MAbs specific for β1[138-141] and β3[142,143] integrins have also been shown to enhance cell adhesion mediated by these receptors. What remains to be determined is the exact intracellular mechanisms by which such conformational changes might occur. It has been proposed that the upregulation of αMβ2 on neutrophils may be caused by a conformational change mediated by the activation-dependent expression of a novel lipid termed integrin-modulating factor-1 (IMF-1).[144] Other mechanisms may also exist, and it is possible that all of the potential mechanisms discussed here may play a role in integrin upregulation. For example, T cell activation may serve to phosphorylate a cytoskeletal protein that initiates a transient cytoskeletal linkage with αLβ2 that results in receptor clustering and conformational changes to a more active state. There is still much to be discovered in this area of integrin biology, including: (1) the exact sequence of intracellular events that result in integrin upregulation; (2) potential differences in intracellular mechanisms between the various activation signals that upregulate integrin activity; and (3) mechanisms responsible for downregulating integrin activity after adhesion has peaked.

V. *IN VIVO* RELEVANCE OF SIGNALING TO INTEGRINS
A. T CELL RECOGNITION

A fundamental requirement of recognition of foreign antigen by T cells is engagement of the CD3/TCR by a peptide antigen presented in the context of an appropriate major histocompatibility complex antigen.[145-147] However, it is now clear that other cell surface molecules on both the T cell and the antigen presenting cell (APC) play important roles in facilitating this interaction.[146,147] The ability of CD3-mediated activation to rapidly increase activity has suggested a multistep model by which αLβ2 initiates and maintains adhesion between a T cell and an appropriate antigen-presenting cell.[74] First, T cells can survey various cells in the extracellular environment by antigen-independent adhesion mediated by both the CD2/LFA-3 pathway and a low but detectable adhesion mediated by αLβ2.[72,73] It is interesting to note that resting memory T cells show greater integrin-mediated adhesion than resting naive T cells,[20,65] which may be in part responsible for the greater responsiveness of memory T cells.[62,64,70] If the CD3/TCR is not engaged by the appropriate Ag/MHC complex, then the adhesion is terminated. However, appropriate activation via the TCR would result in upregulation of αLβ2 activity and increased adhesion between the T cell and the APC. This increased adhesion would allow sufficient time for the full spectrum of signals to be delivered to both of the interacting cells to allow the appropriate immune response to occur. The transience of CD3-mediated upregulation of αLβ2 activity suggests a mechanism by which the T cell and the APC can subsequently detach.

Signaling to integrins occurs not only in the T cell but also in the APC, which also expresses αLβ2.[95,97,98] Thus, the ability of surface immunoglobulin to upregulate αLβ2 activity on human B cells suggests that uptake of antigen by APC may serve to transmit signals that upregulate αLβ2 activity on B cells.[97] Consequently, the increased functional activity of αLβ2 on an APC may allow for more efficient initial interactions between the APC and a passing T cell,[147] most likely by engagement of αLβ2 on the APC with a ligand such as ICAM-3 on the T cell.[12]

The potential role of other T cell molecules that upregulate integrin activity, such as CD7 and CD28, remains undefined. However, the ability of both of these molecules to facilitate T cell proliferative responses[83-85] suggests that one of the functions of the signals transmitted by such molecules is to prolong integrin-mediated adhesion that would normally be transient when the CD3/TCR alone is engaged.[82] Alternatively, engagement of these molecules without prior activation via CD3/TCR might serve to initiate strong antigen-independent adhesion of T cells to an apposing cell.

B. T CELL MIGRATION

The identification of T cell interactions with components of the ECM via β1 integrins has resulted in renewed interest in the role of the ECM in T cell migration.[148] The activation-dependent increase in β1 integrin-mediated T cell adhesion to FN and LN has been speculated to play a role in the arrest of lymphocyte migration upon activation.[20] A recent *in vivo* study demonstrating inhibition of T cell-mediated contact hypersensitivity to the hapten 2,4,6-trinitro-1-chlorobenzene (TNCB) by FN peptides also suggests that integrin-mediated adhesion to the ECM is important in T cell migration to a site of antigenic challenge.[149] This study also found that the TNCB-reactive T cells were found in the small fraction of T cells that bound to FN *in vitro*, suggesting that these cells express active forms of β1 integrins as a result of their response to antigen.

Movement of T cells in the thymus may also be dependent on adhesion to the ECM and thus may be critical to thymic differentiation. Studies of the adhesion of CD4-CD8-thymocytes to stromal cells have implicated the involvement of β1 integrin-mediated adhesion to FN on the surface of the stromal cells;[59] the importance of this interaction for thymic development is suggested by the ability of FN peptides to inhibit the *in vitro* differentiation of these CD4-CD8-double negative cells to CD4 or CD8 single positive thymocytes.

The interaction of T cells with endothelial cells has been shown to be critical to T cell migration into lymphoid tissue and into inflammatory sites.[150,151] A large number of studies have clearly demonstrated the involvement of both αLβ2 and α4β1 in T cell adhesion to the endothelium,[32,33,150,152,153] and T cell activation has been shown to increase adhesion to the endothelium by both of these integrins.[33] Multiple other molecules are also involved in lymphocyte interactions with endothelial cells, and recent models of this process have emphasized that a sequential series of events, an "adhesion cascade", allows a T cell to initiate adhesion to endothelial cells, firmly attach, and then migrate through the endothelial layer into the surrounding tissue.[150,154,155] Integrins such as αLβ2, αMβ2, and α4β1 have been proposed to play a central role in this adhesion cascade by serving as the main adhesive force that mediates strong adhesion to the endothelial cell. Since integrin function is minimal on resting lymphocytes, it has been proposed that an activation signal or trigger must be delivered to the lymphocyte in order for this firm attachment to occur. For neutrophils, cytokines and cell surface molecules such as E-selectin (ELAM-1) have been proposed to serve as such triggers.[154,155] For lymphocytes, two molecules have been proposed to serve as potential activators that upregulate integrin activity upon interaction with endothelium. One of these molecules is CD31, which is expressed at high levels on endothelial cells and has been suggested to mediate adhesion by homophilic interactions.[87] The second molecule is the cytokine MIP-1β, which has been shown to upregulate α4β1-mediated adhesion of CD8+ T cells to VCAM-1 when MIP-1β is immobilized by a proteoglycan.[88] Based on these results, a model has been proposed whereby proteoglycan-immobilized cytokines can serve to upregulate T cell adhesion to vascular endothelial cells.[156] The intriguing aspects of this model remain its inherent flexibility in regulation and its potential to explain subset-specific trafficking patterns of lymphocytes *in vivo*. Therefore, the ability of a lymphocyte to bind to endothelium is determined by a number of factors that include: (1) the phenotypic status of the lymphocyte, particularly with regard to expression of various integrins; (2) the ability of the lymphocyte to respond to the activating cytokine, which presumably is a direct reflection of its state of differentiation and/or activation; (3) differences in the expression of relevant activating cytokines at the site of lymphocyte interaction with endothelium; and (4) heterogeneity in proteoglycans expressed by endothelial cells and the ability of these proteoglycans to "capture" relevant activating cytokines.[156]

C. PERTURBATION OF SIGNALING TO INTEGRINS AND DISEASE

Changes in the level of expression of integrins have been shown to play an important role in the development of disease. Perhaps the most striking examples are the lack of expression of β2 integrins that characterizes leukocyte adhesion deficiency (LAD),[157] and the dramatic effects of changes in β1 integrin expression on tumor development and metastasis.[158-160] Although strong evidence is currently lacking, alterations in the signaling properties that regulate T cell integrin activity may also serve as an important factor in disease development. The strong adhesion of synovial fluid T cells to FN via α4β1[161] suggests that these cells may express constitutively active forms of α4β1 that may allow such cells to remain in the synovial fluid. The retention of specific autoreactive T cells at a site of activation by constitutive integrin activity may represent an important component to the development of autoimmune disease. Furthermore, the implication of the presence of α1β1 and/or α2β1-positive T cells in patients suffering from diseases such as arthritis and AIDS[22-24] is that such cells have altered adhesive capacities that may play a role in the pathogenesis of these diseases. Further work is clearly warranted in our understanding of how disregulation of integrin functional activity affects the development of various immune disorders.

D. ADDITIONAL *IN VIVO* IMPLICATIONS

As discussed earlier, recent studies demonstrating multiple modes of activating T cell integrins, and differential effects depending on the T cell differentiation state (for CD7 activation) or on β1 vs. β2 integrins (for CD31), suggest exquisitely sensitive methods by which integrin activity on specific T cell subsets can be regulated *in vivo*.[82,87] A fundamental question that remains to be answered is whether all integrins expressed on T cells are regulated by T cell activation events. This is particularly critical for those specialized T cell subsets that appear to express rather unique integrins, such as α1β1 in the respiratory tract,[21] and αEβ7 on intraepithelial lymphocytes.[27,56,162,163] It remains possible that some integrins, such as those that are involved in mediating lymphocyte homing (such as α4β7), may be in a constitutively active state.

An additional consideration is whether the integrins that are regulated by activation on peripheral T cells are similarly regulated on T cells that are found resident in tissue. Since it is becoming apparent that tissue-resident lymphocytes differ with regard to expression of integrins[60] when compared to their peripheral counterparts, it is also possible that important differences may exist with regard to the functional activity of the integrins that are expressed on these cells. The analysis of such small but functionally significant lymphocyte subsets is likely to reveal important information on the mechanisms and functional significance of signal transduction to integrin receptors.

VI. SIGNALING FROM T CELL INTEGRINS

In addition to responding to activation signals, evidence is emerging that integrins also contribute to the signaling machinery in T cells by transducing their own signals when appropriately engaged by ligands. Thus, integrins join a multitude of other T cell surface antigens that appear to play an important role in delivering important signals that facilitate T cell responses to antigen.[147]

A. MODULATION OF T CELL PROLIFERATIVE RESPONSES BY INTEGRINS

While T cell activation requires engagement of the antigen-specific CD3/TCR complex, it is clear that complete T cell activation and differentiation also requires the delivery of additional "costimulatory" signals provided by the antigen-presenting cells or other components in the extracellular environment.[145-147] Many *in vitro* studies have demonstrated that integrin ligands can provide such costimulation of T cell activation.

The role of αLβ2 in delivering costimulatory signals to T cells has been the most extensively analyzed. Initial studies focused on the ability of αLβ2 specific MAbs to modulate T cell proliferation induced by MAbs specific for the CD3/TCR that are

(1) T CELL COSTIMULATION

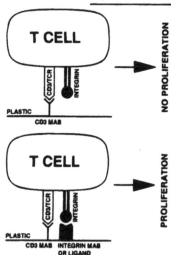

COSTIMULATORY INTERACTIONS:

αLβ2 /ICAM-1	REF. 81,167,168
αLβ2 /ICAM-2	REF. 172
α4β1/VCAM-1	REF. 81,184,185
α4β1/FN	REF. 175-179
α5β1/FN	REF. 174-177
α6β1/LN	REF. 175
α3β1/COLL	REF. 183

(2) INTEGRIN CROSSLINKING BY MAB

*CA++ FLUX	REF. 193,194
*PHOSPHORYLATION	REF. 202-211
*INTRACELLULAR pH CHANGES	REF. 195

(3) INTERACTION WITH ECM SUBSTRATE

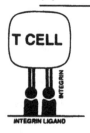

*INDUCTION OF GENE EXPRESSION	REF. 196,201
(AP-1, METALLOPROTEINASE)	
*INDUCTION OF CYTOKINE SECRETION	REF. 199,200

Figure 1 Schematic illustration of three *in vitro* systems used to demonstrate integrin-mediated signal transduction: (1) facilitation of T cell proliferation by integrin ligand or MAb; (2) induction of cellular responses by MAb cross-linking of integrins; (3) induction of cellular responses by engagement of integrins by ligand.

immobilized on a plastic substrate (Figure 1). One study demonstrated that while soluble anti-αL chain MAbs enhanced T cell proliferation, soluble β2 chain MAbs generally inhibited the proliferative response.[164] Using a slightly different approach of coimmobilizing the CD3 and αLβ2 MAbs together on the same plastic substrate, Carrera et al. showed that both αL and β2 chain MAbs could dramatically enhance T cell proliferation.[83] Such MAbs were also shown to induce T cell proliferation in combination with the phorbol ester PMA.[83] Recent studies have further demonstrated that the combination of a unique β2 chain MAb and monocytes was sufficient to induce T cell proliferation,[165] providing additional evidence for signal transduction via αLβ2. Human T cells can also be activated by specific combinations of MAbs recognizing the CD2 molecule,[80] and MAbs specific for αLβ2 or ICAM-1 have also been shown to modulate CD2-mediated T cell proliferation.[166]

Studies using αLβ2 specific MAbs have subsequently been extended with the use of purified ICAM-1 to deliver T cell costimulation. Studies in both the human and mouse have demonstrated that T cell proliferation in response to immobilized CD3 MAb is dramatically enhanced by the coimmobilization of purified ICAM-1.[81,167,168] Furthermore, the expression of ICAM-1 on APCs has been shown in several studies to dramatically enhance antigen presentation by these cells.[168–171] The interaction of T cell αLβ2 with ICAM-2 has also been shown to mediate costimulation, although ICAM-1 delivers stronger costimulation than ICAM-2.[172] In addition, studies using APC that express low levels of ICAM-1 demonstrate that purified ICAM-1 can enhance antigen presentation by binding to LFA-1 expressed on the APC.[173] Thus, αLβ2 expressed on both the T cell and the APC may mediate signal transduction. Of particular importance in future studies will be to assess the role of ICAM-3 in delivering LFA-1-mediated costimulatory signals, since: (1) ICAM-3 is expressed at high levels on lymphoid cells, and (2) the binding of resting lymphocytes to purified αLβ2 has been shown to be almost completely ICAM-3-dependent.[12]

Similar studies have been conducted with ligands for the β1 integrins expressed on T cells. The ability of β1 integrin ligands such as FN to mediate costimulation has recently shed light on a role for the ECM in regulating T cell responses. T cell responses to immobilized CD3 MAb have been shown to be dramatically enhanced by the coimmobilization of FN,[174–177] FN peptides,[178] or β1-specific MAbs.[179] Proliferation in response to the combination of CD3 MAb and FN was inhibitable by either α4β1 or α5β1 MAbs.[174–176] These findings are consistent with earlier studies demonstrating modulatory effects of soluble FN on T cell proliferative responses.[180–182] Similar enhancement of CD3-mediated T cell proliferation was also observed with LN (via α6β1)[175] and by collagen (via both α3β1 and CD26),[183] although other groups have failed to observe costimulation with collagen in similar systems.[175,176] The α4β1 cell surface ligand VCAM-1, like ICAM-1, has also been shown to facilitate CD3-mediated T cell proliferation.[81,184,185] The effects of β1 integrins on CD2-mediated T cell activation have been less extensively analyzed; one report described enhancement by α4β1 MAbs[186] while another reported no effect on CD2 activation by either FN or β1 MAbs.[179]

Studies of other integrins on T cells have also provided evidence for signal transduction. The αvβ3 integrin on mouse γδ T cell clones derived from the epidermis has been shown to mediate a costimulatory signal by binding to FN, FB or VN.[187] A novel integrin termed gp140/95 on murine cytotoxic T cell clones that binds FN, FB, and VN was shown to augment serine esterase release in combination with CD3 MAb.[58] Some αX-specific MAbs were also shown to enhance B cell proliferation in response to PMA,[18] and this proliferation could be inhibited by soluble FB.

In many of these systems, the CD3 MAb and the integrin ligand must be on the same physical surface in order to observe integrin-mediated costimulation. Thus, it has been argued that integrins facilitate T cell responses in these systems not by transducing signals, but by facilitating the adhesion of the T cell with the CD3 MAb. However, there are several lines of evidence to suggest that the effect of integrins in these systems is in fact due to true signaling. First, under appropriate conditions of activation, the integrin-mediated costimulation can be delivered on a physically separate substrate from the CD3 signal. This has been shown for both αLβ2/ICAM-1[171,188] and the interaction of α4β1 and α5β1 with FN.[176] Second, in many of these systems, there is a discordance between the increase in antigen presentation caused by the addition of ICAM-1 to the system and the observed ICAM-1-dependent increase in adhesion.[65,168] Third, there are notable examples of MAbs or adhesion ligands that mediate adhesion but fail to mediate costimulation when coimmobilized with CD3 MAb. These include E-selectin,[81,172] which mediates adhesion of a memory T cell subset[189,190] and the α4β1 MAb HP1/2.[184] Finally, biochemical studies have shown that costimulation of CD3-mediated T cell proliferation by the αLβ1/ICAM-1 interaction results in both prolonged inositol phospholipid hydrolysis and sustained increases in intracellular Ca^{2+}.[191]

B. OTHER EVIDENCE FOR INTEGRIN-MEDIATED SIGNAL TRANSDUCTION

Besides effects on T cell proliferation, there are multiple lines of additional evidence implicating signaling by integrin receptors in immune cells (Figure 1). A unique β1 subunit-specific MAb has been shown to increase the levels of cyclic AMP in T cells that were stimulated via CD3 or CD2.[192] Similarly, antibody engagement of αLβ2 has been reported to either induce or enhance changes in intracellular Ca^{2+}[191,193,194] and to also cause changes in intracellular pH.[195] The interaction of T cells with FN has been shown to induce expression of the AP-1 transcription factor that is involved in regulating expression of the IL-2 gene.[196] αLβ2-deficient lymphocytes isolated from LAD patients have been shown to have poor proliferative responses to T cell receptor-specific MAb[197] and no immunoglobulin production in the presence of activated T cells.[198] The interaction of α2β1 on monocytes with collagen has recently been shown to result in an increased release of the cytokine IL-1.[199] Furthermore, while α5β1-mediated adhesion to FN has no effect by itself on IL-1 production by monocytes, the α5β1/FN interaction appears to potentiate the enhancing effects of collagen in this system.[200] In fibroblasts, induction of metalloproteinase gene expression dependent on the α5β1/FN interaction has also been reported.[201]

Recent studies have focused on the role of activation of tyrosine kinases in integrin-mediated signal transduction. In fibroblasts, platelets, and embryonal carcinoma cells, β1 integrin engagement by MAb cross-linking or adhesion to FN results in tyrosine phosphorylation of a 125-kDa protein.[202-209] This protein has subsequently been shown to be a tyrosine kinase that is colocalized with focal adhesions in human fibroblasts,[209] leading to its designation as focal adhesion kinase (FAK) or pp125[FAK]. The identification of pp125[FAK] as a tyrosine kinase, the localization of pp125[FAK] in the cell to focal adhesions, the reported tyrosine phosphorylation of various cytoskeletal elements, and the known linkage of integrins to the cytoskeleton suggests that pp125[FAK] and its kinase activity may play a crucial role in the formation of integrin-mediated focal contacts with the ECM.[207] In lymphoid cells, similar engagement of the α4β1 integrin has been shown to result in tyrosine phosphorylation of a 105-kDa protein.[210,211] The relationship of pp125[FAK] to this 105-kDa protein remains to be determined. Thus, the emerging picture is one of integrins as important environmental sensors on the cell surface that transmit critical intracellular information regarding the extracellular environment. In that sense, integrins truly "integrate" the outside of the cell with the inside.

VII. *IN VIVO* RELEVANCE OF SIGNALING FROM T CELL INTEGRINS

It is becoming clear from *in vitro* analysis that a number of T cell surface antigens are capable of mediating signal transduction when engaged by a relevant ligand. Integrins clearly fall under this category of T cell signaling molecules. A major question that remains to be resolved is the precise role that integrins and other signaling molecules play in the activation events that occur during T cell recognition of foreign antigen *in vivo*.

It has been speculated that while the primary activation signal is provided by engagement of the antigen-specific TCR, additional signals that synergize with the TCR serve to provide the T cell with important contextual cues that determine the specificity of the resulting T cell response.[65] Various studies, including those investigating integrin-mediated signal transduction as described above, indicate that such signals can be provided by: (1) ligands expressed on the surface of an apposing cell, (2) components of the ECM, and (3) soluble factors such as cytokines. Integrins clearly can play a role in providing such signals via either of the first two mechanisms. The large number of costimulatory signals that have been defined *in vitro* may reflect the fact that many of these costimulatory signals may only be delivered in specific anatomic regions *in vivo*. Such regionalization of costimulation may serve to drive T cell responses in that region towards a specific functional response, which might be manifested by the expression of specific cell surface molecules, such as homing receptors, or the production of a specific spectrum of cytokines that might be particularly effective in combating antigens that are encountered at that site. While the delivery of such costimulatory signals is an important component of T cell activation, the lack of delivery of such signals may in fact result in subsequent nonresponsiveness to future antigenic challenge.[146] Recent studies using *in vivo* treatment with ICAM-1-specific MAb suggest that blocking of the αLβ2/ICAM-1 interaction may play a role in such induction of tolerance.[212] Essential to our further understanding of the role of integrin-mediated signal transduction in T cell function is an elucidation of: (1) the specific combinations of receptors that must be engaged in order to mediate a specific T cell proliferative response or differentiation event, and (2) the precise intracellular signaling events that are initiated by ligand engagement of integrins.

VIII. SUMMARY

Integrins clearly represent an important group of cell surface receptors that mediate the interactions of T cells with the surrounding environment. T cells utilize integrins for a variety of functions, including adhesion to APC, adhesion to the ECM, and migration of specific T cell subpopulations to distinct anatomic sites *in vivo*. A central element by which these various functions are mediated is signal transduction to and from integrin receptors. Thus, integrins are bidirectional signaling structures, both responding to signals delivered by other molecules and delivering their own signals when appropriately engaged. It is likely that future studies of T cell function will continue to benefit from elucidation of the mechanisms that serve to mediate and regulate signaling to and from T cell integrins.

REFERENCES

1. Springer, T. A., Adhesion receptors of the immune system, *Nature*, 346, 425, 1990.
2. Hynes, R. O., Integrins: versatility, modulation, and signaling in cell adhesion, *Cell*, 69, 11, 1992.
3. Shimizu, Y. and Shaw, S., Lymphocyte adhesion mediated by VLA (β1) integrins: functional roles of multiple extracellular matrix and cell surface ligands, *Chem. Immunol.*, 50, 34, 1991.
4. Albelda, S. M. and Buck, C. A., Integrins and other cell adhesion molecules, *FASEB J.*, 4, 2868, 1990.
5. Brandley, B. K., Swiedler, S. J., and Robbins, P. W., Carbohydrate ligands of the LEC cell adhesion molecules, *Cell*, 63, 861, 1990.
6. Hemler, M. E., VLA proteins in the integrin family: structures, functions, and their role on leukocytes, *Annu. Rev. Immunol.*, 8, 365, 1990.
7. Krensky, A. M., Sanchez-Madrid, F., Robbins, E., Nagy, J. A., Springer, T. A., and Burakoff, S. J., The functional significance, distribution, and structure of LFA-1,

LFA-2, and LFA-3: cell surface antigens associated with CTL-target interactions, *J. Immunol.*, 131, 611, 1983.

8. Krensky, A. M., Robbins, E., Springer, T. A., and Burakoff, S. J., LFA-1, LFA-2, and LFA-3 antigens are involved in CTL-target conjugation, *J. Immunol.*, 132, 2180, 1984.

9. Makgoba, M. W., Sanders, M. E., Luce, G. E. G., Dustin, M. L., Springer, T. A., Clark, E. A., Mannoni, P., and Shaw, S., ICAM-1 a ligand for LFA-1-dependent adhesion of B, T and myeloid cells, *Nature*, 331, 86, 1988.

10. Marlin, S. D. and Springer, T. A., Purified intercellular adhesion molecule-1 (ICAM-1) is a ligand for lymphocyte function-associated antigen-1 (LFA-1), *Cell*, 51, 813, 1987.

11. Staunton, D. E., Dustin, M. L., and Springer, T. A., Functional cloning of ICAM-2, a cell adhesion ligand for LFA-1 homologous to ICAM-1, *Nature*, 339, 61, 1989.

12. De Fougerolles, A. R. and Springer, T. A., Intercellular adhesion molecule 3, a third adhesion counter-receptor for lymphocyte function-associated molecule 1 on resting lymphocytes, *J. Exp. Med.*, 175, 185, 1992.

13. Vazeux, R., Hoffman, P. A., Tomita, J. K., Dickinson, E. S., Jasman, R. L., St. John, T., and Gallatin, W. M., Cloning and characterization of a new intercellular adhesion molecule ICAM-R, *Nature*, 360, 485, 1992.

14. Fawcett, J., Holness, C. L. L., Needham, L. A., Turley, H., Gatter, K. C., Mason, D. Y., and Simmons, D. L., Molecular cloning of ICAM-3, a third ligand for LFA-1, constitutively expressed on resting leukocytes, *Nature*, 360, 481, 1992.

15. Yamada, H., Martin, P. J., Bean, M. A., Braun, M. P., Beatty, P. G., Sadamoto, K., and Hansen, J. A., Monoclonal antibody 9.3 and anti-CD 11 antibodies define reciprocal subsets of lymphocytes, *Eur. J. Immunol.*, 15, 1164, 1985.

16. Keizer, G. D., Borst, J., Visser, W., Schwarting, R., De Vries, J. E., and Figdor, C. G., Membrane glycoprotein p150,95 of human cytotoxic T cell clones is involved in conjugate formation with target cells, *J. Immunol*, 138, 3130, 1987.

17. Diamond, M. S., Staunton, D. E., Marlin, S. D., and Springer, T. A., Binding of the integrin Mac-1 (CD11b/CD18) to the third immunoglobulin-like domain of ICAM-1 (CD54) and its regulation by glycosylation, *Cell*, 65, 961, 1991.

18. Postigo, A. A., Corbí, A. L., Sánchez-Madrid, F., and De Landázuri, M. O., Regulated expression and function of CD11c/CD18 integrin on human B lymphocytes. Relation between attachment to fibrinogen and triggering of proliferation through CD11c/CD18, *J. Exp. Med.*, 174, 1313, 1991.

19. Hemler, M. E., Jacobson, J. G., Brenner, M. B., Mann, D., and Strominger, J. L., VLA-1: a T cell surface antigen which defines a novel late stage of human T cell activation, *Eur. J. Immunol.*, 15, 502, 1985.

20. Shimizu, Y., van Seventer, G. A., Horgan, K. J., and Shaw, S., Regulated expression and function of three VLA (β1) integrin receptors on T cells, *Nature*, 345, 250, 1990.

21. Saltini, C., Hemler, M. E., and Crystal, R. G., T lymphocytes compartmentalized on the epithelial surface of the lower respiratory tract express the very late activation antigen complex VLA-1, *Clin. Immunol. Immunopathol.*, 46, 221, 1988.

22. Hemler, M. E., Glass, D., Coblyn, J. S., and Jacobson, J. S., Very late activation antigens on rheumatoid synovial fluid T lymphocytes. Association with stage of T cell activation, *J. Clin. Invest.*, 78, 696, 1986.

23. Cush, J. J. and Lipsky, P. E., Phenotypic analysis of synovial tissue and peripheral blood lymphocytes isolated from patients with rheumatoid arthritis, *Arth. Rheum.*, 31, 1230, 1988.

24. Pantaleo, G., Koenig, S., Baseler, M., Lane, H. C., and Fauci, A. S., Defective clonogenic potential of CD8+ T lymphocytes in patients with AIDS. Expansion in vivo of a nonclonogenic CD3+CD8+DR+CD25-T cell population, *J. Immunol.*, 144, 1696, 1990.

25. Holzmann, B., McIntyre, B. W., and Weissman, I. L., Identification of a murine Peyer's patch-specific lymphocyte homing receptor as an integrin molecule with an α chain homologous to human VLA-4, *Cell*, 56, 37, 1989.

26. Yuan, Q., Jiang, W., Krissansen, G. W., and Watson, J. D., Cloning and sequence analysis of a novel β2-related integrin transcript from T lymphocytes: homology of integrin cysteine-rich repeats to domain III of laminin B chains, *Int. Immunol.*, 2, 1097, 1990.

27. Kilshaw, P. J. and Murant, S. J., Expression and regulation of β7(βp) integrins on mouse lymphocytes: relevance to the mucosal immune system, *Eur. J. Immunol.*, 21, 2591, 1991.

28. Wayner, E. A., Garcia-Pardo, A., Humphries, M. J., McDonald, J. A., and Carter, W. G., Identification and characterization of the T lymphocyte adhesion receptor for an alternative cell attachment domain (CS-1) in plasma fibronectin, *J. Cell Biol.*, 109, 1321, 1989.

29. Elices, M. J. and Hemler, M. E., The human integrin VLA-2 is a collagen receptor on some cells and a collagen/laminin receptor on others, *Proc. Natl. Acad. Sci. U.S.A.*, 86, 9906, 1989.

30. Goldman, R., Harvey, J., and Hogg, N., VLA-2 is the integrin used as a collagen receptor by leukocytes, *Eur. J. Immunol.*, 22, 1109, 1992.

31. Ennis, E., Isberg, R. R., and Shimizu, Y., Very late antigen-4-dependent adhesion and costimulation of resting human T cells by the bacterial β1 integrin ligand invasin, *J. Exp. Med.*, 177, 207, 1993.

32. Elices, M. J., Osborn, L., Takada, Y., Crouse, C., Luhowskyj, S., Hemler, M. E., and Lobb, R. R., VCAM-1 on activated endothelium interacts with the leukocyte integrin VLA-4 at a site distinct from the VLA-4/fibronectin binding site, *Cell*, 60, 577, 1990.

33. Shimizu, Y., Newman, W., Graber, N., Horgan, K. J., Beall, L. D., Gopal, T. V., van Seventer, G. A., and Shaw, S., Four molecular pathways of T cell adhesion to endothelial cells: roles of LFA-1, VCAM-1 and ELAM-1 and changes in pathway hierarchy under different activation conditions, *J. Cell Biol.*, 113, 1203, 1991.

34. Issekutz, T. B., Effect of antigen challenge on lymph node lymphocyte adhesion to vascular endothelial cells and the role of VLA-4 in the rat, *Cell. Immunol.*, 138, 300, 1991.

35. Issekutz, T. B., Inhibition of in vivo lymphocyte migration to inflammation and homing to lymphoid tissues by the TA-2 monoclonal antibody. A likely role for VLA-4 in vivo, *J. Immunol.*, 147, 4178, 1991.

36. Freedman, A. S., Munro, J. M., Rice, G. E., Bevilacqua, M. P., Morimoto, C., McIntyre, B. W., Rhynhart, K., Pober, J. S., and Nadler, L. M., Adhesion of human B cells to germinal centers in vitro involves VLA-4 and INCAM-110, *Science*, 249, 1030, 1990.

37. Rosen, G. D., Sanes, J. R., LaChance, R., Cunningham, J. M., Roman, J., and Dean, D. C., Roles for the integrin VLA-4 and its counter receptor VCAM-1 in myogenesis, *Cell*, 69, 1107, 1992.

38. Bednarczyk, J. L. and McIntyre, B. W., A monoclonal antibody to VLA-4 α-chain (CDw49d) induces homotypic lymphocyte aggregation, *J. Immunol.*, 144, 777, 1990.

39. Campanero, M. R., Pulido, R., Ursa, M. A., Rodriguez-Moya, M., De Landazuri, M. O., and Sanchez-Madrid, F., An alternative leukocyte homotypic adhesion mechanism, LFA-1/ICAM-1 independent, triggered through the human VLA-4 integrin, *J. Cell Biol.*, 110, 2157, 1990.

40. Ryan, D. H., Nuccie, B. L., Abboud, C. N., and Winslow, J. M., Vascular cell adhesion molecule-1 and the integrin VLA-4 mediate adhesion of human B cell precursors to cultured bone marrow adherent cells, *J. Clin. Invest.*, 88, 995, 1991.

41. Vonderheide, R. H. and Springer, T. A., Lymphocyte adhesion through very late antigen 4: evidence for a novel binding site in the alternatively spliced domain of vascular cell adhesion molecule 1 and an additional α4 integrin counter-receptor on stimulated endothelium, *J. Exp. Med.*, 175, 1433, 1992.

42. Takada, Y., Elices, M. J., Crouse, C., and Hemler, M. E., The primary structure of the α4 subunit of VLA-4—homology to other integrins and a possible cell-cell adhesion function, *EMBO J.*, 8, 1361, 1989.

43. Mackay, C. R. and Imhof, B. A., Cell adhesion in the immune system, *Immunol. Today*, 14, 99, 1993.

44. Rüegg, C., Postigo, A. A., Sikorski, E. E., Butcher, E. C., Pytela, R., and Erle, D. J., Role of integrin α4β7/α4βP in lymphocyte adherence to fibronectin and VCAM-1 and in homotypic cell clustering, *J. Cell Biol.*, 117, 179, 1992.

45. Chan, B. M. C., Elices, M. J., Murphy, E., and Hemler, M. E., Adhesion to vascular cell adhesion molecule 1 and fibronectin. Comparison of α⁴β₁ (VLA-4) and α⁴β₇ on the human B cell line JY, *J. Biol. Chem.*, 267, 8366, 1992.

46. Mosher, D. F., Physiology of thrombospondin, *Annu. Rev. Med.*, 41, 85, 1990.

47. Yabkowitz, R., Dixit, V. M., Guo, N., Roberts, D. D., and Shimizu, Y., Activated T cell adhesion to thrombospondin is mediated by the α4β1 (VLA-4) and α5β1 (VLA-5) integrins, *J. Immunol.*, 151, 149, 1993.

48. Isberg, R. R. and Leong, J. M., Multiple β1 chain integrins are receptors for invasin, a protein that promotes bacterial penetration into mammalian cells, *Cell*, 60, 861, 1990.

49. Bergelson, J. M., Shepley, M. P., Chan, B. M. C., Hemler, M. E., and Finberg, R. W., Identification of the integrin VLA-2 as a receptor for echovirus 1, *Science*, 255, 1718, 1992.

50. Fernández, M. A., Muñoz-Fernández, M. A., and Fresno, M., Involvement of β₁ integrins in the binding and entry of *Trypanosoma cruzi* into human macrophages, *Eur. J. Immunol.*, 23, 552, 1993.

51. Moulder, K., Roberts, K., Shevach, E. M., and Coligan, J. E., The mouse vitronectin receptor is a T cell activation antigen, *J. Exp. Med.*, 173, 343, 1991.

52. D'Souza, S. E., Ginsberg, M. H., and Plow, E. F., Arginyl-glycyl-aspartic acid (RGD): a cell adhesion motif, *Trends Biochem. Sci.*, 16, 246, 1991.

53. Klingemann, H. -G. and Dedhar, S., Distribution of integrins on human peripheral blood mononuclear cells, *Blood*, 74, 1348, 1989.

54. Parker, C. M., Cepek, K. L., Russell, G. J., Shaw, S. K., Posnett, D. N., Schwarting, R., and Brenner, M. B., A family of β₇ integrins on human mucosal lymphocytes, *Proc. Natl. Acad. Sci. U.S.A.*, 89, 1924, 1992.

55. Cerf-Bensussan, N., Begue, B., Gagnon, J., and Meo, T., The human intraepithelial lymphocyte marker HML-1 is an integrin consisting of a β7 subunit associated with a distinctive α chain. *Eur. J. Immunol.*, 22, 273, 1992.

56. Micklem, K. J., Dong, Y., Willis, A., Pulford, K. A., Visser, L., Dürkop, H., Poppema, S., Stein, H., and Mason, D. Y., HML-1 antigen on mucosa-associated T cells, activated cells, and hairy leukemic cells is a new integrin containing the β7 subunit, *Am. J. Pathol.*, 139, 1297, 1991.

57. Giunta, M., Favre, A., Ramarli, D., Grossi, C. E., and Corte, G., A novel integrin involved in thymocyte-thymic epithelial cell interactions, *J. Exp. Med.*, 173, 1537, 1991.

58. Takahashi, K., Nakamura, T., Adachi, H., Yagita, H., and Okumura, K., Antigen-independent T cell activation mediated by a very late activation antigen-like extracellular matrix receptor, *Eur. J. Immunol.*, 21, 1559, 1991.

59. Sawada, M., Nagamine, J., Takeda, K., Utsumi, K., Kosugi, A., Tatsumi, Y., Hamaoka, T., Miyake, K., Nakajima, K., Watanabe, T., Sakakibara, S., and Fujiwara, H., Expression of VLA-4 on thymocytes: maturation stage-associated transition and its correlation with their capacity to adhere to thymic stromal cells, *J. Immunol.*, 149, 3517, 1992.

60. Postigo, A. A., Pulido, R., Campanero, M. R., Acevedo, A., García-Pardo, A., Corbi, A. L., Sanchez-Madrid, F., and De Landazuri, M. O., Differential expression of VLA-4 integrin by resident and peripheral blood B lymphocytes. Acquisition of functionally active α4β1-fibronectin receptors upon B cell activation, *Eur. J. Immunol.*, 21, 2437, 1991.

61. Morimoto, C., Letvin, N. L., Boyd, A. W., Hagan, M., Brown, H. M., Kornacki, M. M., and Schlossman, S. F., The isolation and characterization of the human helper inducer T cell subset, *J. Immunol.*, 134, 3762, 1985.

62. Sanders, M. E., Makgoba, M. W., and Shaw, S., Human naive and memory T cells: reinterpretation of helper-inducer and suppressor-inducer subsets, *Immunol. Today*, 9, 195, 1988.

63. Beverley, P. C., Immunological memory in T cells, *Curr. Opin. Immunol.*, 3, 355, 1991.

64. Akbar, A. N., Salmon, M., and Janossy, G., The synergy between naive and memory T cells during activation, *Immunol. Today*, 12, 184, 1991.

65. Shimizu, Y., van Seventer, G. A., Horgan, K. J., and Shaw, S., Roles of adhesion molecules in T cell recognition: fundamental similarities between four integrins on resting human T cells (LFA-1, VLA-4, VLA-5, VLA-6) in expression, binding, and costimulation, *Immunol. Rev.*, 114, 109, 1990.

66. Sanders, M. E., Makgoba, M. W., Sharrow, S. O., Stephany, D., Springer, T. A., Young, H. A., and Shaw, S., Human memory T lymphocytes express increased levels of three cell adhesion molecules (LFA-3, CD2, and LFA-1) and three other molecules (UCHL1, CDw29, and Pgp-1) and have enhanced IFN-γ production, *J. Immunol.*, 140, 1401, 1988.

67. Cavender, D. E., Haskard, D. O., Maliakkai, D., and Ziff, M., Separation and characterization of human T cell subsets with varying degrees of adhesiveness for endothelial cells (EC). *Cell Immunol.*, 117, 111, 1988.

68. Damle, N. K. and Doyle, L. V., Ability of human T lymphocytes to adhere to vascular endothelial cells and to augment endothelial permeability to macromolecules is linked to their state of post-thymic maturation, *J. Immunol.*, 144, 1233, 1990.

69. Pitzalis, C., Kingsley, G., Haskard, D., and Panayi, G., The preferential accumulation of helper-inducer T lymphocytes in inflammatory lesions: evidence for regulation by selective endothelial and homotypic adhesion, *Eur. J. Immunol.*, 18, 1397, 1988.

70. Sanders, M. E., Makgoba, M. W., June, C. H., Young, H. A., and Shaw, S., Enhanced responsiveness of human memory T cells to CD2 and CD3 receptor-mediated activation, *Eur. J. Immunol.*, 19, 803, 1989.

71. Horgan, K. J., van Seventer, G. A., Shimizu, Y., and Shaw, S., Hyporesponsiveness of naive (CD45RA+) human T cells to multiple receptor-mediated stimuli but augmentation of responses by costimuli, *Eur. J. Immunol.*, 20, 1111, 1990.

72. Shaw, S., Luce, G. E. G., Quinones, R., Gress, R. E., Springer, T. A., and Sanders, M. E., Two antigen-independent adhesion pathways used by human cytotoxic T cell clones, *Nature*, 323, 262, 1986.

73. Spits, H., Van Schooten, W., Keizer, H., Van Seventer, G., Van de Rijn, M., Terhorst, C., and De Vries, J. E., Alloantigen recognition is preceded by nonspecific adhesion of cytotoxic T cells and target cells, *Science*, 232, 403, 1986.

74. Dustin, M. L. and Springer, T. A., T-cell receptor cross-linking transiently stimulates adhesiveness through LFA-1, *Nature*, 341, 619, 1989.

75. Patarroyo, M., Yogeeswaran, G., Biberfeld, P., Klein, E., and Klein, G., Morphological changes, cell aggregation and cell membrane alterations caused by phorbol 12, 13-dibutyrate in human blood lymphocytes, *Int. J. Cancer*, 30, 707, 1982.

76. Rothlein, R. and Springer, T. A., The requirement for LFA-1 in homotypic leukocyte adhesion stimulated by phorbol ester, *J. Exp. Med.*, 163, 1132, 1986.

77. Kurki, P., Vartio, T., and Virtanen, I., Mitogen stimulation promotes human T lymphocyte adhesion to fibronectin, *Scand. J. Immunol.*, 26, 645, 1987.

78. Moingeon, P. E., Lucich, J. L., Stebbins, C. C., Recny, M. A., Wallner, B. P., Koyasu, S., and Reinherz, E. L., Complementary roles for CD2 and LFA-1 adhesion pathways during T cell activation, *Eur. J. Immunol.*, 21, 605, 1991.

79. van Kooyk, Y., van de Wiel-van Kemenade, P., Weder, P., Kuijpers, T. W., and Figdor, C. G., Enhancement of LFA-1-mediated cell adhesion by triggering through CD2 or CD3 on T lymphocytes, *Nature*, 342, 811, 1989.

80. Meuer, S. C., Hussey, R. E., Fabbi, M., Fox, D., Acuto, O., Fitzgerald, K. A., Hodgdon, J. C., Protentis, J. P., Schlossman, S. F., and Reinherz, E. L., An alternative pathway of T-cell activation: a functional role for the 50 kd T11 sheep erythrocyte receptor protein, *Cell*, 36, 897, 1984.

81. van Seventer, G. A., Newman, W., Shimizu, Y., Nutman, T. B., Tanaka, Y., Horgan, K. J., Gopal, T. V., Ennis, E., O'Sullivan, D., Grey, H., and Shaw, S., Analysis of T-cell stimulation by superantigen plus MHC class II molecules or by CD3 mAb: costimulation by purified adhesion ligands VCAM-1, ICAM-1 but not ELAM-1, *J. Exp. Med.*, 174, 901, 1991.

82. Shimizu, Y., van Seventer, G. A., Ennis, E., Newman, W., Horgan, K. J., and Shaw, S., Crosslinking of the T cell-specific accessory molecules CD7 and CD28 modulates T cell adhesion, *J. Exp. Med.*, 175, 577, 1992.

83. Carrera, A. C., Rincon, M., Sanchez-Madrid, F., Lopez-Botet, M., and De Landazuri, M. O., Triggering of co-mitogenic signals in T cell proliferation by anti-LFA-1 (CD18, CD11a), LFA-3 and CD7 monoclonal antibodies, *J. Immunol.*, 141, 1919, 1988.

84. Ledbetter, J. A., June, C. H., Grosmaire, L. S., and Rabinovitch, P. S., Crosslinking of surface antigens causes mobilization of intracellular ionized calcium in T lymphocytes, *Proc. Natl. Acad. Sci. U.S.A.*, 84, 1384, 1987.

85. Schwartz, R. H., Costimulation of T lymphocytes: the role of CD28, CTLA-4, and B7/BB1 in interleukin-2 production and immunotherapy, *Cell*, 71, 1065, 1992.

86. Albelda, S. M., Muller, W. A., Buck, C. A., and Newman, P. J., Molecular and cellular properties of PECAM-1 (endoCAM/CD31): a novel vascular cell-cell adhesion molecule, *J. Cell Biol.*, 114, 1059, 1991.

87. Tanaka, Y., Albelda, S. M., Horgan, K. J., van Seventer, G. A., Shimizu, Y., Newman, W., Hallam, J., Newman, P. J., Buck, C. A., and Shaw, S., CD31 expressed on distinctive T cell subsets is a preferential amplifier of $\beta 1$ integrin-mediated adhesion, *J. Exp. Med.*, 176, 245, 1992.

88. Tanaka, Y., Adams, D. H., Hubscher, S., Hirano, H., Siebenlist, U., and Shaw, S., T-cell adhesion induced by proteoglycan-immobilized cytokine MIP-1β, *Nature*, 361, 79, 1993.

89. Axelsson, B., Youseffi-Etemad, R., Hammarstrom, S., and Perlman, P., Induction of aggregation and enhancement of proliferation and IL-2 secretion in human T cells by antibodies to CD43, *J. Immunol.*, 141, 2912, 1988.

90. Koopman, G., van Kooyk, Y., De Graaff, M., Meyer, C. J. L. M., Figdor, C. G., and Pals, S. T., Triggering of the CD44 antigen on T lymphocytes promotes T cell adhesion through the LFA-1 pathway, *J. Immunol.*, 145, 3589, 1990.

91. Belitsos, P. C., Hildreth, J. E. K., and August, J. T., Homotypic cell aggregation induced by anti-CD44(Pgp-1) monoclonal antibodies and related to CD44(Pgp-1) expression, *J. Immunol.*, 144, 1661, 1990.

92. Keizer, G. D., Visser, W., Vliem, M., and Figdor, C. G., A monoclonal antibody (NK1-L16) directed against a unique epitope on the α-chain of human leukocyte function-associated antigen 1 induces homotypic cell-cell interactions, *J. Immunol.*, 140, 1393, 1988.

93. Pulido, R., Elices, M. J., Campanero, M. R., Osborn, L., Schiffer, S., García-Pardo, A., Lobb, R., Hemler, M. E., and Sánchez-Madrid, F., Functional evidence for three distinct and independently inhibitable adhesion activities mediated by the human integrin VLA-4. Correlation with distinct α4 epitopes, *J. Biol. Chem.*, 266, 10241, 1991.

94. Caixia, S., Stewart, S., Wayner, E., Carter, W., and Wilkins, J., Antibodies to different members of the β1 (CD29) integrins induce homotypic and heterotypic cellular aggregation, *Cell. Immunol.*, 138, 216, 1991.

95. Mourad, W., Geha, R. S., and Chatila, T., Engagement of major histocompatibility complex class II molecules induces sustained, lymphocyte function-associated molecule 1-dependent cell adhesion, *J. Exp. Med.*, 172, 1513, 1990.

96. Lauener, R. P., Geha, R. S., and Vercelli, D., Engagement of the monocyte surface antigen CD14 induces lymphocyte function-associated antigen-1/intercellular adhesion molecule-1-dependent homotypic adhesion, *J. Immunol.*, 145, 1390, 1990.

97. Dang, L. H. and Rock, K. L., Stimulation of B lymphocytes through surface Ig receptors induces LFA-1 and ICAM-1-dependent adhesion, *J. Immunol.*, 146, 3273, 1991.

98. Kansas, G. S. and Tedder, T. F., Transmembrane signals generated through MHC class II, CD19, CD20, CD39, and CD40 antigens induce LFA-1-dependent and independent adhesion in human B cells through a tyrosine kinase-dependent pathway, *J. Immunol.*, 147, 4094, 1991.

99. Smith, S. H., Rigley, K. P., and Callard, R. E., Activation of human B cells through the CD19 surface antigen results in homotypic adhesion by LFA-1-dependent and -independent mechanisms, *Immunology*, 73, 293, 1991.

100. Kansas, G. S., Wood, G. S., and Tedder, T. F., Expression, distribution, and biochemistry of human CD39: role in activation-associated homotypic adhesion of lymphocytes, *J. Immunol.*, 146, 2235, 1991.

101. Barrett, T. B., Shu, G., and Clark, E. A., CD40 signaling activates CD11a/CD18 (LFA-1)-mediated adhesion in B cells, *J. Immunol.*, 146, 1722, 1991.

102. Mazerolles, F., Hauss, P., Barbat, C., Figdor, C. G., and Fischer, A., Regulation of LFA-1-mediated T cell adhesion by CD4, *Eur. J. Immunol.*, 21, 887, 1991.

103. Mazerolles, F., Auffray, C., and Fischer, A., Down regulation of T-cell adhesion by CD4, *Hum. Immunol.*, 31, 40, 1991.

104. Hershkoviz, R., Miron, S., Cohen, I. R., Miller, A., and Lider, O., T lymphocyte adhesion to the fibronectin and laminin components of the extracellular matrix is regulated by the CD4 molecule, *Eur. J. Immunol.*, 22, 7, 1992.

105. Chan, B. M. C., Wong, J. G. P., Rao, A., and Hemler, M. E., T cell receptor-dependent, antigen-specific stimulation of a murine T cell clone induces a transient, VLA protein-mediated binding to extracellular matrix, *J. Immunol.*, 147, 398, 1991.

106. Harding, C. V. and Unanue, E. R., Modulation of antigen presentation and peptide-MHC-specific, LFA-1-dependent T cell-macrophage adhesion, *J. Immunol.*, 147, 767, 1991.

107. Kansas, G. S., Cambier, J. C., and Tedder, T. F., CD4 binding to major histocompatibility complex class II antigens induces LFA-1-dependent and -independent homotypic adhesion of B lymphocytes, *Eur. J. Immunol.*, 22, 147, 1992.

108. Selvaraj, P., Plunkett, M. L., Dustin, M., Sanders, M. E., Shaw, S., and Springer, T. A., The T lymphocyte glycoprotein CD2 binds the cell surface ligand LFA-3, *Nature*, 326, 400, 1987.

109. Linsley, P. S., Clark, E. A., and Ledbetter, J. A., T-cell antigen CD28 mediates adhesion with B cells by interacting with activation antigen B7/BB-1, *Proc. Natl. Acad. Sci. U.S.A.*, 87, 5031, 1990.

110. Mercurio, A. M. and Shaw, L. M., Macrophage interactions with laminin: PMA selectively induces the adherence and spreading of mouse macrophages on a laminin substratum, *J. Cell Biol.*, 107, 1873, 1988.

111. Shaw, L. M. and Mercurio, A. M., Interferon gamma and lipopolysaccharide promote macrophage adherence to basement membrane glycoproteins, *J. Exp. Med.*, 169, 303, 1989.

112. Shaw, L. M., Messier, J. M., and Mercurio, A. M., The activation dependent adhesion of macrophages to laminin involves cytoskeletal anchoring and phosphorylation of the $\alpha6\beta1$ integrin, *J. Cell Biol.*, 110, 2167, 1990.

113. Tamatani, T., Kotani, M., Tanaka, T., and Miyasaka, M., Molecular mechanisms underlying lymphocyte recirculation. II. Differential regulation of LFA-1 in the interaction between lymphocytes and high endothelial cells, *Eur. J. Immunol.*, 21, 855, 1991.

114. Masumoto, A. and Hemler, M. E., Multiple activation states of VLA-4. Mechanistic differences between adhesion to CS1/fibronectin and to vascular cell adhesion molecule-1, *J. Biol. Chem.*, 268, 228, 1993.

115. Chatila, T. A., Geha, R. S., and Arnaout, M. A., Constitutive and stimulus-induced phosphorylation of CD11/CD18 leukocyte adhesion molecules, *J. Cell Biol.*, 109, 3435, 1989.

116. Buyon, J. P., Slade, S. G., Reibman, J., Abramson, S. B., Philips, M. R., Weissmann, G., and Winchester, R., Constitutive and induced phosphorylation of the α- and β-chains of the CD11/CD18 leukocyte integrin family: relationship to adhesion-dependent functions, *J. Immunol.*, 144, 191, 1990.

117. Valmu, L., Autero, M., Siljander, P., Patarroyo, M., and Gahmberg, C. G., Phosphorylation of the β-subunit of CD11/CD18 integrins by protein kinase C correlates with leukocyte adhesion, *Eur. J. Immunol.*, 21, 2857, 1991.

118. Hibbs, M. L., Jakes, S., Stacker, S. A., Wallace, R. W., and Springer, T. A., The cytoplasmic domain of the integrin lymphocyte function-associated antigen 1 β subunit: sites required for binding to intercellular adhesion molecule 1 and the phorbol ester-stimulated phosphorylation site, *J. Exp. Med.*, 174, 1227, 1991.

119. Larson, R. S., Hibbs, M. L., and Springer, T. A., The leukocyte integrin LFA-1 reconstituted by cDNA transfection in a nonhematopoietic cell line is functionally active and not transiently regulated. *Cell Regul.*, 1, 359, 1990.

120. Hibbs, M. L., Xu, H., Stacker, S. A., and Springer, T. A., Regulation of adhesion to ICAM-1 by the cytoplasmic domain of LFA-1 integrin β subunit, *Science*, 251, 1611, 1991.

121. Kupfer, A. and Singer, S. J., The specific interaction of helper T cells and antigen-presenting B cells. IV. Membrane and cytoskeletal reorganizations in the bound T cell as a function of antigen dose, *J. Exp. Med.*, 170, 1697, 1989.

122. Horwitz, A., Duggan, K., Buck, C., Beckerle, M. C., and Burridge, D., Interaction of plasma membrane fibronectin receptor with talin-a transmembrane linkage, *Nature*, 320, 531, 1986.

123. Otey, C. A., Pavalko, F. M., and Burridge, K., An interaction between α-actinin and the β1 integrin subunit in vitro, *J. Cell Biol.*, 111, 721, 1990.

124. Detmers, P. A., Wright, S. D., Olsen, E., Kimball, B., and Cohn, Z. A., Aggregation of complement receptors on human neutrophils in the absence of ligand, *J. Cell Biol.*, 105, 1137, 1987.

125. Figdor, C. G., van Kooyk, Y., and Keizer, G. D., On the mode of action of LFA-1, *Immunol. Today*, 11, 277, 1990.

126. Figdor, C. G. and van Kooyk, Y., Regulation of cell adhesion. in *Adhesion. Its role in Inflammatory Disease*, Harlan, J. M. and Liu, D. Y., Eds., W. H. Freeman, New York, 1992, 151.

127. Dransfield, I., Regulation of leukocyte integrin function, *Chem. Immunol.*, 50, 13, 1991.

128. van Kooyk, Y., Weder, P., Hogervorst, F., VerHoeven, A. J., Te Velde, A. A., Borst, J., Keizer, G. D., and Figdor, C. G., Activation of LFA-1 through a Ca^{2+}-dependent epitope stimulates lymphocyte adhesion, *J. Cell Biol.*, 112, 345, 1991.

129. Dransfield, I. and Hogg, N., Regulated expression of Mg^{2+} binding epitope on leukocyte integrin α subunits, *EMBO J.*, 8, 3759, 1989.

130. Dransfield, I., Cabañas, C., Craig, A., and Hogg, N., Divalent cation regulation of the function of the leukocyte integrin LFA-1, *J. Cell Biol.*, 116, 219, 1992.

131. Gailit, J. and Ruoslahti, E., Regulation of the fibronectin receptor affinity by divalent cations, *J. Biol. Chem.*, 263, 12927, 1988.

132. Hautanen, A., Gailit, J., Mann, D. M., and Ruoslahti, E., Effects of modifications of the RGD sequence and its context on recognition by the fibronectin receptor, *J. Biol. Chem.*, 264, 1437, 1989.

133. Bohnsack, J. F. and Zhou, X. -N., Divalent cation substitution reveals CD18- and very late antigen-dependent pathways that mediate human neutrophil adherence to fibronectin. *J. Immunol.*, 149, 1340, 1992.

134. Kirchhofer, D., Gailit, J., Ruoslahti, E., Grzesiak, J. J., and Pierschbacher, M. D., Cation-dependent changes in the binding specificity of the platelet receptor gpIIb/IIIa, *J. Biol. Chem.*, 265, 18525, 1990.

135. Elices, M. J., Urry, L. A., and Hemler, M. E., Receptor functions for the integrin VLA-3: fibronectin, collagen, and laminin binding are differentially influenced by ARG-GLY-ASP peptide and divalent cations, *J. Cell Biol.*, 112, 169, 1991.

136. Sonnenberg, A., Modderman, P. W., and Hogervorst, F., Laminin receptor on platelets is the integrin VLA-6, *Nature*, 336, 487, 1988.

137. Grzesiak, J. J., Davis, G. E., Kirchhofer, D., and Pierschbacher, M. D., Regulation of $α_2β_1$-mediated fibroblast migration on type I collagen by shifts in the concentrations of extracellular Mg^{2+} and Ca^{2+}, *J. Cell Biol.*, 117, 1109, 1992.

138. Chan, B. M. C. and Hemler, M. E., Multiple functional forms of the integrin VLA-2 can be derived from a single $α^2$ cDNA clone: interconversion of forms induced by an anti-$β_1$ antibody, *J. Cell Biol.*, 120, 537, 1993.

139. Kovach, N. L., Carlos, T. M., Yee, E., and Harlan, J. M., A monoclonal antibody to $β_1$ integrin (CD29) stimulates VLA-dependent adherence of leukocytes to human umbilical vein endothelial cells and matrix components, *J. Cell Biol.*, 116, 499, 1992.

140. Arroyo, A. G., Sanchez-Mateos, P., Campanero, M. R., Martin-Padura, I., Dejana, E., and Sanchez-Madrid, F., Regulation of the VLA integrin-ligand interactions through the β1 subunit, *J. Cell Biol.*, 117, 659, 1992.

141. Van de Wiel-van Kemenade, E., van Kooyk, Y., De Boer, A. J., Huijbens, R. J. F., Weder, P., Van de Kasteele, W., Melief, C. J. M., and Figdor, C. G., Adhesion of T and B lymphocytes to extracellular matrix and endothelial cells can be regulated through the β subunit of VLA, *J. Cell Biol.*, 117, 461, 1992.

142. Kouns, W. C., Wall, C. D., White, M. M., Fox, C. F., and Jennings, L. K., A conformation-dependent epitope of human platelet glycoprotein IIIa, *J. Biol. Chem.*, 265, 20594, 1990.

143. O'Toole, T. E., Loftus, J. C., Du, X., Glass, A. A., Ruggieri, Z. M., Shattil, S. J., Plow, E. F., and Ginsberg, M. H., Affinity modulation of the αIIbβ3 integrin (platelet gpIIb-IIIa) is an intrinsic property of the receptor, *Cell Regul.*, 1, 883, 1990.

144. Hermanowski-Vosatka, A., van Strijp, J. A. G., Swiggard, W. J., and Wright, S. D., Integrin modulating factor-1: a lipid that alters the function of leukocyte integrins, *Cell*, 68, 341, 1992.

145. Geppert, T. D. and Lipsky, P. E., Antigen presentation at the inflammatory site, *CRC Crit. Rev. Immunol.*, 9, 313, 1989.

146. Schwartz, R. H., A cell culture model for T lymphocyte clonal anergy, *Science*, 248, 1349, 1990.

147. van Seventer, G. A., Shimizu, Y., and Shaw, S., Roles of multiple accessory molecules in T-cell activation, *Curr. Opin. Immunol.*, 3, 294, 1991.

148. Shimizu, Y. and Shaw, S., Lymphocyte interactions with extracellular matrix, *FASEB J.*, 5, 2292, 1991.

149. Ferguson, T. A., Mizutani, H., and Kupper, T. S., Two integrin-binding peptides abrogate T cell-mediated immune responses *in vivo*, *Proc. Natl. Acad. Sci. U.S.A.*, 88, 8072, 1991.

150. Shimizu, Y., Newman, W., Tanaka, Y., and Shaw, S., Lymphocyte interactions with endothelial cells, *Immunol. Today*, 13, 106, 1992.

151. Yednock, T. A. and Rosen, S. D., Lymphocyte homing, *Adv. Immunol.*, 44, 313, 1989.

152. Dustin, M. L. and Springer, T. A., Lymphocyte function associated antigen-1 (LFA-1) interaction with intercellular adhesion molecule-1 (ICAM-1) is one of at least three mechanisms for T lymphocyte adhesion to cultured endothelial cells, *J. Cell Biol.*, 107, 321, 1988.

153. Hamann, A., Jablonski-Westrich, D., Duijvestijn, A., Butcher, E. C., Baisch, H., Harder, R., and Thiele, H. G., Evidence for an accessory role of LFA-1 in lymphocyte-high endothelium interaction during homing, *J. Immunol.*, 140, 693, 1988.

154. Kishimoto, T. K., A dynamic model for neutrophil localization to inflammatory sites, *J. NIH Res.*, 3, 75, 1991.

155. Butcher, E. C., Leukocyte-endothelial cell recognition: three (or more) steps to specificity and diversity, *Cell*, 67, 1033, 1991.

156. Tanaka, Y., Adams, D. H., and Shaw, S., Proteoglycans on endothelial cells present adhesion-inducing cytokines to leukocytes, *Immunol. Today*, 14, 111, 1993.

157. Arnaout, M. A., Leukocyte adhesion molecules deficiency: its structural basis, pathophysiology and implications for modulating the inflammatory response, *Immunol. Rev.*, 114, 145, 1990.

158. Giancotti, F. G. and Ruoslahti, E., Elevated levels of the $\alpha5\beta1$ fibronectin receptor suppress the transformed phenotype of chinese hamster ovary cells, *Cell*, 60, 849, 1990.

159. Ruoslahti, E., Control of cell motility and tumour invasion by extracellular matrix interactions, *Br. J. Cancer*, 66, 239, 1992.

160. Chan, B. M. C., Matsuura, N., Takada, Y., Zetter, B. R., and Hemler, M. E., In vitro and in vivo consequences of VLA-2 expression on rhabdomyosarcoma cells, *Science*, 251, 1600, 1991.

161. Laffón, A., García-Vicuña, R., Humbría, A., Postigo, A. A., Corbí, A. L., De Landázuri, M. O., and Sánchez-Madrid, F., Upregulated expression and function of VLA-4 fibronectin receptors on human activated T cells in rheumatoid arthritis, *J. Clin. Invest.*, 88, 546, 1991.

162. Kilshaw, P. J. and Murant, S. J., A new surface antigen on intraepithelial lymphocytes on the intestine, *Eur. J. Immunol.*, 20, 2201, 1990.

163. Yuan, Q., Jiang, W., Hollander, D., Leung, E., Watson, J. D., and Krissansen, G. W., Identity between the novel integrin β_7 subunit and an antigen found highly expressed on intraepithelial lymphocytes in the small intestine, *Biochem. Biophys. Res. Commun.*, 176, 1443, 1991.

164. van Noesel, C., Miedema, F., Brouwer, M., De Rie, M. A., Aarden, L. A., and van Lier, R. A., Regulatory properties of LFA-1 α and β chains in human lymphocyte activation, *Nature*, 333, 850, 1988.

165. David, V., Leca, G., Corvaia, N., Le Deist, F., Boumsell, L., and Bensussan, A., Proliferation of resting lymphocytes is induced by triggering T cells through an epitope common to the three CD18/CD11 leukocyte adhesion molecules, *Cell. Immunol.*, 136, 519, 1991.

166. Cerdan, C., Lipcey, C., Lopez, M., Nunes, J., Pierres, A., Mawas, C., and Olive, D., Monoclonal antibodies against LFA-1 or its ligand ICAM-1 accelerate CD2 (T11.1+T11.2)-mediated T cell proliferation, *Cell. Immunol.*, 123, 344, 1989.

167. van Seventer, G. A., Shimizu, Y., Horgan, K. J., and Shaw, S., The LFA-1 ligand ICAM-1 provides an important costimulatory signal for T cell receptor-mediated activation of resting T cells, *J. Immunol.*, 144, 4579, 1990.

168. Kuhlman, P., Moy, V. T., Lollo, B. A., and Brian, A. A., The accessory function of murine intercellular adhesion molecule-1 in T lymphocyte activation: contributions of adhesion and co-activation, *J. Immunol.*, 146, 1773, 1991.

169. Altmann, D. M., Hogg, N., Trowsdale, J., and Wilkinson, D., Cotransfection of ICAM-1 and HLA-DR reconstitutes human antigen presenting cell function in mouse L cells. *Nature*, 338, 512, 1989.

170. Siu, G., Hedrick, S. M., and Brian, A. A., Isolation of the murine intercellular adhesion molecule 1 (ICAM-1) gene. ICAM-1 enhances antigen-specific T cell activation, *J. Immunol.*, 143, 3813, 1989.

171. Nickoloff, B. J., Mitra, R. S., Green, J., Zheng, X. -G., Shimizu, Y., Thompson, C., and Turka, L. A., Accessory cell function of keratinocytes for superantigens: dependence on LFA-1/ICAM-1 interaction, *J. Immunol.*, 150, 2148, 1993.

172. Damle, N. K., Klussman, K., and Aruffo, A., Intercellular adhesion molecule-2, a second counter-receptor for CD11a/CD18 (leukocyte function-associated antigen-1), provides a costimulatory signal for T-cell receptor-initiated activation of human T cells, *J. Immunol.*, 148, 665, 1992.

173. Moy, V. T. and Brian, A. A., Signaling by lymphocyte function-associated antigen 1 (LFA-1) in B cells: enhanced antigen presentation after stimulation through LFA-1, *J. Exp. Med.*, 175, 1, 1992.

174. Matsuyama, T., Yamada, A., Kay, J., Yamada, K. M., Akiyama, S. K., Schlossman, S. F., and Morimoto, C., Activation of CD4 cells by fibronectin and anti-CD3 antibody. A synergistic effect mediated by the VLA-5 fibronectin receptor complex, *J. Exp. Med.*, 170, 1133, 1989.

175. Shimizu, Y., van Seventer, G. A., Horgan, K. J., and Shaw, S., Costimulation of proliferative responses of resting CD4+ T cells by the interaction of VLA-4 and VLA-5 with fibronectin or VLA-6 with laminin, *J. Immunol.*, 145, 59, 1990.

176. Davis, L. S., Oppenheimer-Marks, N., Bednarczyk, J. L., McIntyre, B. W., and Lipsky, P. E., Fibronectin promotes proliferation of naive and memory T cells by signalling through both VLA-4 and VLA-5 integrin molecules, *J. Immunol.*, 145, 785, 1990.

177. Cardarelli, P. M., Yamagata, S., Scholz, W., Moscinski, M. A., and Morgan, E. L., Fibronectin augments anti-CD3-mediated IL-2 receptor (CD25) expression on human peripheral blood lymphocytes, *Cell. Immunol.*, 135, 105, 1991.

178. Nojima, Y., Humphries, M. J., Mould, A. P., Komoriya, A., Yamada, K. M., Schlossman, S. F., and Morimoto, C., VLA-4 mediates CD3-dependent CD4+ T cell activation via the CS1 alternatively spliced domain of fibronectin, *J. Exp. Med.*, 172, 1185, 1990.

179. Yamada, A., Nojima, Y., Sugita, K., Dang, N. H., Schlossman, S. F., and Morimoto, C., Cross-linking of VLA/CD29 molecule has a co-mitogenic effect with anti-CD3 on CD4 cell activation in serum-free culture system, *Eur. J. Immunol.*, 21, 319, 1991.

180. Klingemann, H. G., Dedhar, S., Kohn, F. R., and Phillips, G. L., Fibronectin increases lymphocyte proliferation by mediating adhesion between immunoreactive cells, *Transplantation*, 42, 412, 1986.

181. Klingemann, H. G., Tsoi, M. S., and Storb, R., Fibronectin restores defective in vitro proliferation of patients' lymphocytes after marrow grafting, *Transplantation*, 42, 412, 1986.

182. Klingemann, H. G. and Kohn, F. R., Involvement of fibronectin and its receptor in human lymphocyte proliferation, *J. Leukocyte Biol.*, 50, 464, 1991.

183. Dang, N. H., Torimoto, Y., Schlossman, S. F., and Morimoto, C., Human CD4 helper T cell activation: functional involvement of two distinct collagen receptors, 1F7 and VLA integrin family, *J. Exp. Med.*, 172, 649, 1990.

184. Burkly, L. C., Jakubowski, A., Newman, B. M., Rosa, M. D., Chi-Rosso, G., and Lobb, R. R., Signaling by vascular cell adhesion molecule-1 (VCAM-1) through VLA-4 promotes CD3-dependent T cell proliferation, *Eur. J. Immunol.*, 21, 2871, 1991.

185. Damle, N. K. and Aruffo, A., Vascular cell adhesion molecule 1 induces T-cell antigen receptor-dependent activation of $CD4^+$ T lymphocytes, *Proc. Natl. Acad. Sci. U.S.A.*, 88, 6403, 1991.

186. Blue, M. -L., Conrad, P., Davis, G., and Kelley, K. A., Enhancement of CD2-mediated T cell activation by the interaction of VLA-4 with fibronectin, *Cell. Immunol.*, 138, 238, 1991.

187. Roberts, K., Yokoyama, W. M., Kehn, P. J., and Shevach, E. M., The vitronectin receptor serves as an accessory molecule for the activation of a subset of γδ T cells, *J. Exp. Med.*, 173, 231, 1991.

188. van Seventer, G. A., Shimizu, Y., Horgan, K. J., Luce, G. E. G., Webb, D., and Shaw, S., Remote T-cell costimulation via LFA-1/ICAM-1 and CD2/LFA-3: demonstration with immobilized ligand/mAb and implication in monocyte-mediated costimulation, *Eur. J. Immunol.*, 21, 1711, 1991.

189. Picker, L. J., Kishimoto, T. K., Smith, C. W., Warnock, R. A., and Butcher, E. C., ELAM-1 is an adhesion molecule for skin-homing T cells, *Nature*, 349, 796, 1991.

190. Shimizu, Y., Shaw, S., Graber, N., Gopal, T. V., Horgan, K. J., van Seventer, G. A., and Newman, W., Activation-independent binding of human memory T cells to adhesion molecule ELAM-1, *Nature*, 349, 799, 1991.

191. van Seventer, G. A., Bonvini, E., Yamada, H., Conti, A., Stringfellow, S., June, C. H., and Shaw, S., Costimulation of T cell receptor/CD3-mediated activation of resting human $CD4^+$ T cells by leukocyte function-associated antigen-1 ligand intercellular cell adhesion molecule-1 involves prolonged inositol phospholipid hydrolysis and sustained increase of intracellular Ca^{2+} levels, *J. Immunol.*, 149, 3872, 1992.

192. Groux, H., Huet, S., Valentin, H., Pham, D., and Bernard, A., Suppressor effects and cyclic AMP accumulation by the CD29 molecule of CD4+ lymphocytes, *Nature*, 339, 152, 1989.

193. Wacholtz, M. C., Patel, S. S., and Lipsky, P. E., Leukocyte function-associated antigen 1 is an activation molecule for human T cells, *J. Exp. Med.*, 170, 431, 1989.

194. Pardi, R., Bender, J. R., Dettori, C., Giannazza, E., and Engleman, E. G., Heterogeneous distribution and transmembrane signaling properties of lymphocyte function-associated antigen (LFA-1) in human lymphocyte subsets, *J. Immunol.*, 143, 3157, 1989.

195. Schwartz, M. A., Ingber, D. E., Lawrence, M., Springer, T. A., and Lechene, C., Multiple integrins share the ability to induce elevation of intracellular pH, *Exp. Cell Res.*, 195, 533, 1991.

196. Yamada, A., Nikaido, T., Nojima, Y., Schlossman, S. F., and Morimoto, C., Activation of human CD4 T lymphocytes. Interaction of fibronectin with VLA-5 receptor on CD4 cells induces the AP-1 transcription factor, *J. Immunol.*, 146, 53, 1991.

197. Knobloch, C., Diamantstein, T., Flegel, W. A., and Friedrich, W., Stimulation of human T cells via anti-T cell receptor monoclonal antibody BMA031: distinct cellular events involving interleukin-2 receptor and lymphocyte function antigen 1, *Cell. Immunol.*, 138, 150, 1991.

198. Miedema, F., Tetteroo, P. A., Terpstra, F. G., Keizer, G., Roos, M., Weening, R. S., Weemaes, C. M. R., Roos, D., and Melief, C. J. M., Immunologic studies with LFA-1 and Mo1-deficient lymphocytes from a patient with recurrent bacterial infections, *J. Immunol.*, 134, 3075, 1985.

199. Pacifici, R., Carano, A., Santoro, S. A., Rifas, L., Jeffrey, J. J., Malone, J. D., McCracken, R., and Avioli, L. V., Bone matrix constituents stimulate interleukin-1 release from human blood mononuclear cells, *J. Clin. Invest.*, 87, 221, 1991.

200. Pacifici, R., Basilico, C., Roman, J., Zutter, M. M., Santoro, S. A., and McCracken, R., Collagen-induced release of interleukin 1 from human blood mononuclear cells. Potentiation by fibronectin binding to the $\alpha_5\beta_1$ integrin, *J. Clin. Invest.*, 89, 61, 1992.

201. Werb, Z., Tremble, P. M., Behrendtsen, O., Crowley, E., and Damsky, C. H., Signal transduction through the fibronectin receptor induces collagenase and stromelysin gene expression, *J. Cell Biol.*, 109, 877, 1989.

202. Guan, J. -L., Trevithick, J. E., and Hynes, R. O., Fibronectin/integrin interaction induces tyrosine phosphorylation of a 120-kDa protein, *Cell Regul.*, 2, 951, 1991.

203. Guan, J. -L. and Shalloway, D., Regulation of focal adhesion-associated protein tyrosine kinase by both cellular adhesion and oncogenic transformation, *Nature*, 358, 690, 1992.

204. Zachary, I. and Rozengurt, E., Focal adhesion kinase (p125[FAK]): a point of convergence in the action of neuropeptides, integrins, and oncogenes, *Cell*, 71, 891, 1992.

205. Shattil, S. J. and Brugge, J. S., Protein tyrosine phosphorylation and the adhesive functions of platelets, *Curr. Biol.*, 3, 869, 1991.

206. Lipfert, L., Haimovich, B., Schaller, M. D., Cobb, B. S., Parsons, J. T., and Brugge, J. S., Integrin-dependent phosphorylation and activation of the protein tyrosine kinase pp125[FAK] in platelets, *J. Cell Biol.*, 119, 905, 1992.

207. Burridge, K., Turner, C. E., and Romer, L. H., Tyrosine phosphorylation of paxillin and pp125[FAK] accompanies cell adhesion to extracellular matrix: a role in cytoskeletal assembly, *J. Cell Biol.*, 119, 893, 1992.

208. Hanks, S. K., Calalb, M. B., Harper, M. C., and Patel, S. K., Focal adhesion protein-tyrosine kinase phosphorylated in response to cell attachment to fibronectin, *Proc. Natl. Acad. Sci. U.S.A.*, 89, 8487, 1992.

209. Schaller, M. D., Borgman, C. A., Cobb, B. S., Vines, R. R., Reynolds, A. B., and Parsons, J. T., pp125[FAK] a structurally distinctive protein-tyrosine kinase associated with focal adhesions, *Proc. Natl. Acad. Sci. U.S.A.*, 89, 5192, 1992.

210. Nojima, Y., Rothstein, D. M., Sugita, K., Schlossman, S. F., and Morimoto, C., Ligation of VLA-4 on T cells stimulates tyrosine phosphorylation of a 105-kD protein, *J. Exp. Med.*, 175, 1045, 1992.

211. Freedman, A. S., Rhynhart, K., Nojima, Y., Svahn, J., Eliseo, L., Benjamin, C. D., Morimoto, C., and Vivier, E., Stimulation of protein tyrosine phosphorylation in human B cells after ligation of the $\beta 1$ integrin VLA-4, *J. Immunol.*, 150, 1645, 1993.

212. Charlton, B., Guymer, R. H., Slattery, R. M., and Mandel, T. E., Intercellular adhesion molecule (ICAM-1) inhibition can induce tolerance *in vivo, Immunol. Cell Biol.*, 69, 89, 1991.

213. Kramer, R. H. and Marks, N., Identification of integrin collagen receptors on human melanoma cells, *J. Biol. Chem.*, 264, 4684, 1989.

214. Ignatius, M. J. and Reichardt, L. F., Identification of a neuronal laminin receptor: an Mr 200/120 kD integrin heterodimer that binds laminin in a divalent cation-dependent manner, *Neuron*, 1, 713, 1988.

215. Wayner, E. A. and Carter, W. G., Identification of multiple cell adhesion receptors for collagen and fibronectin in human fibrosarcoma cells possessing unique α and common β subunits, *J. Cell Biol.*, 105, 1873, 1987.

216. Wright, S. D., Weitz, J. I., Huang, A. J., Levin, S. M., Silverstein, S. C., and Loike, J. D., Complement receptor type three (CD11b/CD18) of human polymorphonuclear leukocytes recognizes fibrinogen, *Proc. Natl. Acad. Sci. U.S.A.*, 85, 7734, 1988.

217. Stacker, S. A. and Springer, T. A., Leukocyte integrin p150,95 (CD11c/CD18) functions as an adhesion molecule binding to a counter-receptor on stimulated endothelium, *J. Immunol.*, 146, 648, 1991.

218. Loike, J. D., Sodeik, B., Cao, L., Leucona, S., Weitz, J. I., Detmers, P. A., Wright, S. D., and Silverstein, S. C., CD11c/CD18 on neutrophils recognizes a domain at the N terminus of the Aα chain of fibrinogen, *Proc. Natl. Acad. Sci. U.S.A.*, 88, 1044, 1991.

Chapter 6

Integrins as Signal Transducing Receptors

Martin A. Schwartz

CONTENTS

I. INTRODUCTION

Integrin-mediated contact of cells with extracellular matrices promotes not only adhesion, spreading, and cytoskeletal organization, but also regulates a variety of cell functions. Adhesion has been shown to affect activation of platelets and leukocytes, differentiation and gene expression in many cell types, and growth of all anchorage-dependent cells. Because effects of ECM on cell shape or cytoskeletal organization often parallel effects on specific cell functions, because effects of adhesion can often be inhibited by cytochalasins, and because for a long time the only known intracellular effector of integrin action was the cytoskeleton, the regulatory actions of integrins have generally attributed to the cytoskeleton (for reviews see References 1 to 4).

More recent data, however, show that integrins can regulate intracellular second messengers. Such messengers provide another potential means by which ECM can control cell functions. In many cases, these effects can be distinguished from changes in cell shape or the cytoskeleton. The purpose of this article is to review this area, concentrating on recent work on regulation of specific signaling pathways. The control by ECM of cell functions, such as growth and differentiation, constitutes a much broader area, and is beyond the scope of this review.

II. GENE EXPRESSION

Several groups have reported that adhesion of cells to ECM or treatment with antibodies to integrins stimulates expression of specific genes. In some cases, these effects were quite rapid, preceding longer term changes in cell spreading, growth, or differentiation. The rapid onset of the responses argues that they were primary responses to adhesion, rather than secondary consequences of changes in growth or differentiation.

Dike and Farmer[5] found that in fibroblasts made quiescent by keeping them in suspension, mRNA levels for a number of growth-related genes were low. Upon replating on fibronectin (FN), expression of genes for *fos* and *myc* was rapidly stimulated, in the absence of soluble growth factors or serum. In the case of *fos*, nuclear run-off assays showed increased transcription 5 min after adhesion. Subsequent work demonstrated that collagen genes were also sensitive to adhesion, independent of soluble mitogens or cell growth.[6] These results strongly argued for rapid signaling following contact of cells

0-8493-4711-4/94/$0.00+$.50
© 1994 by CRC Press, Inc.

with ECM proteins. The rapidity of the response also argues against cell shape as a critical variable, since cells have not spread noticeably at these early times.

Werb and co-workers have studied expression of the protease genes stromelysin and collagenase in synovial fibroblasts. They had earlier shown that well-spread cells in culture had low levels of protease gene expression, and that putting cells under non-adhesive conditions or disrupting the cytoskeleton with cytochalasin induced protease gene expression 8 to 24 h later.[7,8] They subsequently found that these genes can be induced without detectable changes in cell shape or cytoskeletal organization by plating cells on a 120-kDa cell binding fragment of FN that lacks a heparin-binding domain, or on anti-integrin α5β1 IgG.[9] Furthermore, treating spread cells on FN with antibodies to α5β1 also induced protease gene expression in the absence of shape changes; antibody-induced clustering of integrins appeared to be the critical event in these experiments. Subsequent work[10] has provided evidence that rapid stimulation of c-*fos* expression mediates this effect on protease genes. The results are therefore in general agreement with those of Dike and Farmer. The protease system, however, exhibits considerable complexity, and several features remain to be explained. Chief among them is the observation that cells on intact FN, or on a complex, FN-containing matrix, have low levels of protease messages, whereas cells on a 120-kDa cell binding fragment or on an anti-integrin IgG have high levels. This result is puzzling, since intact FN also engages integrin α5β1. It could be that regions outside the cell binding region of FN induce inhibitory signals; alternately, it could be that these other regions modulate either the interaction between FN and α5β1 or the signal from this interaction. At present, the answer is unknown.

Evidence for effects of integrins on expression of a variety of genes has been observed in several other systems. In monocytes, contact of cells with FN triggers expression of a number of immediate-early genes, including interleukin Iβ and a protein that suppresses transcription-promoting activity of NF-κb.[11,12] Gene expression in these cells can also be triggered by antibody cross-linking of integrins β1 or α4, but not β2. Other instances in which adhesion or anti-integrin antibodies appear to regulate gene expression include alkaline phosphatase genes in osteosarcoma cells,[13] PDGF, c-*jun*, c-*fos*, and EGF2 in monocytes,[14] and casein genes in mammary epithelial cells,[15] though little is known about the mechanisms of signal transduction in these systems.

III. TYROSINE PHOSPHORYLATION

Several members of the *src* family of nonreceptor tyrosine kinases, as well as much of the cellular phosphotyrosine, localize by immunohistochemistry to sites of cell-ECM and cell-cell contact (reviewed in Reference 2). Though the significance of these observations remains unknown, there has been considerable speculation that c-*src* or other nonreceptor tyrosine kinases might be activated by adhesion.

The first evidence suggesting that integrins might regulate tyrosine phosphorylation was obtained in platelets,[16] where treating thrombin-activated platelets with RGD peptides to block fibrinogen binding to integrin αIIbβ3 (GPIIb-IIIa) inhibited phosphorylation on tyrosine of proteins of mol wt 100, 108, and 126 kDa. This phosphorylation coincided with platelet aggregation; however, the authors claimed that tyrosine phosphorylation was independent of cell-cell aggregation *per se*. They suggested that phosphorylation was a relatively direct consequence of binding of fibrinogen to its receptor, rather than a consequence of the cell-cell interactions that accompany platelet aggregation. Golden and Brugge[17] confirmed many of these results, but found that phosphorylation of similar molecular weight proteins on tyrosine did require aggregation. They argued convincingly that low levels of aggregation present in the washed platelet preparations most likely accounted for the "aggregation-independent" phosphorylation observed by Ferrell and

Martin. Indeed, Brugge's laboratory subsequently found that several nonreceptor tyrosine kinases associate tightly with CD4, a receptor involved in platelet aggregation,[18] whereas neither group could identify physical associations between kinases and integrin αIIbβ3.[16,18] It has recently been observed, however, that adhesion of cells to immobilized fibrinogen, which does not involve cell-cell interactions, also triggers tyrosine phosphorylation.[19] Furthermore, neither focal adhesion kinase (FAK) phosphorylation nor c-*src* association with the cytoskeleton is observed in platelets lacking integrin αIIbβ3.[20] These results argue more directly that αIIbβ3 is linked to tyrosine kinases.

Other integrins clearly appear to be functionally linked to tyrosine phosphorylation. Adhesion of platelets to collagen, mediated most likely by integrin α2β1 (GP Ia-IIa), triggers rapid phosphorylation of several proteins on tyrosine.[21] Adhesion of 3T3 cells to fibronectin or an anti-integrin β1 IgG, but not polylysine, induces rapid and reversible phosphorylation of a 125-kDa protein.[22] Treating A431 cells with antibodies to integrins β1 or α3 followed by a second antibody to induce clustering triggers tyrosine phosphorylation of the same protein.[23] This protein has been identified as the 120-kDa substrate of v-*src* identified by Parsons' laboratory,[24] and has been named FAK, for focal adhesion kinase. It has also been shown that phosphorylation of FAK, either by v-*src* or in response to adhesion, increases its own tyrosine kinase activity towards exogenous substrates.[25]

Guan et al.[22] reported that adhesion of 3T3 cells to the 120-kDa cell binding fragment of FN induced FAK phosphorylation much less than intact FN. Since cells also spread poorly on the 120-kDa fragment, they suggested that phosphorylation could be a prerequisite for spreading. They also suggested that regions of FN outside the integrin binding domain might interact with other cell surface proteins to enhance phosphorylation, perhaps by bringing multiple proteins together inside the cell. This conclusion, however, appears to be contradicted by their result that adhesion of cells to anti-β1 integrin IgG did trigger tyrosine phosphorylation and cell spreading.

Burridge et al.[26] failed to observe a difference in tyrosine phosphorylation triggered by FN and the 120-kDa fragment with rat embryo fibroblasts, indicating that the difference is not universal. They did find, however, that inhibiting tyrosine phosphorylation with the tyrosine kinase inhibitor herbimycin A blocked not spreading, but formation of focal adhesions and stress fibers on FN. In all, there are significant discrepancies that may be accounted for by differences between cell types, but in both systems there is evidence that FAK phosphorylation may promote cytoskeletal rearrangements. No data is currently available concerning the role of FAK or tyrosine phosphorylation in regulating other cell functions.

IV. PROTEIN KINASE C

In a number of instances where ECM proteins regulate cell functions, exogenous activators of protein kinase C (PKC) can replace the need for adhesion. In neurons, extension of neurites on laminin is blocked by PKC inhibitors, and is stimulated by PKC activators on low, but not high, coating densities of laminin.[27] This result was interpreted to suggest that PKC promotes neurite extension and that laminin, at high density, activates PKC maximally and therefore does not require exogenous activators. In neutrophils and monocytes, the ability to respond to cytokines is greatly enhanced by adhesion. The enhancement is blocked by antibodies to integrin αMβ2 (Mac-1, CD11b/CD18), but phorbol esters can activate in the absence of adhesion.[28] The ability of monocytes to phagocytose C3bi-coated particles is similarly dependent on adhesion, but can be triggered by phorbol esters in the absence of adhesion.[29] Though other models can be proposed, one explanation for these results is that adhesion leads to or is required for activation of PKC.

PKC also seems to be involved in the formation of focal contacts. Woods and Couchman[30] found that 3T3 cells do not form focal adhesions on the 120-kDa cell-binding fragment of FN that lacks the heparin-binding region. Addition of the isolated heparin-binding fragment restores the focal contacts and stress fibers, and exogenous activation of PKC allows their formation even on the cell-binding fragment of FN. Furthermore, inhibition of PKC blocks focal contacts on intact FN. These results are thus consistent with the notion that adhesion to FN leads to activation of PKC, and that the heparin-binding region is important for this event.

In HeLa cells, PKC was found to be required for cell spreading and to be involved in an interesting regulatory circuit with arachidonic acid.[31,32] The authors observed a rapid release of arachidonic acid, production of diacylglycerol, and translocation of PKC to the plasma membrane upon contact of HeLa cells with gelatin or collagen. These effects preceded cell spreading. Pharmacological inhibition of either arachidonic acid release, or of arachidonic acid metabolism through the lipoxygenase pathway, inhibited DAG production and cell spreading. Spreading could also be blocked by calphostin C, a specific PKC inhibitor. Finally, the inhibition of spreading by blockade of phospholipase A_2 or the lipoxygenase could be reversed by exogenous activation of PKC with a phorbol ester. Taken together, these data suggest that initial cell adhesion triggers arachidonic acid release, and that a lipoxygenase metabolite of arachidonate activates PKC, which then triggers cell spreading. It was also observed that PKC activation augmented the release of arachidonic acid, suggesting a positive feedback circuit in which production of diacylglycerol and arachidonate each stimulates the other to amplify an initial signal.

One question that remains unresolved in the HeLa cell system is the identity of the adhesion receptors. On the one hand, spreading of HeLa cells on gelatin or collagen requires divalent cations and was inhibited by RGD peptides, results that are characteristic of integrins. On the other hand, affinity chromatography of HeLa cell extracts on gelatin identified a number of proteins at molecular weights distinct from integrins. Until these proteins are more fully characterized, the identity of the receptors that mediate the interactions and signaling must be regarded as unknown.

V. INTRACELLULAR CALCIUM

Changes in levels of intracellular free calcium ($[Ca^{2+}]_i$) are provoked by a wide variety of soluble agonists that act via a number of different pathways. Several groups have now reported that adhesion of cells to ECM or antibodies to integrins can affect $[Ca^{2+}]_i$.

In neutrophils, irregularly timed spikes in $[Ca^{2+}]_i$ can be observed in cells adhered to plastic, but not cells in suspension, and these spikes are blocked by antibodies to $\alpha M \beta 2$.[33,34] Cross-linking integrin $\alpha M \beta 2$ on these cells also triggered a large $[Ca^{2+}]_i$ transient,[35] leading the authors to suggest that formation of adhesions followed by detachment are responsible for the rise and fall of $[Ca^{2+}]_i$ in each spike. Cross-linking integrin $\alpha L \beta 2$ (LFA-1, CD11a/CD18) on lymphocytes also induced a calcium transient.[36]

These calcium spikes have been shown to be required for migration of neutrophils, since blockade by removal of extracellular calcium or by buffering intracellular calcium inhibited migration of these cells on vitronectin or serum-coated surfaces.[37] Inhibition of the calcium/calmodulin-activated phosphatase calcineurin, either with inhibitory peptides or with the immunosuppressant FK506, also blocked neutrophil migration.[38] In all of these instances, the cells adopted a highly spread morphology and appeared unable to detach from their substratum contacts in order to migrate. Addition of an RGD peptide to loosen attachment restored migration. These results suggest a model in which calcium spikes activate calcineurin, leading to dephosphorylation of a substrate and detachment from vitronectin.

In platelets, addition of RGD peptides to block Fg binding partially blocks calcium entry.[39] In the same study, partially purified preparations of integrin αIIbβ3 had calcium channel activity when reconstituted into liposomes, from which the authors concluded that αIIbβ3 functions as a calcium channel. This conclusion is likely to be an overinterpretation, since a contaminating calcium channel could easily account for the result. It has, however, also been shown that addition of ligand (Fg or RGD peptides) blocks the channel. This result is difficult to understand, since in intact platelets, ligands activate the channels. The results do, however, support the idea that the calcium channels are likely to be closely associated with the integrin. Calcium channels that are regulated by ligands of integrin αIIbβ3 have also been identified electrophysiologically in reconstituted platelet membranes.[40]

Integrin αIIbβ3 expressed in a tumor cell line is able to trigger highly periodic calcium transients.[41] Inhibiting these transients, either by removing extracellular calcium or buffering intracellular calcium blocked phosphorylation on tyrosine of FAK. This result was interpreted to indicate that the calcium transients are required for activation of FAK. This conclusion, however, should be regarded with caution (see below).

An integrin-mediated rise in intracellular calcium has also been observed in endothelial cells.[42] In this system, no spikes were observed, but individual cells underwent an abrupt increase in $[Ca^{2+}]_i$ at some point during cell spreading, followed by a plateau phase during which $[Ca^{2+}]_i$ remained elevated. This rise was completely dependent on extracellular Ca^{2+}, and appeared to be due to a calcium channel with novel properties. Of some interest was the result that the Ca^{2+} rise appeared to be restricted to αv-containing integrins.[43] Vitronectin stimulated Ca^{2+} entry through integrin αvβ3; fibronectin also did so via αv integrins, but did not do so through α5β1. Collagen or integrin α2β1 failed to stimulate measurable Ca^{2+} entry at all. Since both α5β1 and collagen or α2β1 do trigger FAK phosphorylation, these results indicate that a rise in $[Ca^{2+}]_i$ cannot be a general requirement for tyrosine kinase activation by integrins.

The function of these Ca^{2+} signals in various cell types is not fully characterized, but an interesting correlation exists with cell migration.[44] When migration towards vitronectin or collagen was compared using a transfilter migration assay in a modified Boyden chamber, it was observed that movement towards vitronectin was completely blocked by removing Ca^{2+} from the medium, whereas migration towards collagen was unaffected. Neither attachment nor spreading were significantly or differentially altered by low Ca^{2+} on these substrata. Thus, the Ca^{2+} rise may be important for endothelial cell migration under these conditions. Further work, however, will be needed to firmly establish this connection.

One other system in which integrins impact $[Ca^{2+}]_i$ levels concerns calcium mobilization by PDGF in fibroblasts. Whereas adherent cells release Ca^{2+} from internal stores in response to PDGF, suspended cells fail to do so.[45] Attachment to FN for a few minutes restores the ability to respond to PDGF, before cell spreading occurs. These results demonstrate that some step in the pathway linking the PDGF receptor to calcium release must require adhesion. They therefore show that synergy between adhesion and a soluble growth factor can be observed at the level of an early signaling event.

VI. INTRACELLULAR pH

Intracellular pH (pHi) appears to function in growth control in eukaryotic cells. pHi is low in quiescent, serum-deprived cells and is increased by treatment with serum or growth factors (for a review see Reference 46). This change in pHi is due primarily to activation of the Na-H antiporter, which can occur via Ca^{2+}/calmodulin-dependent kinase, protein kinase C, or cAMP-dependent kinase. Blockade of the antiporter inhibits cell

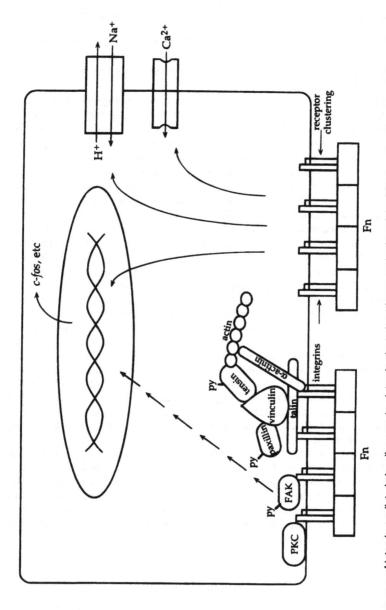

Figure 1 Summary of integrin-mediated signaling events. Integrin clustering triggers activation of the FAK tyrosine kinase, and other unknown mediators, that lead to activation of the Na+/H+ antiporter, opening of calcium channels, and expression of c-fos and other genes.

growth in a number of systems. Thus, pHi is an excellent marker for activation of the intracellular signaling pathways that regulate growth.

Integrins can also regulate the Na-H antiporter and pHi. Blocking the binding of Fg to integrin αIIbβ3 was shown to diminish activation of the antiporter in platelets, and to partially inhibit platelet activation.[47] Adhesion to FN also stimulates the antiporter and elevates pHi in fibroblasts,[48-50] endothelial cells,[51] lymphocytes,[52] hepatocytes, and smooth muscle cells (M. Schwartz and D. Ingber, unpublished data), suggesting that the phenomenon is widespread if not universal. Since inhibition of the antiporter frequently inhibits cell growth,[46,51] these results suggested that the decrease in antiporter activity could account for some of the diminished growth rate of cells lacking adhesion. In support of this conclusion, it has also been observed that cells transformed by several plasma membrane oncogenes have a higher pHi in suspension in direct proportion to their ability to grow in suspension.[53] Furthermore, transfection of cells with a proton pump from yeast elevated pHi and conferred the ability to grow in suspension culture.[54] Taken together, these results argue that changes in pHi or proton transport activity mediated at least some of the stimulation of cell growth by adhesion. Results from experiments in which pHi was manipulated by changing the pH of the medium indicated that pHi *per se* was less important than the transport of protons.[51,54] Thus, the mechanisms by which these membrane transporters affect cell growth remain unknown.

In fibroblasts and endothelial cells, local clustering of integrin α5β1 was sufficient to activate the antiporter and elevate pHi.[55] Experiments in which clustering was induced with soluble antibodies revealed an interesting twist: activation by soluble antibodies was highly transient, apparently due to internalization of receptor-antibody complexes. Only when ligands were immobilized on a solid surface was the signal sustained. These results illustrate the importance of proteins being incorporated into an insoluble extracellular matrix, where ligand internalization is prevented and the response is prolonged. The importance of the matrix is also revealed by the result that DNA synthesis is not stimulated by soluble ligands, or by ligands bound to particles that are too small to support cell spreading, even when they induce a sustained elevation of pHi. It seems most likely that integrins regulate cell growth (and probably other functions) by two distinct mechanisms: one via receptor clustering and effects on signaling pathways, the other via changes in cell shape and cytoskeletal organization.

Analysis of the signaling pathways responsible for the activation of the antiporter by ECM proteins revealed differences between cell lines. In capillary endothelial cells, the effect was independent of soluble growth factors and independent of protein kinase C (see Reference 51; and M. Schwartz, unpublished data). In the 10T1/2 fibroblast line, however, it was shown that serum or PDGF could elevate pHi via a PKC-dependent pathway in adhered cells, but could not do so in poorly adhered cells.[56] It was also shown that poorly adhered cells could still respond to phorbol ester, indicating that PKC itself and subsequent steps in the pathway were functional in nonadhered cells. Thus, coupling of PDGF receptor to PKC, presumably via inositol lipid breakdown, required adhesion. Data from pHi measurements were, therefore, in general agreement with previous studies suggesting an effect of adhesion on PKC, and with data indicating that PDGF binding was uncoupled from calcium mobilization in nonadhered cells.

VII. INOSITOL LIPID METABOLISM

Two papers have reported ostensibly direct effects of ECM on inositol lipid breakdown. One short note reported that attachment of a transformed strain of BHK cells to FN resulted in an increase in levels of intracellular water-soluble inositol phosphates.[57] A second group analyzed growth and lipid metabolism in kidney epithelial cells.[58] These cells grew much better on a complex ECM than on tissue culture plastic, and had substantially higher levels of diacylglycerol and inositol phosphates on the complex ECM. Both groups concluded that adhesion to ECM proteins triggered activation of phospholipase C (PLC).

Investigation of why the coupling of PDGF receptor to PKC activation required adhesion showed that addition of PDGF induced release of water soluble inositol phosphates in adhered but not suspended cells.[59] This result shows that inositol lipid breakdown is impaired in suspended cells, and confirms the hypothesis formulated on the basis of pHi measurements. Analysis of proteins phosphorylated on tyrosine revealed that addition of PDGF triggered phosphorylation of PLC and other proteins on tyrosine equally well in adherent and suspended cells. This result showed that PDGF receptors are still present and functional in suspended cells, and that PLC is present and able to be phosphorylated.

The lack of inositol lipid hydrolysis is most likely related to synthesis of phosphatidyl inositol bis phosphate (PIP_2). Detaching cells from the substratum led to a decline in total PIP_2 levels to about 20% of the original value, and reattachment to fibronectin stimulated an increase to the initial levels.[59] Incorporation of $^{32}PO_4$ into lipids showed that, in the complete absence of serum or growth factors, adhesion to FN rapidly (<2 min) stimulated the synthesis of PIP_2. This increase in PIP_2 was accompanied by a decrease in PIP, suggesting that a PIP kinase was activated. Under identical conditions, no release of water soluble inositol phosphates could be observed, ruling out inositol lipid breakdown as the primary event. These results support a model in which integrins control inositol lipid synthesis, while soluble growth factors control lipid breakdown. Stimulation of both events is required for optimum release of second messengers.

This model has the potential to explain many instances in which soluble factors and ECM are synergistic. It is also attractive in light of the systems (discussed above) in which exogenous activators of PKC bypass the need for adhesion. Finally, the effect of adhesion on synthesis of PIP_2 may be important for the control of cytoskeletal organization, since PIP_2 can modulate the function of several actin binding proteins.

One question that remains unanswered is the relationship between experiments showing that adhesion induces inositol lipid breakdown and those demonstrating increased inositol lipid synthesis. Under conditions where PLC is activated to some degree, enhanced synthesis should lead to enhanced breakdown, so that adhesion might appear to "directly" trigger lipid hydrolysis. Alternatively, integrins might trigger either synthesis or breakdown depending upon the cell line or the specific integrin.

VIII. SUMMARY AND PERSPECTIVES

It is noteworthy in the preceding discussion that several issues appear as recurring themes. The first such theme is receptor clustering. Changes in gene expression, tyrosine phosphorylation, intracellular calcium, and intracellular pH can be triggered by clustering of integrins. It, therefore, seems reasonable that clustering might actually trigger a single event, which could then lead to secondary changes. In this light, it is tempting to speculate that activation of one or more of the nonreceptor tyrosine kinases is the initial event, since receptor tyrosine kinases are activated by clustering, due to cross-phosphorylation. Receptor clustering could, alternatively, create a new binding site to localize and activate the nonreceptor kinases. In any case, tyrosine phosphorylation could then trigger activation of an inositol lipid kinase or other events.

The second theme concerns the heparin-binding domain of FN. It seems clear that this region can modulate the response of cells to the RGD-containing cell binding domain of FN, whether the assay is gene expression, cell spreading, or tyrosine phosphorylation. Whether it does so by directly signaling the cell or by modulating the RGD-integrin interaction and whether binding of heparin sulfate proteoglycans on the cell surface is responsible for these effects is unknown.

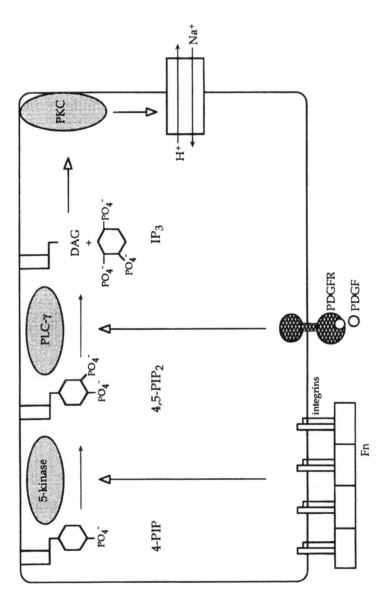

Figure 2 Cooperation between integrins and growth factor receptors. Integrin clustering triggers activation of a PIP 5-kinase, leading to increased synthesis and accumulation of 4,5-PIP$_2$. Binding of PDGF by the PDGF receptor triggers activation of phospholipase Cγ. Both events are required for maximal 4,5-PIP$_2$ hydrolysis and generation of second messengers.

The third theme concerns protein kinase C and inositol lipid metabolism. PKC could be activated if integrins triggered phospholipase C. Alternatively, if PLC had significant basal activity, activation of a PIP kinase to increase the pool of PLC-sensitive PIP_2 could lead to increased lipid hydrolysis and PKC activation. Thus, adhesion alone would promote inositol lipid hydrolysis by an indirect mechanism. It could well be that both of these mechanisms operate under different circumstances. Whatever the mechanism, PKC activation could trigger activation of leukocytes, focal adhesion formation, gene expression, and other events induced by ECM.

One point that is very clear is that much remains to be done before these apparent connections will be fully understood and the mechanism of signal transduction established for any of these signaling events. Furthermore, there were, at last count, 13 α and 8 β integrin subunits that can combine to form more than 20 integrin dimers. These subunits have significant differences in their C-terminal cytoplasmic domains, raising the possibility that they could generate distinct intracellular signals. Results from $[Ca^{2+}]_i$ measurements support this view. I think it is likely that signaling by different integrins will show considerable diversity, and will be an exciting area of research for many years.

REFERENCES

1. Bissell, M., How does extracellular matrix direct gene expression?, *J. Theor. Biol.*, 99, 31, 1981.
2. Burridge, K., Substrate adhesions in normal and transformed fibroblasts: organization and regulation of cytoskeletal, membrane and extracellular matrix components at focal contacts, *Cancer Rev.*, 4, 18, 1986.
3. Hemler, M. E., VLA proteins in the integrin family: structures, functions and their role on leukocytes, *Annu. Rev. Immunol.*, 8, 365, 1990.
4. Ruoslahti, E., Fibronectin and its receptors, *Annu. Rev. Biochem.*, 57, 375, 1988.
5. Dike, L. E. and Farmer, S. R., Cell adhesion induces expression of growth-associated genes in suspension-arrested fibroblasts, *Proc. Natl. Acad. Sci. U.S.A.*, 85, 6792, 1988.
6. Dhawna, J. and Farmer, S. R., Regulation of $\alpha1(I)$ collagen gene expression in response to cell adhesion in Swiss 3T3 cells, *J. Biol. Chem.*, 265, 9015, 1990.
7. Aggeler, J., Frisch, S. M., and Werb, Z., Changes in cell shape correlate with collagenase gene expression in rabbit synovial fibroblasts, *J. Cell Biol.*, 98, 1662, 1984.
8. Unemori, E. N. and Werb, Z., Cell shape and actin reorganization as a trigger of procollagenase gene expression, *J. Cell Biol.*, 103, 1021, 1986.
9. Werb, Z., Tremble, P. M., and Damsky, C. D., Signal transduction through the fibronectin receptor induces collagenase and stromelysin gene expression, *J. Cell Biol.*, 109, 877, 1989.
10. Tremble, P. and Werb, Z., Regulation of metalloproteinase gene expression by the fibronectin receptor involves a TPA-responsive element and c-*fos*, *J. Cell Biol.*, 11, 263a.
11. Haskill, S., Beg, A. A., Tompkins, S. M., Moriss, J. S., Yurochko, A. D., Sampson-Johannes, A., Mondal, K., Ralph, P., and Baldwin, A. S., Characterization of an intermediate-early gene induced in adherent monocytes that encodes IκB-like activity, *Cell*, 65, 1281, 1991.
12. Yurochko, A. D., Liu, D. Y., Eierman, D., and Haskill, S., Integrin as a primary signal transduction molecule regulating monocyte immediate-early gene induction, *Proc. Natl. Acad. Sci. U.S.A.*, 89, 9034, 1992.
13. Dedhar, S., Signal transduction via the $\beta1$ integrins is a required intermediate in interleukin 1β induction of alkaline phosphatase activity in human osteosarcoma cells, *Exp. Cell Res.*, 183, 207, 1989.

14. Shaw, R. J., Doherty, D. E., Ritter, A. G., Benedict, S. H., and Clark, R. A. F., Adherence-dependence increase in human monocyte PDGF(B) mRNA is associated with increases in c-fos, c-jun and EGF2 mRNA, *J. Cell Biol.*, 111, 2139, 1990.

15. Steuli, C. H., Bailey, N. and Bissel, M. J., Control of mammary epithelial differentiation: basement membrane induces tissue-specific gene expression in the absence of cell-cell interaction and morphological polarity, *J. Cell Biol.*, 115, 1382, 1991.

16. Ferrell, J. E. and Martin, G. S., Tyrosine-specific phosphorylation is regulated by glycoprotein IIb/IIIa in platelets, *Proc. Natl. Acad. Sci. U.S.A.*, 86, 2234, 1989.

17. Golden, A., Brugge, J. S., and Shattil, S. J., Role of platelet membrane glycoprotein IIb-IIIa in agonist-induced tyrosine phosphorylation of platelet proteins, *J. Cell Biol.*, 111, 3117, 1990.

18. Huang, M. M., Bolen, J. B., Barnwell, J. W., Shattil, S. J., and Brugge, J. S., Membrane glycoprotein IV (CD36) is physically associated with the fyn, lyn and yes protein-tyrosine kinases in human platelets, *Proc. Natl. Acad. Sci. U.S.A.*, 88, 7844, 1991.

19. Lipfert, L., Haimovitch, B., Schaller, M. D., Cobb, B. S., Parsons, J. T., and Brugge, J. S., Integrin dependent phosphorylation anc activation of the protein tyrsoine kinase pp125^{FAK} in platelets, *J. Cell Biol.*, 119, 905, 1992.

20. Clark, E. A. and Brugge, J. S., Redistribution of activated pp60^{c-src} to integrin-dependent cytoskeletal complexes in thrombin stimulated platelets, *Mol. Cell. Biol.*, 13, 1863, 1993.

21. Nakamura, S. I. and Yamamura, H., Thrombin and collagen induce rapid phosphorylation of a common set of cellular proteins on tyrosine in human platelets, *J. Biol. Chem.*, 264, 7089, 1989.

22. Guan, J.-L., Trevithick, J. E., and Hynes, R.O., Fibronectin/integrin interaction induces tyrosine phosphorylation of a 120 kD protein, *Cell Regul.*, 2, 951, 1991.

23. Kornberg, L. J., Earp, H. S., Turner, C. E., Prockop, C., and Juliano, R. L., Signal transduction by integrins: increased protein tyrsoine phosphorylation caused by integrin clustering, *Proc. Natl Acad. Sci. U.S.A.*, 88, 8392, 1991.

24. Schaller, M. D., Borgman, C. A., Cobb, B. S., Vines, R. R., Reynolds, A. B., and Parsons, J. T., pp125 FAK, a structurally unique protein tyrosine kinase associated with focal adhesions, *Proc. Natl. Acad. Sci. U.S.A.*, 89, 5192, 1992.

25. Guan, J.-L. and Shalloway, D., Regulation of pp125FAK both by cellular adhesion and by oncogenic transforamtion, *Nature*, 358, 690, 1992.

26. Burridge, K., Turner, C. E., and Romer, L. H., Tyrosine phosphorylation of paxillin and pp125FAK accompanies cell adhesion to extracellular matrix: a role in cytoskeletal assembly, *J. Cell Biol.*, 119, 893, 1992.

27. Bixby, J. L. and Jhabvala, P., Extracellular matrix molecules and cell adhesion molecules induce neurites through different mechanisms, *J. Cell Biol.*, 111, 2725, 1990.

28. Nathan, C., Srimal, S., Farber, C., Sanchez, E., Kabbash, L., Asch, A., Gailit, J., and Wright, S. D., Cytokine-induced respiratory burst of human neutrophils: dependence on extracellular matrix protein and CD11/CD18 integrins, *J. Cell Biol.*, 109, 1341, 1989.

29. Wright, S. D., Licht, M. R., Craigmyle, L. S., and Silverstein, S. C., Communication between different receptors for different ligands in a single cell: ligation of fibronectin receptors induces a reversible alteration in the function of complement receptors on cultured human monocytes, *J. Cell Biol.*, 99, 336, 1984.

30. Woods, A. and Couchman, J. R., Role of protein kinase C in focal adhesions, *J. Cell Sci.*, 101, 277, 1992.

31. Chun, J.-S. and Jacobson, B. S., Spreading of Hela cells on a collagen substratum requires a second messenger formed by the lipoxygenase metabolism of arachadonic acid released by collagen receptor clustering, *Mol. Biol. Cell*, 3, 481, 1992.

32. Chun, J. S. and Jacobson, B. S., Requirement for diacylglycerol and protein kinase C in Hela cell-substratum adhesion and their feedback amplification of arachidonic acid production for optimum cell spreading, *Mol. Biol. Cell*, 4, 271, 1993.

33. Richter, J., Ng-Sikorski, J., Olsson, J., and Anderson, T., Tumor necrosis factor induced degranulation in adherent neutrophils is dependent on CD11b/CD18 integrin triggered oscillations of free Ca^{2+}, *Proc. Natl. Acad. Sci. U.S.A.*, 87, 9472, 1990.

34. Jaconi, M. E. E., Theler, J. M., Schlegel, W., Appel, R. D., Wright, S. D., and Lew, P. D., Multiple elevations of cytosolic free Ca2+ in human neutrophils: initiation by adherence receptors of the integrin family, *J. Cell Biol.*, 112, 1249, 1991.

35. Ng-Sikorski, J., Andersson, R., Patarroyo, M., and Andersson T., Calcium signaling capacity of the CD11b/CD18 integrin on human neutrophils, *Exp. Cell Res.*, 195, 504, 1991.

36. Pardi, R., Bender, J. R., Dettori, C., Giannazza, E., and Engelman, E. G., Heterogeneous distribution and transmembrane signaling properties of lymphocyte function associated antigen (LFA-1) in human lymphocyte subsets, *J. Immunol.*, 143, 3157, 1989.

37. Marks, P. W., Hendey, B., and Maxfield, F. R., Attachment to fibronectin or vitronectin makes human neutrophil migration sensitive to alterations in cytosolic free calcium, *J. Cell Biol.*, 112, 149, 1991.

38. Hendey, B., Klee, C. B., and Maxfield, F. R., Inhibition of neutrophil chemokinesis on vitronectin by inhibitors of calcineurin, *Science*, 258, 296, 1992.

39. Ryback, M. E. and Renzulli, L. A., Ligand inhibition of the platelet glycoprotein IIb-IIIa complex function as a calcium channel in liposomes, *J. Biol. Chem.*, 264, 14617, 1989.

40. Fujimoto, T., Fujimura, K., and Kuramoto, A., Electrophysiological evidence that glycoprotein IIb-IIIa complex is involved in calcium channel activation on human platelet plasma membrane, *J. Biol. Chem.*, 266, 16370, 1991.

41. Pelletier, A. J., Bodary, S. C., and Levinson, A. D., Signal transduction by the platelet integrin $\alpha_{IIb}\beta_3$: induction of calcium oscillations required for protein-tyrosine phosphorylation and ligand-induced spreading of stably transfected cells, *Mol. Biol. Cell*, 3, 989, 1992.

42. Schwartz, M. A., Spreading of human endothelial cells on fibronectin or vitronectin triggers elevation of intracellular free calcium, *J. Cell Biol.*, 120, 1003, 1993.

43. Schwartz, M. A., Leavesley, D., and Cheresh, D. A., A calcium influx triggered by αv integrins is required for migration towards vitronectin, *Mol. Biol. Cell*, 3, 95a, 1992.

44. Leavesley, D. I., Schwartz, M. A., Rosenfeld, M., and Cheresh, D. A., Integrin β1- and β3-mediated endothelial cell migration is triggered through distinct signaling mechanisms, *J. Cell Biol.*, 121, 163, 1993.

45. Tucker, R. W., Meade-Cobun, K., and Ferris, D., Cell shape and increased free cytosolic calcium induced by growth factors, *Cell Calcium*, 11, 201, 1990.

46. Grinstein, S., Rotin, D., and Mason, M. M., Na^+/H^+ exchange and growth factor induced cytosolic pH changes: role in cellular proliferation, *Biochim. Biophys. Acta*, 988, 73, 1989.

47. Banga, H. S., Simons, E. R., Brass, L. T., and Rittenhouse, S. F., Activation of phospholipases A and C in human platelets exposed to epinephrine: role of glycoproteins IIb/IIIa and dual role of epinephrine, *Proc. Natl. Acad. Sci. U.S.A.*, 83, 9197, 1986.

48. Margolis, L. B., Rozovskaya, I. A., and Cragoe, E., Intracellular pH and cell adhesion to solid substrate, *FEBS Lett.*, 234, 449, 1988.

49. Schwartz, M. A., Both, G., and Lechene, C., The effect of cell spreading on cytoplasmic pH in normal and transformed fibroblasts, *Proc. Natl. Acad. Sci. U.S.A.*, 86, 4525, 1989.

50. Schwartz, M. A., Cragoe, E. J. C., Jr., and Lechene, C. P., Regulation of cytoplasmic pH in spread cells and round cells, *J. Biol. Chem.*, 265, 1327, 1990.

51. Ingber, D. E., Prusty, D., Frangioni, J. J., Cragoe, E. J. C., Jr., Lechene, C. P., and Schwartz, M. A., Control of intracellular pH and growth by fibronectin in capillary endothelial cells, *J. Cell Biol.*, 110, 1803, 1990.

52. Schwartz, M. A., Ingber, D. E., Lawrence, M., Springer, T. A., and Lechene, C., Multiple integrins share the ability to induce elevation of intracellular pH, *Exp. Cell Res.*, 195, 533, 1991.

53. Schwartz, M. A., Rupp, E. E., Frangioni, J. V., and Lechene, C. P., Cytoplasmic pH and anchorage independent growth induced by v-Ki-*ras*, v-*src* or polyoma middle T, *Oncogene*, 5, 55, 1990.

54. Perona, R. and Serrano, R., Increased pH and tumorigenicity of fibroblasts expressing a yeast proton pump, *Nature*, 334, 438, 1988.

55. Schwartz, M. A., Lechene, C. and Ingber, D. E., Insoluble fibronectin activates the Na/H antiporter by clustering and immobilizing integrin $\alpha5\beta1$, independent of cell shape, *Proc. Natl. Acad. Sci. U.S.A.*, 88, 7849, 1991.

56. Schwartz, M. A. and Lechene, C., Adhesion is required for protein kinase C dependent activation of the Na-H antiporter by platelet-derived growth factor, *Proc. Natl. Acad. Sci. U.S.A.*, 89, 6138, 1992.

57. Breuer, D. and Wagener, C., Activation of the phosphatidylinositol lipid cycle in spreading cells, *Exp. Cell Res.*, 182, 659, 1989.

58. Cybulsky, A. V., Bonventre, J. V., Quigg, R. J., Wolfe, L. S., and Salant, D. J., Extracellular matrix regulates proliferation and phospholipid turnover in glomerular epithelial cells, *Am. J. Physiol.*, 259, F326, 1990.

59. McNamee, H. P., Ingber, D. E., and Schwartz, M. A., Adhesion to fibronectin stimulates inositol lipid synthesis and enhances PDGF-induced inositol lipid breakdown, *J. Cell Biol.*, 121, 673, 1993.

Chapter 7

Integrin Receptors and Epiligrin in Cell-Cell and Cell-Substrate Adhesion in the Epidermis

William G. Carter, Susana G. Gil, Banu E. Symington,
Tod A. Brown, Shunji Hattori, and Maureen C. Ryan

CONTENTS

0-8493-4711-4/94/$0.00+$.50

I. CELL ADHESION IN THE EPIDERMIS: BACKGROUND

A. INTRODUCTION

This chapter will outline current knowledge of the role of the integrin receptors in mediating adhesion of epidermal basal cells to the basement membrane (BM) and to other cells in the basal cell layer. In particular we will focus on integrins $\alpha2\beta1$, $\alpha3\beta1$, $\alpha6\beta4$, and the BM ligand epiligrin. Adhesion of basal keratinocytes to the BM generates positional signals that are required for maintenance of the undifferentiated phenotype of basal cells. We will describe evidence that links changes in keratinocyte gene expression to changes in basal cell adhesion.

B. ORGANIZATION AND FUNCTION OF THE EPIDERMIS

The basal cells of the epidermis are involved in at least four functions including normal homeostatic regeneration (proliferation and differentiation)[1] assembly of and adhesion to the BM, immune surveillance,[2,3] and wound repair.[4-6] Skin is composed of three layers, the dermis that contains primarily fibroblasts and macrophages as resident cells, epidermis that forms the protective outer layer of the skin, and BM that separates the epidermis from the dermis (for general reviews on epidermal structure and function see Fuchs).[7,8] The epidermis contains four resident cell types: keratinocytes, comprising 90% of the cells; melanocytes, that also adhere to the BM; Langerhans cells, a dendritic cell, and Merkel cells which play a role in cutaneous sensation. In addition, there are variable populations of infiltrating leukocytes that are involved in immune surveillance making the skin the largest organ of the immune system (for review see Reference 2). The epidermis is a regenerating tissue, and the predominant keratinocyte population is composed of four stratifying cell layers listed in order of increasing degree of differentiation: Stratum Germinativum (basal), Stratum Spinosum (spinous, including suprabasal cells immediately above the basal), Stratum Granulosum (granular), and Stratum Corneum (cornified[9-11]). The basal keratinocytes contain a subpopulation of stem cells that is estimated at 10% of the basal cell population. The stem cells are relatively slow cycling cells that retain their proliferative potential throughout the life of the organism.[10,12-14] The stem cells give rise to transient amplifying cells in both the basal and suprabasal cell layers.

C. PROLIFERATION, DIFFERENTIATION AND WOUND REPAIR IN THE EPIDERMIS

It is possible to distinguish between early differentiation and proliferation near the BM and late differentiation processes that occur in the upper stratified cell layers. Early events are regulated by fibroblast products, while late stage differentiation processes, such as formation of the detergent insoluble cell envelope, occur in the absence of the dermis[9] or attachment to the BM.[15] Basal and suprabasal cells are the most highly proliferative and the least differentiated cell layers. Hemidesmosomes are the major anchoring junctions that attach basal cells to the BM. The attached basal cells express a cytoskeletal network rich in keratin subunits K5 and K14.[7,8] Above the suprabasal cells, spinous cells are postmitotic but metabolically active and are rich in cell-cell adhesions of the desmosomal type. Spinous cells upregulate keratins K1 and K10 and a cell envelope protein, involucrin. Granular cells synthesize filaggrin, a protein involved in aggregation of keratin filaments (tonofilaments). The cornified layer is composed of enucleated and terminally differentiated keratinocytes sealed together with lipids to form an impermeable protective layer. Increased membrane permeability in cornified cells allows an influx of Ca^{2+} that activates transglutaminase and results in cross-linking of cell envelope proteins to form the insoluble envelope.

Analysis of epidermal proliferation, differentiation, and wound repair is possible because basal keratinocytes can be established in culture and induced to proliferate or differentiate.[9,16,17] Many extracellular regulators controlling the balance between growth and terminal differentiation have been identified (for reviews see References 7, 8). Positive growth regulators include EGF, TGFα, low concentrations of retinoic acid (10^{-7} to 10^{-10} *M*), KGF, IL-6 and IL-1α. Negative growth regulators include TGFβ, TPA, serum, Ca^{2+} and detachment from the substratum.[15,18] Withdrawal from the cell cycle is necessary but not sufficient to induce commitment to terminal differentiation.[7,8] For example, TGF-β inhibits myc transcription in keratinocytes resulting in withdrawal from the cell cycle but does not induce differentiation. It has been proposed that gradients of these extracellular factors may play a major role in regulating proliferation and differentiation in the epidermis.[7,8] However, it is not clear at this time how these gradients could be adequately controlled to generate the precise differentiation stages in the stratified epidermis.

In addition to normal proliferation and differentiation, one of the primary functions of the epidermis is wound repair. Wounding of the skin results in fibroblast, platelet, and leukocyte accumulation in the wound site. These cells are involved in the production of cytokines, ECM components, and degradative enzymes that mediate wound activation and repair.[19,20] One of the first steps in wound repair is platelet attachment to collagen in the wound site via integrin α2β1, the prototype collagen receptor.[21,22] This adhesion step induces platelet thrombi and release of PDGF.[23] In turn, PDGF stimulates fibroblast migration into the wound site where the fibroblasts release (1) ECM components, such as tenascin, and SPARC that regulate adhesion (for reviews see Reference 19), (2) cytokines that stimulate keratinocyte migration, and (3) metalloproteinases involved in ECM remodeling. Production of metalloproteinases is stimulated by fibroblast adhesion to fibronectin. TGFα, TGFβ, EGF, and gamma-interferon produced by these infiltrating cells facilitate the migration of keratinocytes into the wound.[6,24,25]

Understanding the mechanisms that regulate wound repair, proliferation, and differentiation in keratinocytes is complicated by the heterogeneity of keratinocytes in culture. Multiple cell subpopulations including proliferating, differentiating, and activated wound keratinocytes can coexist in a single culture dish. As a result, a single extracellular regulator can alter multiple differentiation pathways within the heterogenous culture. For example, retinoids can induce proliferation of stem cells but inhibit expression of keratin K1 and K10 in cells that have committed to differentiate.[26] As a possible solution to the problem of keratinocyte heterogeneity, a methodology has been described for selection of a subpopulation of keratinocytes with the long-term proliferative potential characteristic of stem cells.[27] The approach is based on the selection of keratinocytes with high expression levels of β1 integrin adhesion receptors. However, it is not clear that this technique will distinguish stem cells from keratinocytes that have been activated to upregulate β1 integrins by wound repair stimuli.[6]

Extracellular regulators and cell adhesion have both been identified as regulatory factors of gene expression in various cell systems and in the epidermis in particular (reviewed in Reference 28). Exit from the cell cycle and commitment to terminal differentiation occur as basal cells move into the spinous cell layers.[15] TGF-βs expressed in the spinous cell layer, retinoids that diffuse in from the circulation in the dermis, and Ca^{2+} may each play a specific role in regulating postmitotic commitment to terminal differentiation.[7,8] It is possible that each of these extracellular regulators may have distinct effects on cells, depending on whether the cells are adherent to the BM or to other cells. For example, PDGF binding to the PDGF receptor initiates proliferation only in cells adherent to the substratum.[29] Thus, distinct adhesion systems in the basal and spinous cells provide a stable environment in which relatively unstable gradients of extracellular regulators can generate a precise differentiation cascade.

D. ANCHORING JUNCTIONS AND ADHESION RECEPTORS IN THE EPIDERMIS

1. Integrins and Cadherins in Anchoring Junctions

Cell-cell and cell-substrate adhesion is mediated by two types of anchoring junctions termed (1) adherens junctions including focal adhesions (FAs[30]) and (2) hemidesmosomes (HDs[31]) and desmosomes.[32,33] Each anchoring junction consists of an extracellular ligand or coreceptor, a transmembrane adhesion receptor(s), and an associated cytoskeletal complex. Desmosomes and HDs are stable anchoring junctions and mediate cell-cell and cell-BM adhesions, respectively. They are associated with intermediate filaments.[34] Adherens junctions are linked to actin containing-filaments[35A,35B,36] and are more dynamic and labile then desmosomes or HDs. Specialized forms of adherens junctions can mediate either cell-cell or cell-substrate adhesion.

There are five families of cell adhesion receptors that have been characterized: calcium-independent cell adhesion molecules in the immunoglobulin gene family (CAMs), selectins, calcium-dependent cadherins, integrins, and intrinsic membrane proteoglycans.[37] Of these families, the last three are expressed at high levels in the epidermis, and two of these, integrins and cadherins, are functional components of anchoring junctions and will be discussed further. Although intrinsic membrane proteoglycans, including CD44 antigens[38] and syndecans[39], have been suggested to participate in both cell-cell and cell-substrate adhesion, this role is not clear in the epidermis. CD44 is a hyaluronate binding[40] and collagen binding protein.[41] It is expressed by keratinocytes in the basal and spinous cell layers as a heparan sulfate intrinsic membrane proteoglycan.[42]

2. Cell-Cell Adhesion: Cadherins in Desmosomes and Adherens Junctions

Cadherins are Ca^{2+}-dependent mediators of homotypic cell-cell adhesion and are the primary cell-cell adhesion receptors in the epidermis.[43-45] There are two major classes of cadherins termed classical cadherins and desmosomal cadherins that are components of adherens junctions and desmosomes, respectively.[46,47] The two types of cadherins exhibit a high degree of sequence and functional homology in their extracellular domains but differ in their cytoplasmic sequences that associate with the cytoskeleton.

Epidermal cells form extensive cell-cell adhesions of the desmosomal type[32] that facilitate the epidermal function as a permeability barrier. Desmosomes contain eight major proteins: desmoplakins I and II (265 and 215 kDa), plakoglobin (82 kDa), and band 6 (76 kDa) are all cytoplasmic components. Desmoglein (165 kDa), desmocollins I and II (130 and 115 kDa), and a 22-kDa component are integral membrane adhesive components. Desmoglein[47] and desmocollins I and II[46,48-50] are desmosomal cadherins.

Desmosomes are relatively few in the basal cell layer but are upregulated in the suprabasal and spinous cell layer. Consistently, serum from patients with the autoimmune blistering disorder Pemphigus vulgaris (PV) express autoantibodies that cause detachment of suprabasal cells from the basal cells. Recently, the PV antigen was shown to be identical to desmoglein, a desmosomal cadherin (Amagai et al., 1991). The PV antigen, like the desmosomes, is upregulated in the suprabasal cell layer where it is a target for the autoimmune antibodies (Karpati et al., 1993).

Epidermal cells also express cell-cell adherens junctions that contain classic cadherins. P cadherin is expressed primarily in the basal cell layer while E cadherin is expressed in basal, suprabasal, and spinous cells.[53] The cadherins are associated with a cytoskeletal protein, catenin. Catenin is homologous to vinculin, a component of integrin-containing FAs. Catenin and vinculin link the cell-cell adhesion receptors to the actin-containing cytoskeleton.[36,54,55] In addition to mediating cell-cell adhesion, cadherins have been implicated in the assembly of intercellular communications via gap junctions[56] and in limiting invasiveness of tumor cells.[57]

3. Cell-Substrate Adhesion: Integrins in Hemidesmosomes and Focal Adhesions

The integrins are the primary adhesion receptors present in both FAs and HDs. Further, FAs[30] and HDs[31,32] are the primary mediators of basal keratinocyte adhesion to the substratum. Integrins are heterodimeric adhesion receptors that contain dissimilar α and β subunits (for reviews see References 37, 58 to 61). There are, currently, 13 different α subunits that associate with 8 different β subunits. Normal human skin and cultured human foreskin keratinocytes (HFKs) express high levels of integrins $\alpha 6\beta 4$, $\alpha 3\beta 1$, and $\alpha 2\beta 1$.[62-66] In addition, keratinocytes in culture express low but detectable quantities of $\alpha 1\beta 1$, $\alpha 5\beta 1$, $\alpha 6\beta 1$, $\alpha v\beta 5$, and $\alpha v\beta 6$, in part due to the activation-dependent expression of these receptors in culture and in wounds.[6,67-70] Based on the described receptor-ligand interactions for many cell types, the possible ligand-specificities for the major epidermal integrin receptors are $\alpha 2\beta 1$, collagen[21] and laminin;[71,72] $\alpha 3$, epiligrin,[73-75] fibronectin, collagen,[76] entactin,[77A] and laminin;[21,76,77B] and $\alpha 6\beta 4$, epiligrin[73] and laminin.[78] However, studies with primary human keratinocytes indicate that the preferred ligand-specificities are $\alpha 2\beta 1$, collagen; $\alpha 3\beta 1$, epiligrin; and $\alpha 6\beta 4$, epiligrin or an epiligrin associated component.[64-66,73]

HDs are the primary cell-substrate adhesion structures identifiable in the epidermis at the ultrastructural level, but are more difficult to detect in cell culture.[65] In skin, HDs are identified by electron microscopy as electron dense plaques localized to the cytoplasmic side of the basal membrane of keratinocytes attached to the BM. Immature forms of HDs have been identified in culture[31,79,80] and referred to as stable anchoring contacts (SACs[65]).

In the basal plasma membrane, HDs contain integrin $\alpha 6\beta 4$ as a transmembrane adhesion receptor.[65,67,81-84] In the cytoplasm, HDs contain a 230-kDa form of the bullous pemphigoid antigen (BPA) as a component of the electron dense plaque.[85] Desmoplakin I,[86] a component of desmosomes, is homologous, but not identical to BPA. A unique 180-kDa form of BPA that contains a collagen-like sequence has been localized to the cytoplasmic and extracellular domains of HDs.[87-90] The cytoplasmic plaque also contains a newly identified 200-kDa component;[79,80] HD-1, a high molecular mass protein;[91] and intermediate filaments.[32]

Cell-substrate adhesion is also mediated by FAs.[30,64,92,93] In FAs of keratinocytes, $\beta 1$ integrins are the primary adhesion receptor involved in ligand recognition.[64,65] In contrast to HDs, FAs are readily identified in cell culture but have not been identified in skin. The differences in detectability of HDs and FAs in tissue and culture may be due to differences in stability as well as function. FAs are relatively labile structures that are short-lived and may not be observed in normal epidermis. In contrast, HDs are stable structures that form only after adhesion mediated by FAs.[65] Consistently, when anchored cells are induced to migrate, as in wound repair, $\alpha 6\beta 4$ and BPA relocate from HDs to a diffuse distribution over the cell surface and cytoplasm.[94] The cytoplasmic tail of the integrin $\beta 4$ subunit in HDs is larger (1000 amino acids) than that of the $\beta 1$ integrin (50 amino acids[95]). The unique cytoplasmic sequences of the $\beta 4$ and $\beta 1$ subunits regulate associations of the HDs and FA with keratin and actin filaments through the linker proteins.

Despite the instability of FAs relative to HDs, FAs are formed in cultured cells only after initial adhesion, cell spreading, and migration have ceased. Adhesion structures that mediate initial adhesion, spreading, and migration are not detectable by light microscropy and are not characterized. However, they are probably structurally and functionally related to FAs. The suggested similarity is based on the fact that anti-integrin antibodies that block or activate initial cell adhesion, spreading, and migration react with the same integrin receptors that localize in FAs.[96,97] Although the labile adhesion structures that mediate migration contain integrins and actin filaments, a cascade of regulatory factors may modulate the function of the integrins and actin filaments and distinguish migration adhesion structures from FAs. The role of integrins[98] and actin-containing filaments and associated cytoskeletal components in regulating locomotion has recently been reviewed.[99] Vesicular traffic may provide adhesion receptors in tips of lamella of migrating cells that are subsequently incorporated into adhesion structures and are eventually endocytosed behind the leading edge of the cell.

In FAs, the β1 or β3 subunit of the integrin heterodimer receptors are associated directly with cytoplasmic talin or α-actinin and indirectly with vinculin, paxillin, zyxin, tensin, and a series of regulatory components including the tyrosine kinases, pp60src, and p125[FAK].[19,29,100,101] For example, overexpression of vinculin suppresses cell motility.[102] Phosphorylation of tyrosine residues in tensin, zyxin, paxillin, and p125[FAK] has been implicated as a regulatory signal for the assembly/disassembly of FAs.

4. The Basement Membrane

Gipson et al.[103] suggested that formation of HDs by basal cells may be controlled by the components of the BM. The unique organization of the epidermal BM in proximity to HDs supports its possible regulatory role in HD formation.[33,104–106] In turn the synthesis and assembly of the BM is a cooperative effort of epidermal and dermal cells.[107] In the epidermal BM, type VII collagen-containing anchoring fibrils in the dermis extend into the lamina densa of the BM adjacent to HDs.[108–111] Anchoring filaments extend from the anchoring fibrils across the lamina lucida, to the HDs. These BM structures may function as nucleation points for formation of FAs and HDs via integrin adhesion receptors. The composition of the anchoring filaments is being established and includes a number of glycoproteins: epiligrin,[73] kalinin,[112] a 125-kDa BM component,[84] and the high molecular mass antigen complex recognized by the GB3 monoclonal antibody (MAb), termed nicein.[113,114] Recently, epiligrin[115] and kalinin[116] were shown to be related to nicein.[117] It is likely that epiligrin, kalinin, and nicein may be related to a unique truncated laminin chain, termed B2t, that was recently cloned and sequenced.[118] The laminin-associated glycoproteins described by Frenette et al.[119] and the 140-/100-kDa antigen recognized by MAb AE26[120] may also be related or identical to one or more subunits of epiligrin.

Variant forms of laminin[121–123] are also present in the lamina lucida of the epidermal BM.[118,124A] Additional components of the BM include entactin/nidogen, heparan sulfate proteoglycan, and type IV collagen.[105,123] Epiligrin (Reference 73 and see later discussions), laminin variants,[124B–127] collagen,[64] and entactin/nidogen[77A] have all been shown to interact with integrins *in vitro* and mediate basal cell adhesion to the BM *in vivo*. This suggests that basal cell adhesion to the BM may involve one or more different receptors, ligands, and anchoring junctions.

The accumulated data suggest that the receptors and ligands involved in basal keratinocyte adhesion may be involved in regulation of basal cell differentiation, proliferation, and wound healing. Extracellular regulators (cytokines and ECM components) produced by dermal cells contribute to the regulation of the epidermal basal cells both in normal regeneration and wound repair. The following sections will evaluate the components and mechanisms involved in cell-cell and cell-substrate adhesion in normal basal epidermis, and the role that adhesion may play in regulation of proliferation, differentiation, and wound repair in the basal cell layer.

II. EPILIGRIN, AN EPIDERMAL BASEMENT MEMBRANE LIGAND FOR INTEGRINS α3β1 AND α6β4.
A. DIFFERENTIAL AND POLARIZED EXPRESSION OF INTEGRINS

In normal skin, expression of the major integrins, α6β4, α3β1, and α2β1, is restricted to the proliferating basal/suprabasal cell layers (Figure 1, normal skin). All three integrins

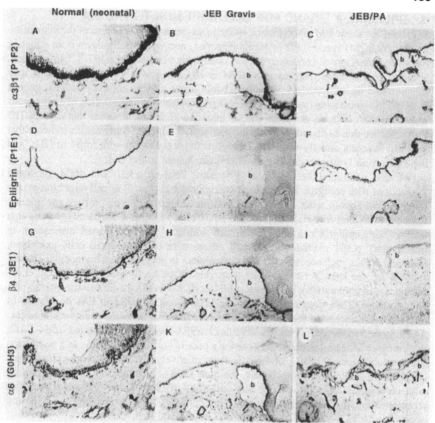

Normal (neonatal)   JEB Gravis   JEB/PA

α3β1 (P1F2)   Epiligrin (P1E1)   β4 (3E1)   α6 (GOH3)

Figure 1 Immunoperoxidase staining of serial cryostat skin sections from a healthy neonate (A,D,G,J), a JEB gravis patient (B,E,H,K), and a JEB gravis patient with pyloric atresia (JEB/PA, panels C,F,I,L) with indicated MAbs for α3β1 (A to C), epiligrin (D to F), β4 subunit (G to I), and α6 subunit (J to L). Original magnification = 250X. Sections from the two individuals with JEB display both areas of epidermal/dermal attachment and detachment with small blisters (b). The patient with JEB and JEB/PA exhibited no detectable HDs by electron microscopy. The basement membrane region associated with the dermal side is demarcated by arrows. The individual with JEB alone exhibits an absence of epiligrin (E), which has previously been associated with the JEB gravis (or Herlitz) subgroup. In contrast, the fetus diagnosed with JEB/PA has normal epiligrin expression (F), but an absence of detectable β4 subunit (I) and reduced α6 subunit (L) expression at the basal surface of the basal epidermal cells.

are downregulated in the differentiated spinous cells.[62,64,65,69] During suspension-induced terminal differentiation of keratinocytes, the α5β1 integrin is downregulated in two stages; first, the ability of the receptor to bind fibronectin declines, and second, mRNA levels reduce and the quantity of receptor declines. The decline reflects both inhibition of subunit transcription and inhibition of maturation and intracellular transport of subunits.[128] Expression of α6β4 is restricted to the basal plasma membrane,[63-65] except in developing epidermis prior to formation of HDs.[129] In contrast, α3β1 is expressed on the basal, lateral, and apical membranes, while α2β1 is expressed primarily on the apical and lateral membrane.[62,64,65,68] This suggests that α3β1, in contrast to both α2β1 and α6β4, functions in both cell-cell and cell-BM adhesion in normal epidermis.

B. EPILIGRIN, A LIGAND FOR CELL ADHESION TO THE BM

We identified a new BM glycoprotein, epiligrin, as the major component of the extracellular matrix (ECM) synthesized by HFKs (Figure 1, normal skin). Epiligrin is an effective ligand for adhesion of keratinocytes and other cells via integrin $\alpha3\beta1$.[73-75] In addition, epiligrin codistributes with integrin $\alpha6\beta4$ in HD-like SACs,[73] suggesting that it may function as a ligand for $\alpha6\beta4$. In culture, $\alpha6\beta4$ forms a stable complex with exogenous HFK-ECM,[65] indicating that exogenous epiligrin controls the organization of endogenous $\alpha6\beta4$. Further, both an acquired and an inherited blistering disorder that disrupt HD formation are due to disorders involving epiligrin.[115] Together these results indicate that epiligrin interacts directly with $\alpha3\beta1$ and directly or indirectly with $\alpha6\beta4$ in HDs, and this interaction is instrumental in basal cell attachment to the BM.

We have examined the functional differences between $\alpha3\beta1$- and $\alpha6\beta4$- mediated interactions with epiligrin. In short-term cell adhesion assays[73,74] or cell migration assays (Nguyen and Carter, work in progress), cell interactions with epiligrin are specifically inhibited with anti-$\alpha3\beta1$ (P1B5) or anti-$\beta1$ (P4C10) MAbs. Consistently, MAbs that interact with epiligrin can also inhibit cell adhesion (Gil and Carter, manuscript in preparation). $\alpha3\beta1$ mediates initial cell adhesion to epiligrin in all cells examined, including HFKs, melanocytes (Gil and Carter, work in progress) and thymocytes (Reference 74 and see later section), and $\alpha3$-transfected K-562 cells (Reference 75; Gil et al., manuscript in preparation). In contrast, $\alpha6\beta4$ in HDs does not mediate initial interactions with epiligrin. Time course studies on the localization of $\alpha3\beta1$ in FAs in relation to $\alpha6\beta4$ in SACs indicated that $\alpha3\beta1$ was involved in initial adhesion events while $\alpha6\beta4$-SACs formed in the quiescent cell colonies.[65] $\alpha3\beta1$-FAs usually encircled $\alpha6\beta4$-SACs in a complex adhesion structure, suggesting a possible role for $\alpha3\beta1$-FAs as a nucleation site for formation of HDs. $\alpha6\beta4$-SACs were relatively stable to detergents and urea, suggesting a nonmotile, or anchoring, function for SACs. Inhibition of HFK adhesion with combined anti-$\alpha3\beta1$ (P1B5) and anti-$\alpha6\beta4$ (G0H3) MAbs indicated that both receptors were functional in adhesion to the epiligrin complex in long-term adhesion assays.[65]

Epiligrin is the major component of the ECM assembled by HFKs and contains three disulfide-bonded, noncollagenous, glycoprotein subunits E170, E145, and E135, based on molecular mass in kDas. Two additional glycoproteins, E200 and E100, are variably associated with epiligrin and are a precursor to E170 and a degradation product of the E145 subunit, respectively. We prepared MAb P1E1 that reacts with E170, and was used for the initial immunopurification of epiligrin from HFK culture medium and for localization studies. Immunolocalization in tissue identified epiligrin in the dermal-epidermal junction of epithelial cells in organs primarily of endodermal/ectodermal origin. At the ultrastructural level, epiligrin localized to the lamina lucida of BMs concentrated adjacent to HDs.

C. MIGRATION VIA $\alpha3\beta1$ AND DEPOSITION OF EPILIGRIN IN BMs

The above results suggest that adhesion to epiligrin might be inhibitory for keratinocyte migration since it would induce formation of anchoring HDs. We have examined epiligrin as a substrate for keratinocyte migration and have found that HFK-ECM surfaces generated extensive adhesion, spreading, and migration. This migration was blocked with antibodies against $\alpha3$, $\beta1$, or epiligrin, but not antibodies against other receptors or ligands (Nguyen and Carter, work in progress). We have previously published that epiligrin is deposited on adhesion surfaces even when adhesion and migration are mediated by collagen or fibronectin. The epiligrin appears as a trail deposited behind the migrating cells.[73] Consistently, when migrating keratinocytes are incubated with MAb anti-$\alpha3\beta1$ (P1B5), the deposition of epiligrin is inhibited. These results suggest that cell adhesion and migration via $\alpha3\beta1$ are linked to the deposition of epiligrin in the BM. We suggest that this functional linkage may play a role in assembly of epiligrin in the BM. In related results, Kim et al.[130] observed that keratinocyte migration on collagen could be inhibited with antibodies to α2 as has been reported for HT-1080 cells.[96] However, keratinocyte migration on collagen was stimulated with antibodies to α3. They suggested that α3 interactions with epiligrin inhibited migration. Our results would suggest that inhibition of α3 with P1B5 inhibits deposition of epiligrin and may result in increased migration on collagen.

D. EXPRESSION OF EPILIGRIN AND β4 RELATE TO INHERITED AND ACQUIRED ADHESION DISORDERS AND MALIGNANCIES

We have identified both inherited and acquired epidermal adhesion disorders that may be caused by defects in the adhesive function of epiligrin and β4. Results from these studies support the proposed functions for epiligrin and β4 as physiologically significant components of HDs and as mediators of basal cell adhesion to the BM. During the course of our studies, it became apparent that epiligrin was related or identical to at least two other BM proteins,[115] nicein[113,114] and kalinin.[112,116] Each of these protein complexes had been identified as a result of different research directions: epiligrin as an adhesive ligand for α3β1 and α6β4,[73] kalinin as a component of the anchoring filaments associated with HDs,[112] and nicein/BM600/GB3 antigen as a protein antigen that was absent from the BM of patients with the gravis form of junctional epidermolysis bullosa (JEB[131]). JEB gravis is an inherited autosomal recessive blistering disorder of the epidermis. At the ultrastructural level, patients with this disease have abnormal or reduced numbers of HDs[106,132] with clinical findings of blister formation with mild mechanical trauma to the skin. Blister cleavage occurs within the lamina lucida of the BM. Consistently, epiligrin expression is altered or absent in almost all cases of the JEB gravis, but was present in the milder forms of JEB, EB simplex, or dystrophic EB (see Figure 1, JEB; Reference 115; Gil et al., 1993, manuscript in preparation).

We examined skin samples from a fetus diagnosed with JEB gravis, due to the absence of detectable HDs, combined with pyloric atresia, or incomplete formation of the pyloris.[133] Surprisingly, skin samples from this patient expressed normal levels of epiligrin (Figure 1, JEB/PA). Consistent with this result, three other cases of pyloric atresia with JEB were also found to be positive for nicein/epiligrin.[134-136] This suggests that this clinical phenotype may define a pleiotropic effect of a single gene or defects in two closely linked genes. Further analyis of this skin sample identified normal α3 and α6 expression but absence of detectable β4 (Figure 1, JEB/PA; Brown et al., manuscript in preparation). This phenotype of β4⁻, epiligrin⁺ may define a new variant of JEB that is distinct from JEB gravis which is epiligrin⁻, β4⁺. Further, these results testify to the physiological significance of β4 in formation of HDs. Defects in either epiligrin or β4 result in reduced HD formation and epidermal blistering at the lamina lucida.

In collaboration with Dr. Kim Yancey (NIH, Bethesda, MD), we have also identified patients with an acquired mucosal subepidermal blistering disease who have IgG anti-BM autoantibodies that bind the lamina lucida/lamina densa interface of epidermal BM. These autoantibodies react with epiligrin synthesized by keratinocytes in culture.[115] Further, the autoantibodies against epiligrin do not react with the BM of patients with JEB gravis. These studies show that anti-epiligrin autoantibodies are a specific marker for a novel autoimmune blistering disease and that the epidermal basement membrane antigen absent in patients with JEB gravis is epiligrin.

We have reported that oncogenic transformation of HFKs downregulates expression of epiligrin and β4.[137] In collaboration with C. B. Zachary, Dept. of Dermatology, University of Minnesota, we have examined the expression of epiligrin in basal cell carcinoma.[138] Basal cell carcinoma is unique, being locally invasive but failing to metastasize. We have found a consistent reduction in the expression of epiligrin and integrin α6β4 in invading basal cell carcinoma. As we previously proposed[73] and consistent with results in organotypic cultures,[137] reduction in adhesion components responsible for formation of HDs, epiligrin, and β4 may facilitate local invasion of the dermis by tumor cells.

E. IDENTIFICATION OF EE-LAMININ AND ITS RELATION TO EPILIGRIN

Immunoprecipitation of the epiligrin heterotrimer (E170, E145/E100, E135) from conditioned culture medium of HFKs with MAb P1E1 also precipitated proteins of 230 and 190 kDa.[73] The 230/190 proteins were described as "laminin-like" because they could be removed by precipitation with anti-laminin antibodies that did not precipitate epiligrin. Thus, epiligrin was defined as the heterotrimer (E170, E145/E100, E135) that was immunoprecipitated with P1E1 after preclearing with antilaminin.[73] We suggested that the laminin-like complex is either bound to epiligrin and/or cross-reacts with the P1E1 MAb. For simplicity in the following discussion, we will refer to the 230/190 laminin-like complex as EE-laminin because it is expressed in BM of both epidermis (E) and microvascular endothelium (E). Conceivably, EE-laminin may be related to K-laminin described by Burgeson et al.[124A] K-laminin is a laminin isoform that is copurified with kalinin and may be immunologically related to kalinin. However, neither antibody nor cDNA probes specific for K-laminin or EE-laminin are currently available to evaluate the relationship.

We have identified seven new MAbs that react with epiligrin (Reference 74; Gil and Carter, manuscript in preparation). These MAbs have been categorized as specific for epiligrin or epiligrin/EE-laminin based on differences in results from immunoprecipitation and tissue staining. The characteristics of these MAbs are summarized in Table 1 and are as follows

1. All seven MAbs react with epiligrin as determined by preclearing with anti-epiligrin P1E1 MAb (Figure 2)
2. Three of the MAbs inhibit melanocyte, keratinocyte, T cell, and fibroblast adhesion to epiligrin but not collagen or fibronectin
3. Epiligrin/EE-laminin-specific MAbs, including the original P1E1 MAb, precipitate both the 190-/230-kDa EE-laminin and the three subunits of epiligrin (E170, E145/E100, E135; Figure 3)
4. Epiligrin-specific MAbs precipitate epiligrin but not EE-laminin
5. Epiligrin-specific MAbs stain only epidermal BM while epiligrin/EE-laminin-specific MAbs stain epidermal and endothelial BM in the microvasculature
6. Epitopes recognized by epiligrin-specific MAbs are completely absent from epidermal BM of patients with JEB gravis. Epitopes recognized by epiligrin/EE-laminin-specific MAbs are reduced, but clearly detectable in epidermal and endothelial BM of patients with JEB gravis

These results indicate that EE-laminin is present in BMs of the epidermis and microvascular endothelium, while epiligrin is expressed only in epidermal BMs. Further, epiligrin, but not EE-laminin, is absent in patients with JEB gravis. These results also indicate that epiligrin and EE-laminin share multiple common epitopes recognized by a number of MAbs. Thus, the E170 subunit of epiligrin and at least one subunit of EE-laminin are immunologically related.

We have identified and sequenced multiple cDNA clones encoding the 170-kDa subunit of epiligrin (E170). Consistent with the immunological data above, partial sequence analysis of the cDNA clones for E170 revealed a correlation with domains IIIa and II of human laminin A chain. Domain IIIa encodes a cysteine-rich region containing EGF-like repeats.[124C] Sequence alignment of E170 and human laminin A

Table 1 Characteristics of new anti-epiligrin monoclonal antibodies

Antibody	HFK-ECM[a]	Reaction of the antibody with									
		Tissue[b]				Epiligrin[c]		EE-Lam[d]	Attachment Inhibition[e]		
		Normal		JEB							
		Epi	End	Epi	End	Blot	Rip	Rip	Epi	Col	Fn
Epiligrin/EE-Laminin											
P1E1	+	+	±	–	±	–	+	+	–	–	–
G3-3-SG	+	+	+	+	+	–	+	+	–	–	–
D3-4-SG	+	+	+	+	+	–	+	+	±	–	–
P3H9-2-EW	+	+	+	+	+	–	+	+	+	–	–
Epiligrin											
P3C10-A6	+	+	–	–	–	+	+	–	–	–	–
C2-9-SG	+	+	–	–	–	–	+	–	++	–	–
B4-6-SG	+	+	–	–	–	+	+	–	–	–	–
P3E4-EW	+	+	–	–	–	+	–	–	–	–	–
GB3	+	+	–	–	–	–	+	–	–	–	–

[a] Immunofluorescence microscopy of ECM from cultured HFKs;

[b] Immunoperoxidase staining of normal and JEB gravis skin (sample: SA 8-14-92); epidermal (Epi); and endothelial (End) BM;

[c] Immunoblot (Blot) or immunoprecipitation (Rip) reaction with epiligrin after affinity purification from HFK culture supernatant;

[d] None of the antibodies react with human placental laminin. However, epiligrin/EE-laminin antibodies coprecipitate a laminin-like protein of 190/230, referred to as EE-laminin (EE-Lam);

[e] MAbs C2-9-SG > P3H9-2-EW > D3-4-SG inhibit cell adhesion to epiligrin (Epi) but not collagen (Col) or fibronectin (Fn).

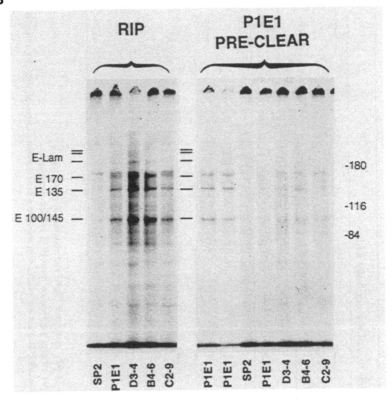

Figure 2 Immunoprecipitation studies with the newly characterized anti-epiligrin monoclonal antibodies. Identification of two epitopes, Epiligrin 1 (epiligrin/EE-Lam) and Epiligrin 2 (epispecific). Primary neonatal keratinocytes were metabolically labeled with [35-S]methionine and the culture supernatant was immunoprecipitated with the following antibodies: (RIP) Control nonimmune-conditioned medium (SP2), original anti-epiligrin (P1E1), new anti-epiligrin 1 (D3-4), and new anti-epiligrin 2 (B4-6 and C2-9). (P1E1 PRE-CLEAR) Sequential preclearing of the culture supernatant with P1E1 prior to immunoprecipitations with the new anti-epiligrin antibodies removed the subunits of epiligrin and EE-Lam, suggesting that the newly identified antibodies react with P1E1 antigen, either epiligrin (B4-6 and C2-9) or epiligrin/EE-Lam (D3-4).

chain in domain IIIa showed conservation of the cysteine residues and 55% sequence homology. In contrast to domain III, the primary amino acid sequence in domain II is not as well conserved between E170 and laminin A chain. However, a structural relationship is maintained in that both proteins encode alpha-helical domains characteristic of a coiled-coil structure previously described for laminin.[123,124C] Complete sequence analysis will be necessary to determine how E170 relates to the other domains present in the laminin A chain and to determine if E170 is derived from the same gene as one of the EE-laminin subunits.

F. MELANOCYTE AND T-LYMPHOCYTE ADHESION TO EPILIGRIN

Human epidermis contains resident melanocytes and infiltrating T lymphocytes that usually localize to the BM. In contrast to keratinocytes, melanocytes and T cells do not express HDs, $\alpha6\beta4$, or epiligrin. We have found that melanocytes (Gil and Carter, work in progress) and T lymphocytes[74] interact with keratinocyte epiligrin via $\alpha3\beta1$.

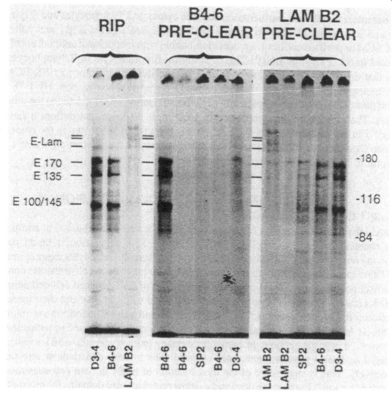

Figure 3 Dissociation of EE-Lam molecules from the epiligrin complex utilizing Epiligrin 1 (epiligrin/EE-Lam) and Epiligrin 2 (epi-specific) antibodies. Immunoprecipitation from the supernatant of [35-S]methionine-labeled HFKs with the indicated antibodies: (RIP) D3-4 (epi/EE-Lam), B4-6 (epiligrin), and Lam B2 antibodies. (B4-6 PRE-CLEAR) Sequential preclearing with anti-epiligrin MAb (B4-6) followed by immunoprecipitation with a control antibody (SP2), anti-Epiligrin 2 (B4-6), and anti-Epiligrin 1 (D3-4) antibodies specifically removed the epiligrin complex. EE-Lam complex can be separated from epiligrin by subsequent immunoprecipitation with D3-4. (LAM B2 PRE-CLEAR) EE-Lam complex was precleared by two consecutive immunoprecipitations with anti-Lam B2 antibody. The conditioned medium was then sequentially immunoprecipitated with control antibody (SP2), anti-Epiligrin 2 (B4-6), and anti-Epiligrin 1 (D3-4). EE-Lam antigen was quantitatively removed with no effect on the epiligrin complex; furthermore, D3-4 still precipitated the three subunits of epiligrin.

Many inflammatory diseases of the skin and the cutaneous T cell lymphomas (CTCLs) are characterized by a superficial perivascular infiltrate of T lymphocytes in the papillary dermis, epidermis, and, in particular, along the dermal-epidermal junction. Alternatively, in cutaneous B cell lymphoma (CBCL) the infiltrating cells are localized to the reticular (or deep) dermis. The close apposition of T cell infiltrates to the dermal/epidermal boundary suggests that the adhesive properties of the basement membrane may contribute to the migration of T cells but not B cells into the epidermis. In skin derived from patients with graft vs. host (GVH) disease, cutaneous-delayed hypersensitivity reactions, and epidermotropic cutaneous T cell lymphoma (CTCL or mycosis fungoides, MF), the infiltrating T cells express $\alpha3\beta1$ and are localized in close proximity to the epiligrin-containing basement membrane. Lymphocytes in cutaneous B cell lymphoma do not express $\alpha3\beta1$ by immunohistochemical techniques and do not associate with the epidermal basement membrane. Furthermore, cultured cytotoxic T lymphocytes and T lymphocyte cell lines established from patients with CTCL also express α3β1 and adhere to HFK ECM or purified epiligrin, a process inhibitable with monoclonal antibodies (MAbs) directed to α3 (P1B5) or β1 (P4C10) or epiligrin. B lymphocyte or T lymphocyte cell lines that do not express α3β1 do not adhere to purified epiligrin or to HFK ECM. In an adhesion assay using tissue sections derived from normal human skin, HUT 78 cells bound throughout the dermis/epidermis while B cell binding was restricted to the reticular dermis. The present findings constitute the first clear evidence that defines a function for α3β1 in hematopoietic cells and strongly suggest that an initial step in the epidermotropism characteristic of cutaneous T cell disorders may be the interaction of α3β1 with epiligrin in the epidermal basement membrane.

III. INTEGRIN α3β1 IN CELL-CELL ADHESION

A. α3β1 IN CELL-CELL ADHESION

Several labs have suggested a role for β1 integrins in cell-cell adhesion in addition to their characterized roles in cell-substrate adhesion. This suggestion is based on the following observations:[64,66,139] (1) In epidermis and culture, keratinocytes express integrin α3β1 that localizes at either lateral/apical cell-cell contacts or in cell-substrate contacts in distinct populations of basal cells. (2) Spontaneous or Ca^{2+}-induced cell-cell adhesion of HFKs results in decreased expression of α2β1 and α3β1 in FAs and their increased expression in cell-cell contacts.[64] (3) Cell-substrate and cell-cell adhesion are inhibited by anti-β1 MAb (P4C10) indicating that a β1 integrin may be involved in both adhesion processes. In cryostat sections of human or monkey palm epidermis, α3β1 localizes to the basal or lateral plasma membranes of basal cells in deep and shallow rete ridges, respectively. This is suggestive of the dual function of α3β1 in both cell-substrate and cell-cell adhesion.[66] Further, the deep and shallow rete ridges are domains for proliferating and nonproliferating epidermal cells, respectively.[11]

Transformed HFKs, termed FEPE1L-8 cells, will form cell-cell adhesions in culture when activated with either anti-α3β1 MAb P1B5 or P1B5 Fab fragments.[66] Other anti-integrins MAbs that we have tested do not have this ability. Further, anti-β1 MAbs (P4C10), anti-α2 (P1H5), and other anti-α3 MAbs (P4E7) block the P1B5-induced cell-cell adhesion, suggesting that another β1 integrin may be a coreceptor for α3β1. In collaboration with Dr. Y. Takada, we have examined α3-mediated cell-cell and cell-substrate adhesion utilizing α3-transfected K562 cells. K562 cells express α5β1, the fibronectin receptor, as their only endogenous integrin. Transfection of cDNAs encoding the integrin α3 subunit into K562 cells results in cell surface expression of α3β1. These α3-K562 cells acquire the ability to adhere to HFK-ECM that can be inhibited with anti-α3 and anti-epiligrin MAbs. Consistently, the α3-K562 cells can also be induced to aggregate with anti-α3 MAb (Symington et al., work in progress). Similar observations have been reported,[75] confirming the role of α3β1 in both cell-epiligrin and cell-cell adhesion. These results suggest that α3β1 functions as a receptor or as an inducer of anti-α3-inducible cell-cell adhesion system.

B. α2β1 AND α3β1 AS CORECEPTORS IN CELL-CELL ADHESION

We have identified α2β1 as a coreceptor for α3β1 in the P1B5-stimulated cell-cell adhesion of keratinocytes.[66] α2β1-coated beads were prepared and bound to P1B5-activated FEPE1L-8 cells. No other integrin-coated beads or antibody induced the binding of the beads. Further, α2β1-coated beads bound to α3β1-coated culture dishes or beads. Thus, purified α2β1 will directly bind purified α3β1. We suggest that this *in vitro*-stimulated α3-dependent cell-cell adhesion reflects a normal differentiation process of increased intercellular adhesion that occurs as cells detach from epiligrin in the BM and move into the suprabasal cell layer. Both the physiological function of α3-dependent cell-cell adhesion and the coreceptor(s) for α3 are still under investigation. For example, our experiments have not ruled out the possible role of other β1 integrins as coreceptors in the α3-dependent cell-cell adhesion. A report has suggested that α3β1 may interact with itself.[140] Weitzman et al.[75] suggested that antibody binding to α3 may activate an adhesion system that does not include α3. Wheelock and Jensen[141] reported that anti-E cadherin inhibited the Ca^{2+}-induced cell-cell adhesion of keratinocytes and the localization of P-cadherin, β1 integrins, vinculin, and desmoplakin to cell-cell contacts. This suggests that E-cadherin may control the localization of α3β1 in cell-cell contacts. From another standpoint, Watt's lab[142] presented evidence against a major role for integrins in calcium-dependent intercellular adhesion. Clearly, α3 function in cell-cell adhesion will continue to be a controversial subject.

C. EPILIGRIN REGULATES α3β1-DEPENDENT CELL-CELL ADHESION

The dual roles of α3β1 in cell-substrate adhesion and cell-cell adhesion has suggested that epiligrin may regulate α3β1 function in cell-cell adhesion. Consistently, we have observed that addition of epiligrin to cells inhibits α3β1 function in cell-cell adhesion.[143] Neither fibronectin nor collagen have a similar inhibitory effect. Since epiligrin is an endogenous ligand for α3β1 in the epidermal BM, this suggests that keratinocyte adhesion to epiligrin via α3β1 may inhibit α3-dependent cell-cell adhesion. In contrast, keratinocyte adhesion to epiligrin via α6β4 may stimulate α3 function in cell-cell adhesion.

IV. CELL ADHESION IN REGULATION OF DIFFERENTIATION IN BASAL CELLS

A. ADHESION IN REGULATION OF PROLIFERATION, DIFFERENTIATION, AND WOUND REPAIR

As discussed in the introduction, both cell adhesion and extracellular factors (TGFα, TGFβ, retinoids, etc.) regulate proliferation, differentiation, and migration in cultured keratinocytes. Cell adhesion to the ECM can regulate the phenotype of adherent cells by three possible mechanisms:[144] (1) by changing cell shape and tensile stress, mediated by the cytoskeleton; (2) by accumulating and presenting cytokines and growth factors; and (3) by changing second messengers through signal transduction that regulate proliferation and gene expression. For reviews on the role of cell adhesion in regulating signal transduction, differentiation, and proliferation, the reader is referred to References 19, 28, 29, and 144. ECM-cell interactions have been shown to stimulate transcription of differentiation-dependent genes, suggesting that the regulatory elements of these genes respond to transcription factors that are regulated by the ECM.[145] In skin, keratinocytes express both TGFα and TGFβ as autocrine cytokines that may contribute to the regulation of proliferation and differentiation.[7,8] In addition, it has been shown that mitogenic stimulation by PDGF[29] and FGF[146] may require cell adhesion.

Barondon and Green[147] evaluated keratinocyte growth in "mega colonies" and concluded that proliferation was restricted to the migrating keratinocytes at the outer edge of the colonies, identifying a correlation between migration and proliferation. BM has been shown to induce tissue-specific gene expression.[18] In addition, interactions of α5β1 with fibronectin inhibit terminal differentiation and facilitate maintenance of the migratory phenotype of the basal keratinocytes.[148] Similarly, keratinocytes adhesion to collagen has been shown to promote proliferation of keratinocytes in the absence of growth factors.[149]

Cell adhesion and cytokines also contribute to activation-dependent processes associated with wound repair in the epidermis. As a result of injury to the skin, keratinocytes become highly motile and proliferative, an activation process that facilitates wound

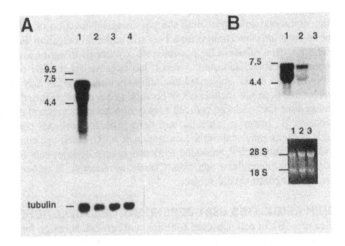

Figure 4 (A) Northern blot of total cellular RNA isolated from human foreskin keratinocytes (lane 1), FEP18-11 (lane 2), FEPE1L-8 (lane 3), and human foreskin fibroblasts (lane 4). FEP18-11 and FEPE1L-8 are human foreskin keratinocytes that have been transformed with human papilloma virus (Kaur and McDougall, 1988; Kaur et al., 1989). Hybridization with the Ep-1 probe detects two transcripts, approximately 5 and 6 kbases in size, which are significantly decreased in the cell lines that have been transformed with papilloma virus (lanes 2 and 3). No signal is detectable in human foreskin fibroblast RNA (lane 4) which was included as a negative control. Positive hybridization to tubulin mRNA shows that equal amounts of RNA were loaded in lanes 1 to 4. (B) Northern blot anlysis of RNA isolated from proliferating keratinocytes (lane 1), keratinocytes treated with 1.3 mM of Ca^{2+} (lane 2), and confluent keratinocytes (lane 3). The reduced signals in lanes 2 and 3 show that the Ep-1 transcript is significantly decreased by factors that cause keratinocyte differentiation. Photograph of 28S and 18S RNA shows that equal amounts of RNA were loaded in lanes 1 to 3.

closure (for review see References 5, 6). The signals that regulate keratinocyte activation are poorly understood. However, maximal activation requires loss of contact with the BM in addition to serum factors. Keratinocyte migration across the wound bed is facilitated by the accumulation of fibronectin in the wound site and upregulation of fibronectin and integrin $\alpha5\beta1$, the fibronectin receptor.[6] The activation is also accompanied by down-regulation of HDs as the cells become migratory and resynthesize the BM.[94] In the following section we will summarize results from our lab that evaluate the role of cell adhesion in regulating normal regeneration and wound repair in the epidermis.

B. REGULATION OF EPILIGRIN EXPRESSION: EFFECTS OF CELL DENSITY, WOUNDING, AND ONCOGENIC TRANSFORMATION

Northern blot analysis of HFK mRNA with a cDNA probe (Figure 4A; Ep-1 transcript) specific for epiligrin identified 2 mRNAs of 5 and 6 kbases. These sizes are compatible with the expected mRNAs coding for the E170 subunit of epiligrin. It is possible that the two mRNA transcripts correspond to the E170 subunit of epiligrin and the immunologically related subunit of EE-laminin. Northern blot analysis of HFK mRNA and mRNA from HPV transformed HFKs with a cDNA probe specific for epiligrin identified reduced amounts of mRNA in HPV-transformed HFKs (FEPE1L-8 cells). This confirms and extends our published report that transformation by HPV causes decreased expression of HFK ECM.[64,137]

We have observed that epiligrin is deposited in the BM localized to the leading edge of the migratory tongue of epithelium 24, 48, and 72 h after wounding in normal human volunteers (Ryan and Carter, manuscript in preparation). In these wounds, epiligrin and α3β1 were polarized to the basal surface of cells in the migrating epithelial tongue. *In situ* hybridization with antisense cRNA probes specific for epiligrin identified elevated epiligrin mRNA in the wound site. These data suggest that α3β1 and epiligrin may be of primary utility during epidermal migration and wound repair, possibly in relation to the deposition and assembly of epiligrin in the BM. In contrast, laminin and α6β4 were reduced in cells at the leading edge of the migratory epidermal tongue, consistent with the role of HDs as stable adhesion structures for nonmigrating cells. In related results, Northern blot analysis of mRNA from HFKs grown at low density vs. confluence using the Ep-1 transcript identified a dramatic increase in epiligrin expression in sparse cells (Figure 4B). This indicates that epiligrin mRNA is upregulated at low cell densities at the wound edge. Further, detachment from the substratum caused an inhibition in expression of epiligrin mRNA. Together, these results indicate that expression of epiligrin is upregulated by wounding or reduced cell-cell contact and downregulated by detachment from the substratum. It should be noted that the decreased expression of epiligrin in transformed cells contrasts markedly with the observed increase in epiligrin in wounds. Thus, the migration of wound cells may be self-limiting by the production of a restricting BM. In contrast, the migration of transformed cells in a wound environment would have have no such limitation.

C. INHIBITION OF CELL ADHESION VIA α3β1-EPILIGRIN INDUCES DIFFERENTIATION AND EXPRESSION OF INVOLUCRIN

We have investigated the role of α3β1-epiligrin interactions in cell adhesion in regulating expression of involucrin as an indicator of normal epidermal differentiation.[143] Involucrin is a component of the cross-linked cell envelope that is normally synthesized in the spinous cell layers. Incubation of keratinocytes with anti-α3β1 (P1B5) inhibits cell adhesion to epiligrin, stops migration, and induces cell-cell adhesion. The resulting rounded and aggregated cells initiate terminal differentiation including increased involucrin expression. We suggest that adhesion of basal keratinocytes to epiligrin inhibits terminal differentiation. Further, these results suggest that induction of α3-mediated cell-cell adhesion is a normal early step in a differentiation cascade that is initiated by detachment of basal cells from epiligrin in the BM. In related observations, it has been shown that keratinocyte interaction with fibronectin via α5β1 inhibits differentiation and expression of involucrin.[148] Fibronectin and α5β1 are expressed in wound epidermis but not in normal regenerating epidermis.[6] Therefore, it is most likely that the inhibitory effects of α5β1-fibronectin interactions are functional in wound repair while the inhibitory effects of epiligrin are functional in normal regeneration.

D. REGULATION OF INTEGRIN EXPRESSION BY DERMAL FACTORS

The above results have suggested that adhesion of basal keratinocyte to epiligrin inhibits differentiation in normal regenerating epidermis. We have also examined the role of dermal factors in regulating the organization and expression of epidermal integrins and CD44 intrinsic membrane proteoglycan in HFKs and FEPE1L-8 cells.[137] In monolayer cultures, integrins α2β1, α3β1, and α6β4 were expressed and functional in both cells; however, decreased epiligrin synthesis and deposition was detected in the FEPE1L-8 cells. Organotypic cultures were used in this study to provide an *in vitro* model system for the ordered stratification and differentiation of cells. In organotypic culture, epidermal cells were plated on a collagen gel containing either primary human dermal fibroblasts (HFFs), Swiss 3T3 fibroblasts, or no fibroblasts and then raised to an air-medium interface. Unlike HFKs, FEPE1L-8s exhibited, (1) disorganized stratification and limited differentiation capacity, (2) invasion into the collagen gel, and (3) unregulated expression of α3β1 and α2β1 and underexpression of α6β4. Ordered stratification and spatial regulation of integrin expression could be induced in the FEPE1L-8s by substituting Swiss 3T3 fibroblasts in the collagen gel. Data from cultures grown in the absence of fibroblasts indicated that the primary human fibroblasts induce the transformed HFKs to invade into the collagen gel. We concluded that stromal cells play an important role in the regulation of integrin expression in the basal cell layer.

E. CELL ADHESION IN REGULATION OF CYTOSKELETAL PROTEINS
1. Identification of New Cytoskeletal Components
We have identified cytoskeletal components of adhesion structures whose expression and organization is regulated by cell-cell and cell-epiligrin adhesion in basal keratinocytes. These responsive components may participate in signal transduction or regulation of differentiation that is dependent on cell adhesion. We have used production of MAbs to identify unique proteins or organization of proteins in cells grown under different adhesion conditions. Using this approach we have identified the P1H8 Ag that associates with HDs and the R3A12 Ag(s) that associate with both FAs and cell-cell adhesions.

2. The P1H8 Ag: A New Component of HDs
We have identified a protein recognized by MAb P1H8, termed P1H8 Ag, whose localization in the dermal-epidermal junction is regulated by keratinocyte adhesion to epiligrin (Brown and Carter, manuscript in preparation). The P1H8 Ag is expressed in most cells in culture and tissue, while epiligrin is restricted to the epithelial BM. MAb P1H8 immunoblots three components in sequential cell extracts of adherent HFKs: a 170-kDa protein in Triton X-100 detergent extracts, a 135-kDa component present in subsequent urea/NaCl extracts, and a 36- to 45-kDa protein released into the conditioned culture medium. The 170-kDa component is expressed in almost all cells examined including keratinocytes, fibroblasts, and leukocytes. In culture, cell adhesion to epiligrin results in P1H8 Ag localization to immature HDs. This suggests that adhesion to epiligrin controls the subcellular localization of the P1H8 Ag. P1H8 Ag precedes BPA in localization to HDs. In split skin, the P1H8 Ag stays with the epidermal roof while the epiligrin stays with the BM floor, indicating that the P1H8 Ag and epiligrin have distinct subcellular locations. Immunoelectron microscopy has localized the P1H8 Ag to the cytoplasmic side of the electron dense plaque of HDs. The epiligrin-dependent mobilization of the P1H8 Ag as cells move from the basal through the suprabasal to the spinous layers is a unique marker for the early morphogenic processes of the basal epidermis.

3. The R3A12 Ag: A Component of FAs and Cell-Cell Adherens Junctions
We have identified two new cytoskeletal proteins recognized by MAb R3A12, and termed the 35- and 60-kDa Ags. The R3A12 Ags localize in actin-containing adherens junctions at cell-substrate and cell-cell contacts (Hattori and Carter, manuscript in preparation). MAb R3A12 was prepared and identified antigen(s) that overlap in distribution with β1 integrins, talin, and actin-containing stress fibers in FAs. The localization of the R3A12 Ag to FAs was observed in fibroblasts (Figure 5) and HFKs on surfaces coated with collagen, fibronectin, and epiligrin. This indicates that the localization in FAs was not ligand or integrin specific. However, the R3A12 Ag did not localize in all FAs in contrast to talin. This indicates that the R3A12 Ag was not a required structural component of all FAs and that its localization in FAs was dependent on a characteristic of a subpopulation of FAs. MAb R3A12 immunoblotted a 35-kDa cytoskeletal protein (35-kDa Ag) in extracts of dermal fibroblasts (Figure 5). A similar R3A12 30-kDa Ag was also identified in cultures of HFKs grown at low cell density. However, a 2nd 60-kDa antigen was identified in keratinocytes grown at confluence, suggesting that the 60-kDa Ag was upregulated at cell confluence and that it functioned in cell-cell adhesion.

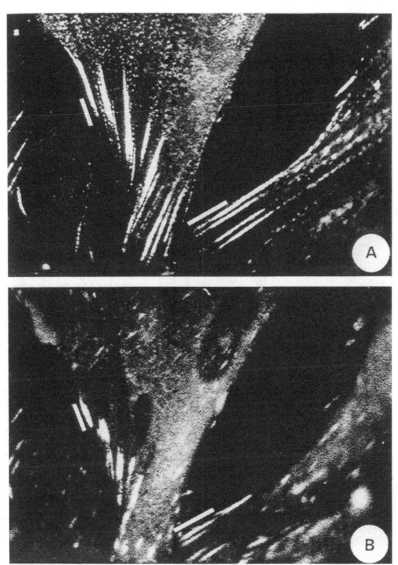

Figure 5 Localization of the 35-kDa R3A12 Ag in relation to talin in FAs of human foreskin fibroblasts (HFFs). HFFs were grown for 24 h on glass coverslips coated with collagen. Cells were fixed in cold 90% methanol, and double-stained with rat MAb R3A12 using FITC-goat anti-rat Ig as a secondary Ab (panel A) and with polyclonal anti-talin Ab using rhodamine-goat anti-rabbit Ig as a secondary Ab (Panel B). The R3A12 antibody is concentrated at the origins of actin stress fibers in and adjacent to FAs. Talin identifies the FA. White bars are located in the same position in panels A and B for location of R3A12 Ag relative to talin.

Consistently, R3A12 was also localized to intercellular contacts in confluent cells (Figure 6). In epidermis, R3A12 localized in the basal cell layer and was downregulated in the suprabasal cells consistent with it performing a differentiation-dependent adhesion function in basal cells. Induction of differentiation in keratinocytes by inhibition of cell-substrate adhesion decreased the expression of both the 35- and 60-kDa R3A12 Ags.

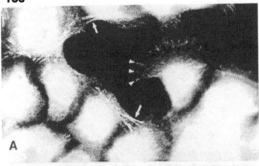

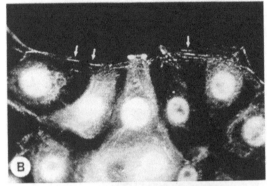

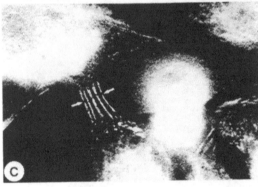

Figure 6 Localization of the 35-kDa R3A12 Ag in FAs and the 60-kDa R3A12 Ag in cell-cell adherens junctions. Human foreskin keratinocytes were grown for 24 h at high cell density on glass slides and stained with R3A12 MAb and FITC-goat anti-rat Ig as a secondary Ab. (A) The 35-kDa R3A12 Ag localized in FAs as indicated with arrowheads. (A-B) The 60-kDA Ag that is expressed at cell confluence, localized to the cell-cell contacts along stress fibers (arrows) that were perpendicular to the junction of the plasma membranes. (C) Arrows designate a break in the continuous staining of R3A12 Ag that runs from one cell into the adjacent cell. The break identifies the location of the two plasma membranes of two contacting cells.

The organization of R3A12 in FAs and cell-cell contacts could be disrupted with cytochalasin D, confirming its association with actin stress fibers. Further, MAbs to the $\beta 1$ subunit of integrins, but not anti-cadherins, selectively disrupted actin stress fibers and R3A12 Ag at cell-cell contacts. Thus, the 35- and 60-kDa R3A12 Ags are 2 new cytoskeletal components that associate with integrin-containing FAs and cell-cell adhesions, respectively. The expression of the R3A12 antigens is regulated by changes in both cell-substrate and cell-cell adhesion of basal keratinocytes. Upregulation in expression of the 60-kDa R3A12 Ag is an early differentation marker that is responsive to changes in cell adhesion in the basal cell layer.

V. OVERVIEW: CELL ADHESION AS A REGULATOR OF DIFFERENTIATION, AND WOUND REPAIR IN EPIDERMAL BASAL CELLS

We have examined the composition, interactions, and functions of epidermal adhesion structures composed of integrin $\alpha 3 \beta 1$ complexed with epiligrin ($\alpha 3 \beta 1$-epiligrin), $\alpha 6 \beta 4$-

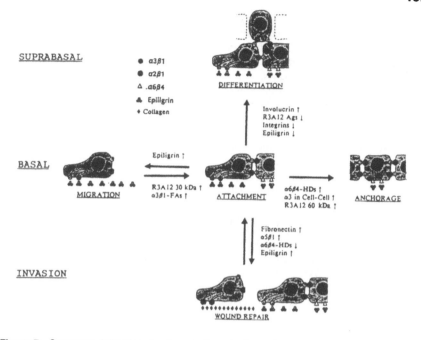

Figure 7 Summary: Adhesion structures in three regions of the epidermal-dermal junction.

epiligrin in HDs, $\alpha2\beta1$-collagen, and $\alpha3\beta1$ in cell-cell adhesions in culture and tissue. Our current understanding of the characteristics of these adhesion structures is summarized diagrammatically for three regions of the epidermal/dermal junction (Figure 7): the basal cell zone, the suprabasal cell layer, and the invasion area of BM and dermis affected by wound cells and/or transformed cells. We have also indicated specific gene products whose expression correlate with and are affected by changes in adhesion structures. Future work will determine if these different adhesion processes are necessary and sufficient to regulate the expression of gene products characteristic of normal regeneration and wound repair in basal and suprabasal cells.

BASAL LAYER: Basal keratinocytes (and melanocytes and T cells) attach and spread on epiligrin in the BM via $\alpha3\beta1$ that associates with actin-containing stress fibers. Initial attachment, spreading, and migration (see MIGRATION) precede formation of $\alpha3\beta1$ in FAs (ATTACHMENT). The formation of $\alpha3\beta1$-FAs functions as a nucleation site for the reorganization of $\alpha6\beta4$ into stable HDs (ANCHORAGE). Concurrent with the formation of the HDs, $\alpha3\beta1$ is upregulated in cell-cell contacts and functions in cell-cell adhesion with $\alpha2\beta1$ as a coreceptor. In culture, these progressive changes in adhesion structures correlate with decreasing migration, increased cell-cell contact, and increased stability of the cell-substrate adhesions. We suggest that regulation of adhesion to epiligrin in the basement membrane via $\alpha3\beta1$ and $\alpha6\beta4$ plays a major role in regulation of differentiation and activation in wound repair. Epiligrin expression is upregulated ($\uparrow$) in migrating and proliferating cells at the edge of wounds or colonies in culture or tissue (Ryan and Carter, manuscript in preparation). Conversely, increased cell density results in decreased synthesis of epiligrin ($\downarrow$), formation of stable $\alpha3\beta1$-FAs followed by $\alpha6\beta4$-HDs, increased expression of the 60-kDa form of the R3A12 antigen at cell-cell contacts, and expression $\alpha3\beta1$ and $\alpha2\beta1$ at cell-cell contacts.

SUPRABASAL LAYER: In culture, detachment (DETACHMENT) of basal cells from the BM with inhibitory antibodies to $\alpha3\beta1$ induces $\alpha3\beta1$-mediated cell-cell adhesion and expression of involucrin, a component of the granular cell layer (*DIFFERENTIA-TION*). This suggests that cell adhesion to epiligrin inhibits epidermal differentiation while detachment induces $\alpha3\beta1$ function in cell-cell adhesion as a normal differentiation step. Detachment of basal cells results in downregulation of integrins, epiligrin, and R3A12 35-and 60-kDa antigens.

INVASION ZONE: At a point of injury to the skin, basal keratinocytes that normally adhere to BM epiligrin are exposed to dermal connective tissue, including collagen ($\blacklozenge$). We propose that $\alpha2\beta1$-collagen interactions contribute to the *ACTIVATION* of basal keratinocytes involved in wound repair. Wounding (*WOUND REPAIR*) results in increased keratinocyte migration, expression of epiligrin and fibronectin, utilization of $\beta1$ integrins ($\alpha2$, $\alpha3$, $\alpha5$) in keratinocyte migration, and, later, after closure is complete, the expression of $\alpha3\beta1$ and $\alpha2\beta1$ at cell-cell contacts. In contrast, $\beta4$ function in stable HDs is decreased in the migrating cells.

ACKNOWLEDGMENTS

The authors would like to acknowledge collaborations with Elizabeth A. Wayner (University of Minnesota, Minneapolis, MN) and Yoshikazu Takada (The Scripps Research Institute, La Jolla, CA).

This work was supported by grants from the National Institutes of Health, RO1-CA49259 (WGC), PO1-AR21557 (WGC), F32-GM15085 (MCR) and K11-HL02216 (BES) and the American Cancer Society, CB-59I (WGC).

REFERENCES

1. Watt, F. M., Cell culture models of differentiation, *FASEB J.*, 5, 287, 1991.
2. Walsh, L. J., Lavker, R. M., and Murphy, G. F., Biology of disease. Determinants of immune cell trafficking in the skin, *Lab. Invest.*, 63, 592, 1990.
3. Nickoloff, B., Karabin, G., Barker, J., Griffiths, C., Sarma, V., Mitra, R., Elder, J., Kunkel, S., and Dixit, V., Cytokine networks: immunobiology surfaces, *J. NIH Res.*, 3, 71, 1991.
4. Toda, K-I., Tuan, T-L., Brown, P. J., and Grinnell, F., Fibronectin receptors of human keratinocytes and their expression during cell culture, *J. Cell Biol.*, 105, 3097, 1987.
5. Clark, R. A. F., Fibronectin matrix deposition and fibronectin receptor expression in healing and normal skin, *J. Invest. Derm. Suppl.*, 94, 128s, 1990.
6. Grinnell, F., Wound repair, keratinocyte activation and integrin modulation, *J. Cell Sci.*, 101, 1, 1992.
7. Fuchs, E., Epidermal differentiation: the bare essentials, *J. Cell Biol.*, 111, 2807, 1990.
8. Fuchs, E., Epidermal differentiation, *Curr. Opin. Cell Biol.*, 2, 1028, 1990.
9. Green, H., Terminal differentiation of cultured human epidermal cells, *Cell*, 11, 405, 1977.
10. Potten, C. S., and Morris, R. J., Epithelial stem cells *in vivo*, *J. Cell Sci. Suppl.*, 10, 45, 1988.
11. Sun, T., Eichner, R., Nelson, W. G., Scheffer, B. A., Tseng, C. G., Weiss, R. A., Jarvinen, M., and Woodcock-Mitchell, J., Keratin classes: molecular markers for different types of epithelial differentiation, *J. Invest. Dermatol.*, 81, 109s, 1983.
12. Wolpert, L., Stem cells: a problem in asymmetry, *J. Cell Sci. Suppl.*, 10, 1–9, 1988.
13. Hall, P. A., and Watt, F. A., Stem cells: the generation and maintenance of cellular diversity, *Development*, 106, 619, 1989.
14. Lavker, R. M. and Sun, T-T., Epidermal stem cells. *J. Invest. Derm.*, 81, 121s, 1983.
15. Watt, F. M., Selective migration of terminally differentiating cells from the basal layer of cultured human epidermis, *J. Cell Biol.*, 98, 16, 1984.

16. Rheinwald, J. G., Serial cultivation of normal human epidermal keratinocytes, *Meth. Cell Biol.*, 21A, 229, 1980.

17. Boyce, S. T. and R. G. Ham., Cultivation, frozen storage, and clonal growth of normal human epidermal keratinocytes in serum free media, *J. Tiss. Cult. Meth.*, 9(2), 83, 1985.

18. Streuli, C. H., Bailey, N., and Bissell, M. J., Control of mammary epithelial differentiation: basement membrane induces tissue-specific gene expression in the absence of cell-cell interaction and morphological polarity, *J. Cell Biol.*, 115, 1383, 1991.

19. Dambsky, C. H. and Werb, Z., Signal transduction by integrin receptors for extracellular matrix: cooperative processing of extracellular information, *Curr. Opin. Cell Biol.*, 4, 772, 1992.

20. Deuel, T. F., Kawahara, R. S., Mustoe, T. A., and Pierce, G. F., Growth factors and wound healing: platelet-derived growth factor as a model cytokine, 42, 567, 1991.

21. Wayner, E. A. and Carter, W. G., Identification of multiple cell adhesion receptors for type VI collagen and fibronectin in human fibrosarcoma cells possessing unique α and common β subunits, *J. Cell Biol.*, 105, 1873, 1987.

22. Kunicki, T. J., Nugent D. J., Staats, S. J., Orchekowski, R. P., Wayner, E. A., and Carter, W. G., The human fibroblast class II extracellular matrix receptor mediates platelet adhesion to collagen and is identical to the platelet glycoprotein Ia-IIa complex, *J. Biol. Chem.*, 263, 4516, 1988.

23. Sakariassen, K. S., Joss, R., Muggli, R., Kuhn, H., Tschopp, T. B., Sage, H., and Baumgartner, H. R., Collagen type III induced ex vivo thrombogenesis in humans. Role of platelets and leukocytes in depostion of fibrin, *Arteriosclerosis*, 10, 276, 1990.

24. Schultz, G., Rotatori, D. S., and Clark, W., EGF and TGF-α in wound healing and repair, *J. Cell. Biochem.*, 45, 346, 1990.

25. Mansbridge, J. N. and Knapp, A. M., Charges in keratinocyte maturation during wound healing, *J. Invest. Dermotol.*, 89, 253, 1987.

26. Jetten, A., Multi-stage program of differentiation in human epidermal keratinocytes: regulation by retinoids, *J. Invest. Dermatol.*, 95, 44s, 1990.

27. Jones, P. H. and Watt, F. M., Separation of human epidermal stem cells from transit amplifying cells on the basis of differences in integrin function and expression, *Cell*, 73, 713, 1993.

28. Adams, J. C. and Watt, F. M., Regulation of development and differentiation by the extracellular matrix, *Development*, 117, 1183, 1993.

29. Schwartz, M. A., Transmembrane signalling by integrins, *Trends Cell Biol.*, 2, 304, 1992.

30. Turner, C. E. and Burridge, K., Transmembrane molecular assemblies in cell-extracellular matrix interactions, *Curr. Opin. Cell Biol.*, 3, 849, 1991.

31. Jones, J. C. R., Yokoo, K. M., and Goldman, R. D., Is the hemidesmosome a half desmosome? An immunological comparison of mammalian desmosomes and hemidesmosomes, *Cell Motil. Cytoskeleton*, 6, 560, 1986.

32. Schwarz, M. A., Owaribe, K., Kartenbeck, J., and Franke, W. W., Desmosomes and hemidesmosomes: constitutive molecular components, *Annu. Rev. Cell Biol.*, 6, 461, 1990.

33. Legan, P. K., Collins, J. E., and Garrod., D. R., The molecular biology of desmosomes and hemidesmosomes: 'what's in a name?', *Bioessays*, 14, 385, 1992.

34. Jones, J. C. R. and Green, K. J., Intermediate filament-plasma membrane interactions, *Curr. Opin. Cell Biol.*, 3, 127, 1991.

35A. Volk, T. and Geiger, B., A-cam: A 135-kD receptor of intercellular adherens junction. I. Immunoelectron microscopic localization and biochemical studies, *J. Cell Biol.*, 103, 1441, 1986.

35B. Volk, T. and Geiger, B., A-cam: A 135-kD receptor of intercellular adherens junctions.

II. Antibody-mediated modulation of junction formation, *J. Cell Biol.*, 103, 1451, 1986.

36. Magee, A. I. and Buxton, R. S., Transmembrane molecular assemblies regulated by the greater cadherin family, *Curr. Opin. Cell Biol.*, 3, 854, 1991.

37. Albelda, S. M. and Buck, C. A., Integrins and other cell adhesion molecules, *FASEB*, 4, 2868, 1990.

38. Haynes, B. F., Telen, M. J., Hale, L. P., and Denning, S. M., CD44—A molecule involved in leukocyte adherence and T-cell activation, *Immunol. Today*, 10, 423, 1989.

39. Bernfield, M., Kokenyesi, R., Kato, M., Hinkes, M. T., Spring, J., Gallo, R. L., and Lose, E. J., Biology of the syndecans: a family of transmembrane heparan sulfate proteoglycans, *Annu. Rev. Cell Biol.*, 8, 365, 1992.

40. Underhill, C., The hyaluronan receptor, *J. Cell Sci.*, 103, 293, 1992.

41. Carter, W. G. and Wayner, E. A., Characterization of a collagen-binding, phosphorylated, transmembrane glycoprotein expressed in nucleated human cells, *J. Biol. Chem.*, 263, 4193, 1988.

42. Brown, T. A., Bouchard, T., St. John, T., Wayner, E., and Carter, W. G, Human keratinocytes express a new CD44 core protein (CD44E) as a heparan-sulfate intrinsic membrane proteoglycan with additional exons, *J. Cell Biol.*, 113, 207, 1991.

43. Takeichi, M., The cadherins: cell-cell adhesion molecules controlling animal morphogenesis, *Development*, 102, 639, 1988.

44. Takeichi, M., Cadherin cell adhesion receptors as a morphogenetic regulator, *Science*, 251, 1451, 1991.

45. Geiger, B. and Ayalon, O., Cadherins, *Annu. Rev. Cell Biol.*, 8, 307, 1992.

46. Collins, J. E., Legan, P. K., Kenny, T. P., MacGarvie, J., Holton, J. L., and Garrod, D. R., Cloning and sequence analysis of desmosomal glycoproteins 2 and 3 (desmocollins): cadherin-like desmosomal adhesion molecules with heterogeneous cytoplasmic domains, *J. Cell Biol.*, 113, 381, 1991.

47. Goodwin, L., Hill, J. E., Raynor, K., Raszi, L., Manabe, M., and Cowin, P., Desmoglein shows extensive homology to the cadherin family of cell adhesion molecules, *Biochem. Biophys. Res. Commun.*, 173, 1224, 1990.

48. Mechanic, S., Raynor, K., Hill, J. E., and Cowin, P., Desmocollins form a subset of the cadherin family of cell adhesion molecules, *Proc. Natl. Acad. Sci. U.S.A.*, 88, 4476, 1991.

49. Holton, J. L., Kenny, T. P., Legan, P. K., Collins, J. E., Keen, J. N., Sharma, R., and Garrod, D. R., Desmosomal glycoproteins 2 and 3 (desmocollins) show N-terminal similarity to calcium-dependent cell-cell adhesion molecules, *J. Cell Sci.*, 97, 239, 1990.

50. Parker, A. E., Wheeler, G. N., Arnemann, J., Pidsley, S. C., Ataliotis, P., Thomas, C. L., Rees, D. A., Magee, A. I., and Buxton, R. S., Desmosomal glycoproteins II and III, *J. Biol. Chem.*, 266, 10438, 1991.

51. Amagai, M., Klaus-Kovtum, V., and Stanley, J. R., Autoantibodies against a novel epithelial cadherin in pemphigus vulgaris, a disease of cell adhesion, *Cell*, 67, 869, 1991.

52. Karpati, S., Amagai, M., Prussick, R., Cehrs, K., and Stanley, J. R., Pemphigus vulgaris antigen, a desmoglein type of cadherin, is localized with keratinocyte desmosomes, *J. Cell Biol.*, 122, 409, 1993.

53. Shimoyama, Y., Hirohashi, S., Hirano, S., Noguchi, M., Shimosato, Y., Takeichi, M., and Abe, O., Cadherin cell-adhesion molecules in human epithelial tissues and carcinomas, *Cancer Res.*, 49, 2128, 1989.

54. Nagafuchi, A., Takeichi, M., and Tsukita, S., The 102 kd cadherin-associated protein: similarity to vinculin and posttranscriptional regulation of expression, *Cell*, 65, 849, 1991.

171

55. Tsukita, S., Tsukita, S., Nagafuchi, A., and Yonemura, S., Molecular linkage between cadherins and actin filaments in cell-cell adherens junctions, *Curr. Opin. Cell Biol.*, 4, 834, 1992.
56. Jongen, W. M. F., Fitzgerald, D. J., Asamoto, M., Piccoli, C., Slaga, T. J., Gros, D., Takeichi, M., and Yamasaki, H., Regulation of connexin 43-mediated gap junctional intercellular communication by Ca^{2+} in mouse epidermal cells is controlled by E-cadherin, *J. Cell Biol.*, 114, 545, 1991.
57. Begrens, J., Vakaet, L., Friis, R., Winterhager, E., Van Roy, F., Mareel, M. M., and Birchmeier, W., Loss of epithelial differentiation and gain of invasiveness correlates with tyrosine phosphorylation of the e-cadherin/β-catenin complex in cells transformed with a temperature-sensitive v-SRC gene, *J. Cell Biol.*, 120, 757, 1993.
58. Akiyama, S. K., Nagata, K., and Yamada, K. M., Cell surface receptors for extracellular matrix components, *Biochim. Biophys. Acta*, 1031, 91, 1990.
59. Humphries, M. J., The molecular basis and specificity of integrin-ligand interactions, *J. Cell Sci.*, 97, 585, 1990.
60. Ruoslahti, E., Integrins, *J. Clin. Invest.*, 87, 1, 1991.
61. Hynes, R. O., Integrins: versatility, modulation, and signalling in cell adhesion, *Cell*, 69, 11, 1992.
62. Wayner, E. A., Carter, W. G., Piotrowicz, R. S., and Kunicki, T. J., The function of multiple extracellular matrix receptors in mediating cell adhesion to extracellular matrix: preparation of monoclonal antibodies to the fibronectin receptor that specifically inhibit cell adhesion to fibronectin and react with platelet glycoproteins Ic-IIa, *J. Cell Biol.*, 107, 1881, 1988.
63. DeLuca, M., Tamura, R. N., Kajiji, S., Bondanza, S., Rossino, P., Cancedda, R., Marchisio, P. C., and Quaranta, V., Polarized integrin mediates human keratinocyte adhesion to basal lamina, *Proc. Natl. Acad. Sci. U.S.A.*, 87, 6888, 1990.
64. Carter, W. G., Wayner, E. A., Bouchard, T. S., and Kaur, P., The role of integrins α2β1 and α3β1 in cell-cell and cell-substrate adhesion of human epidermal cells, *J. Cell. Biol.*, 110, 1387, 1990.
65. Carter, W. G., Kaur, P., Gil, S. G., Gahr, P. J., and Wayner, E. A., Distinct functions for integrins α3β1 in focal adhesions and α6β4/bullous pemphigoid antigen in a new stable anchoring contact (SAC) of keratinocytes: relation to hemidesmosomes, *J. Cell Biol.*, 111, 3141, 1990.
66. Symington, B. E., Takada, Y., and Carter, W. G., Interaction of α3β1 and α2β1: potential role in keratinocyte intercellular adhesion, *J. Cell Biol.*, 120, 523, 1993.
67. Marchisio, P. C., Bondanza, S., Cremona, O., Cancedda, R., and De Luca, M., Polarized expression of integrin receptors (α6β4, α2β1, α3β1, and αvβ5) and their relationship with the cytoskeleton and basement membrane matrix in cultured human keratinocytes, *J. Cell Biol.*, 112, 761, 1991.
68. Klein, C. E., Steinmayer, T., Mattes, J. M., Kaufmann, R., and Weber, L., Integrins of normal human epidermis: differential expression, synthesis and molecular structure, *Br. J. Dermatol.*, 123, 171, 1990.
69. Adams, J. C., and Watt, F. M., Changes in keratinocyte adhesion during terminal differentiation: reduction in fibronectin binding precedes α5β1 integrin loss from the cell surface, *Cell*, 63, 425, 1990.
70. Adams, J. C., and Watt, F. M., Expression of β1, β3, β4 and β5 integrins by human epidermal keratinocytes and non-differentiating keratinocytes, *J. Cell Biol.*, 115, 829, 1991.
71. Languino, L. R., Gehlsen, K. R., Wayner, E. A., Carter, W. G., Engvall, E., and Ruoslahti, E., Endothelial cells use α2β1 integrin as a laminin receptor, *J. Cell Biol.*, 109, 2455, 1989.
72. Elices, M. J. and Hemler, M. E., The human integrin VLA-2 is a collagen receptor

on some cells and a collagen/laminin receptor on others, *Proc. Natl. Acad. Sci. U.S.A.*, 86, 9906, 1989.

73. Carter, W. G., Ryan, M. C., and Gahr, P. J., Epiligrin, a new cell adhesion ligand for integrin α3β1 in epithelial basement membranes, *Cell*, 65, 599, 1991.

74. Wayner, E. A., Gil, S. G., Murphy, G. F., Wilke, M. S., and Carter, W. G., Epiligrin, a component of epithelial basement membranes, is an adhesive ligand for α3β1 positive T lymphocytes, *J. Cell Biol.*, 121, 1141, 1993.

75. Weitzman, J. B., Pasqualin, R., Takada, Y., and Hemler, M. E., The function and distinctive regulation of the integrin VLA-3 in cell adhesion, spreading and homotypic cell aggregation, *J. Biol. Chem.*, 268, 8651, 1993.

76. Elices, M. J., Urry, L. A., and Hemler, M. E., Receptor functions for the integrin VLA-3: fibronectin, collagen, and laminin binding are differentially influenced by ARG-GLY-ASP peptide and by divalent cations, *J. Cell Biol.*, 112, 169, 1991.

77A. Dedhar, S., Jewell, K., Rojian, M., and Gray, V., The receptor for the basement membrane glycoprotein entactin is the integrin α3/β1*, *J. Biol. Chem.* 267, 18908, 1992.

77B. Gehlsen, K. R., Sriramarao, P., Furcht, L. T., and Skubitz, P. N., A synthetic peptide derived from the carboxy terminus of the laminin A chain represents a binding site for the α3β1 integrin, *J. Cell Biol.*, 117, 449, 1992.

78. Lee, E. C., Lotz, M. M., Steele, G. D., and Mercurio, A. M., The integrin α6β4 is a laminin receptor, *J. Cell Biol.*, 117, 671, 1992.

79. Kurpakus, M. A. and Jones, J. C. R., A novel hemidesmosomal plaque component: tissue distribution and incorporation into assembling hemidesmosomes in an in vitro model, *Exp. Cell Res.*, 194, 139, 1991.

80. Klatte, D. H., Kurpakus, M. A., Grelling, K. A., and Jones, J. C. R., Immunochemical characterization of three components of the hemidesmosome and their expression in cultured epithelial cells, *J. Cell Biol.*, 109, 3377, 1989.

81. Stepp, M. A., Spurr-Michaud, S., Tisdale, A., Elwell, J., and Gipson, I. K., α6β4 integrin heterodimer is a component of hemidesmosomes, *Proc. Natl. Acad. Sci. U.S.A.*, 87, 8970, 1990.

82. Jones, J. C. R., Kurpakus, M. A., Cooper, H. M., and Quaranta, V., A function for the integrin α6β4 in the hemidesmosome, *Cell Regul.*, 2, 427, 1991.

83. Sonnenberg, A., Calafat, J., Janssen, H., Daams, H., van der Raaij-Helmer, L. M. H., Falcioni, R., Kennel, S. J., Aplin, J. D., Baker, J., Loizidou, M., and Garrod, D., Intergrin α6β4 complex is located in hemidesmosomes, suggesting a major role in epidermal cell-basement membrane adhesion, *J. Cell Biol.*, 113, 907, 1991.

84. Kurpakas, M. A., Quaranta, V., and Jones, J. C. R., Surface relocation of alpha6 beta4 integrins and hemidesmosomes in and *in vitro* model of wound healing, *J. Cell Biol.*, 115, 1737, 1991.

85. Tanaka, T., Korman, N. J., Shimizu, H., Eady, R. A. J., Klaus-Kovtun, V., Cehrs, K., and Stanley, J. R., Production of rabbit antibodies against carboxy-terminal epitopes encoded by bullous pemphigoid cDNA, *J. Invest. Dermatol.*, 94, 617, 1990.

86. Tanaka, T., Parry, D. A. D., Klaus-Kovtun, V., Steinert, P. M., and Stanley, J. R., Comparison of molecularly cloned bullous pemphigoid antigen to desmoplakin I confirms that they define a new family of cell adhesion junction plaque proteins, *J. Biol. Chem.*, 266, 12555, 1991.

87. Hopkinson, S. B., Riddelle, K. S., and Jones, J. C. R., Cytoplasmic domain of the 180-kD bullous pemphigoid antigen, a hemidesmosomal component: molecular and cell biologic characterization, *J. Invest. Dermatol.*, 99, 264, 1992.

88. Giudice, G. J., Squiquera, H. L., Elias, P., and Diaz, L. A., Identification of two collagen-like domains within the bullous pemphigoid antigen, BP180, *J. Clin. Invest.*, 87, 734, 1991.

89. Giudice, G. J., Emery, D. J., and Diaz, L. A., Cloning and primary structural analysis of the bullous pemphigoid autoantigen BP180, *J. Invest. Dermatol.*, 99, 243, 1992.

90. Li, K., Giudice, G. J., Tamai, K., Do, H. C., Sawamura, D., Diaz, L. A., and Uitto, J., Cloning of partial cDNA for mouse 180-kDa bullous pemphigoid antigen (BPAG2), a highly conserved collagenous protein of the cutaneous basement membrane zone, *J. Invest. Dermatol.*, 99, 258, 1992.

91. Hieda, Y., Nishizawa, Y., Uematsu, J., and Owaribe, K., Identification of a new hemidesmosomal protein, HD-1: a major, high molecular mass component of isolated hemidesmosomes, *J. Cell Biol.*, 116, 1497, 1992.

92. Griepp, E. B. and Robbins, E. S., Epithelium, in *Cell and Tissue Biology*, 6th ed., Weiss, L., Ed., Urban & Schwarzenberg, Baltimore, 1988.

93. Burridge, K., Fath, K., Kelly, T., Nuckolls, G., and Turner, C., Focal adhesions: transmembrane junctions between the extracellular matrix and the cytoskeleton, *Annu. Rev. Cell Biol.*, 4, 487, 1988.

94. Gipson, I. K., Spurr-Michaud, S., Tisdale, A. Elwell, J., and Stepp, M. A., Redistribution of the hemidesmosome components $\alpha6\beta4$ integrin and bullous pemphigoid antigens during epithelial wound healing, *Exp. Cell Res.*, 207, 86, 1993.

95. Quaranta, V. and Jones, J. C. R, The internal affairs of an integrin, *Trends in Cell Biol.*, 1, 2, 1991.

96. Straus, A. H., Carter, W. G., Wayner, E. A., and Hakomori, S.-I., Mechanism of fibronectin-mediated cell migration: dependence or independence of cell migration susceptibility on RGDS-directed receptor (Integrin)[1], *Exp. Cell Res.*, 183, 126, 1989.

97. Duband, J-L., Dufour, S., Yamada, S. S., Yamada, K. M., and Thiery, J. P., Neural crest cell locomotion induced by antibodies to $\beta1$ integrins, *J. Cell Sci.*, 98, 517, 1991.

98. Bretscher, M. S., Circulating integrins: $\alpha5\beta1$, $\alpha6\beta4$ and mac-1, but not $\alpha3\beta1$ or LFA-1, *EMBO J.*, 11, 405, 1992.

99. Stossel, T. P., On the crawling of animal cells, *Science*, 260, 1087, 1993.

100. Burridge, K., Petch, L. A., and Romer, L. H., Signals from focal adhesions, *Curr. Biol.*, 2, 537, 1992.

101. Juliano, R. S. and Haskill, S., Signal transduction from the extracellular matrix, *J. Cell Biol.*, 120, 577, 1993.

102. Rodriquez-Fernandez, J. L., Geiger, D., Salomon, and Ben-Zev'ev, A., Overexpression of vinculin suppresses cell motility in Balb/c 3T3 cells, *Cell Motil. Cytoskeleton*, 22, 127, 1992.

103. Gipson, I. K., Grill, S. M., Spurr, S. J., and Brennan, S. J., Hemidesmosome formation in vitro, *J. Cell Biol.*, 97, 849, 1983.

104. Eady, R. A. J., The basement membrane: interface between the epithelium and the dermis: structural features, *Arch. Dermatol.*, 124, 709, 1988.

105. Yurchenco, P. D. and Schittny, J. C., Molecular architecture of basement membranes, *FASEB J.*, 4, 1577, 1990.

106. Uitto, J. and Christiano, A. M., Molecular genetics of cutaneous membrane zone, *J. Clin. Invest.*, 90, 687, 1992.

107. Bohnert, A., Hornung, J., Mackenzie, I. C., and Fusenig, N. E., Epithelial-mesenchymal interactions control basement membrane production and differentiation in cultured and transplanted mouse keratinocytes, *Cell Tissue Res.*, 244, 413, 1986.

108. Ellison, J. and Garrod, D. R., Anchoring filaments of the amphibian epidermal-dermal junction traverse the basal lamina entirely from the plasma membrane of hemidesmosomes to the dermis, *J. Cell Sci.*, 72, 163, 1984.

109. Sakai, L. Y., Keene, D. R., Morris, N. P., and Burgeson, R. E., Type VII collagen is a major structural components of anchoring fibrils, *J. Cell Biol.*, 103, 1577, 1986.

110. Keene, D. R., Sakai, L. Y., Lunstrum, G. P., Morris, N. P., and Burgeson, R. E., Type

VII collagen forms an extended network of anchoring fibrils, *J. Cell Biol.*, 104, 611, 1987.

111. Smith, L. T. and Sybert, V. P., Intra-epidermal retention of type VII collagen in a patient with recessive dystrophic epidermolysis bullosa, *J. Invest. Dermatol.*, 94, 261, 1990.

112. Rouselle, P., Lunstrum, G. P., Keene, D. R., and Burgeson, R. E., Kalinin: an epithelium-specific basement membrane adhesion molecule that is a component of anchoring filaments, *J. Cell. Biol.*, 114, 567, 1991.

113. Verrando, P., Hsi, B-L., Yeh, C-J., Pisani, A., Serieys, N., and Ortonne, J-P., Monoclonal antibody GB3, a new probe for the study of human basement membranes and hemidesmosomes.

114. Verrando, P., Pisani, A., and Ortonne, J-P., The new basement membrane antigen recognized by the monoclonal antibody GB3 is a large size glycoprotein: modulation of its expression by retinoic acid, *Biochim. Biophys. Acta*, 942, 45, 1988.

115. Domloge-Hultsch, N., Gammon, W. R., Briggman, R. A., Gil, S. G., Carter, W. G., and Yancey, K. B., Epiligrin, the major human keratinocyte integrin ligand, is a target in both and acquired autoimmune and an inherited subepidermal blistering skin disease, *J. Clin. Invest.*, 90, 1628, 1992.

116. Marchisio, P. C., Cremona, O., Savoa, P., Pellegrini, G., Ortonne, J-P., Verrando, P., Burgeson, R. E., Cancedda, R., and De Luca, M., The basement membrane protein BM-600/nicein codistributes with kalinin and the integrin α6β4 in human cultured keratinocytes, *Exp. Cell Res.*, 205, 205, 1993.

117. Watt, F. M., Kalinin, epiligrin and GB3 antigen: kalinephiligrinin-3?, *Curr. Biol.*, 2, 106, 1992.

118. Kallunki, P., Sainio, K., Eddy, R., Byers, M., Kallunki, T., Sariola, H., Beck, K., Hirvonen, H., Shows, T. B., and Tryggvason, K., A truncated laminin chain homologous to the B2 chain: structure, spatial expression and chromosomal assignment, *J. Cell Biol.*, 119, 679, 1992.

119. Frenette, G. P., Carey, T. E., Varani, J., Schwartz, D. R., Fligiel, S. E. G., Ruddon, R. W., and Peters, B. P., Biosynthesis and secretion of laminin and laminin-associated glycoproteins by nonmalignant and malignant human keratinocytes: comparison of cell lines from primary and secondary tumors in the same patient, *Cancer Res.*, 48, 5193, 1988.

120. Kolega, J. and Manabe, M., Tissue-specific distribution of a novel component and epithelial basement membranes, *Exp. Cell Res.*, 189, 213, 1990.

121. Engvall, E., Earwicker, D., Haaparanta, T., Ruoslahti, E., and Sanes, J. R., Distribution and isolation of four laminin variants; tissue restricted distribution of heterotrimers assembled from five different subunits, *Cell Regul.*, 1, 731, 1990.

122. Engel, J., Laminins and other strange proteins, *Biochemistry*, 31, 10643, 1992.

123. Beck, K., Hunter, I., and Engel, J., Structure and function of laminin: anatomy of a multidomain glycoprotein, *FASEB*, 4, 148, 1990.

124A. Marinkovich, M. P., Lunstrum, G. P., Keene, D. R., and Burgeson, R. E., The dermalepidermal junction of human skin contains a novel laminin variant, *J. Cell Biol.*, 119, 695, 1992.

124B. Yamada, K. M., Adhesive recognition sequences, *J. Biol. Chem.*, 266, 12809, 1991.

124C. Sasaki, M., Kleinman, H. K., Huber, H., Deutzmann, R., and Yamada, Y., Laminin, a multidomain protein, *J. Biol. Chem.*, 263, 16536, 1988.

125. Wilke, M. S. and Skubitz, A. P. N., Human keratinocytes adhere to multiple distinct peptide sequences of laminin, *J. Invest. Dermatol.*, 97, 141, 1991.

126. Sonnenberg, A., Modderman, P. W., and Hogervorst, F., Laminin receptor on platelets is the integrin VLA-6, *Nature*, 336, 487, 1988.

127. Sonnenberg, A., Linders, C. J. T., Modderman, P. W., Damsky, C. H., Aumailley,

M., and Timpl, R., Integrin recognition of different cell-binding fragments of laminin (P1, E3, E8) and evidence that α6β1 but not α6β4 functions as a major receptor for fragment E8, *J. Cell Biol.*, 110, 2145, 1990.

128. Hotchin, N. A. and Watt, F. M., Transcriptional and post-translational regulation of β1 integrin expression during keratinocyte terminal differentiation, *J. Biol. Chem.*, 267, 14852, 1992.

129. Hertle, M. D., Adams, J. C., and Watt, F. M., Integrin expression during human epidermal development in vivo and in vitro, *Development*, 112, 193, 1991.

130. Kim, J. P., Zhang, K., Kramer, R. H., Schall, T. J., and Woodley, D. T., Integrin receptors and RGD sequences in human keratinocyte migration: unique anti-migratory function of α3β1 epiligrin receptor, *J. Invest. Dermatol.*, 98, 764, 1992.

131. Verrando, P., Blanchet-Bardon, C., Pisani, A., Thomas, L., Cambazard, F., Eady, R. A. J., Schofield, O., and Ortonne, J-P., Monoclonal antibody GB3 defines a widespread defect of several basement membranes and a keratinocyte dysfunction in patient with lethal junction epidermolysis bullosa, *Lab. Invest.*, 64, 85, 1991.

132. Fine, J. D., Bauer, E. A., Briggman, R. A., Carter, D. M., Eady, R. A. J., Esterly, N. B., Holbrook, K. A., Hurwitz, S., Johnson, L., Lin, A., Pearson, R., and Sybert, V. P., Revised clinical and laboratory criteria for subtypes of inherited epidermolysis bullosa, *J. Am. Acad. Dermatol.*, 24, 119, 1991.

133. Dolan, C. R., Smith, L. T., and Sybert, V. P., Prenatal detection of epidermolysis bullosa letalis with pyloric atresia in a fetus by abnormal ultrasound and elevated alpha-fetoprotein, *Am. J. Med. Genet.*, 47, 395, 1993.

134. LaCour, J. P., Hoffman, P., Bastiani-Griffet, F., Boutte, P., Pisani A., and Ortonne, J. P., Lethal junctional bullosa with normal expression of BM600 and antro-pyloric atresia: a new variant of junctional epidermolysis bullosa?, *Eur. J. Pediatr.*, 151, 252, 1992.

135. Nazzaro, V., Nicolini, U., De Luca, L., Berti, E., and Caputo, R., Prenatal diagnosis of junctional epidermolysis bullosa associated with pyloric atresia, *J. Med. Genet.*, 27, 244, 1990.

136. Lestringant, G. G., Akel, S. R., and Qayed, K. I., The pyloric atresia-junctional bullosa syndrome, *Arch. Dermatol.*, 128, 1083, 1992.

137. Kaur, P. and Carter, W. G., Integrin expression and differentiation in transformed human epidermal cells is regulated by fibroblasts, *J. Cell Sci.*, 103, 755, 1992.

138. Schofield, O. M. V., Kist, D., Lucas, A., Wayner, E., Carter, W., and Zachary, C. B., Abnormal expression of epiligrin and α6β4 integrin in basal cell carcinoma, submitted.

139. Larjava, H., Peltonen, J., Akiyama, S. K., Gralnick, H. R., Uitto, J., and Yamada, K. M., Novel function for β1 integrins in keratinocyte cell-cell interactions, *J. Cell Biol.*, 110, 1990.

140. Sriramarao, P., Steffner, P., and Gehlsen, K. R., Evidence for a homophilic interaction of the α3β1 integrin in cell-cell adhesion, *J. Biol. Chem.* 268, 22036, 1993.

141. Wheelock, M. A. and Jensen, P. J., Regulation of keratinocyte intercellular junction organization of epidermal morphogenesis by e-cadherin, *J. Cell Biol.*, 117, 415.

142. Tenchini, M. L., Adams, J. C., Gilberty, C., Steel, J., Hudson, D. L., Malcovati, M., and Watt, F. M., Evidence against a major role for integrins in calcium-dependent intercellular adhesion of epidermal keratinoctyes, *Cell Adhesion Commun.*, 1, 55, 1993.

143. Symington, B. E. and Carter, W. G., Epiligrin modulates epidermal differentiation and intercellular adhesion, submitted.

144. Schnaper, H. W. and Kleinman, H. K., Regulation of cell function by extracellular matrix, *Pediatr Neohol.*, 7, 96, 1993.

145. Kubota, S., Tashiro, K., and Yamada, Y., Signaling site of laminin with mitogenic activity, *J. Biol. Chem.*, 267, 4285, 1992.

146. Schubert, D. and Kimura, H., Substratum-growth factor collaborations are required for the mitogenic activities of activin and FGF on embryonal carcinoma cells, *J. Cell Biol.*, 114, 841, 1991.

147. Barrandon, Y. and Green, H., Cell migration is essential for sustained growth of keratinocyte colonies: the roles of transforming growth factor-α and epidermal growth factor, *Cell*, 50, 1131, 1987.

148. Adams, J. C. and Watt, F. M., Fibronectin inhibits the terminal differentiation of human keratinocytes, *Nature*, 340, 307, 1989.

149. Woodley, D. T., Wynn, K. C., and O'Keefe, E. J., Type IV collagen and fibronectin enhance human keratinocyte thymidine incorporation and spreading in the absence of soluble growth factors, *J. Invest. Dermatol.*, 94, 139, 1990.

Chapter 8

The Integrin α6β4 in Epithelial and Carcinoma Cells

Vito Quaranta, Ginetta Collo, Carla Rozzo, Lisa Starr,
Guido Gaietta, and Richard N. Tamura

CONTENTS

I. INTRODUCTION

An important issue in current cell adhesion research is to achieve an understanding of the role of integrins in tumorigenesis. It is now increasingly accepted that many integrins can be considered tumor-associated antigens, since their expression is characteristically altered in transformed cells.[1] While integrin precise mechanisms of action in the process of cell transformation are not well understood, several general concepts have begun to emerge from the work of our laboratory as well as others: (1) carcinoma cells are almost invariably endowed with repertoires encompassing numerous integrins, which presumably negotiate complex and versatile interactions with the extracellular matrix; (2) promising correlations exist between certain integrins and tumorigenesis, based on the functions of such integrins in normal cells; and (3) integrin "fingerprinting" approximates the adhesive profile of carcinoma cells, but functional and/or mechanistic analyses are needed to fully appreciate its significance in the context of tumor cell invasion and metastasis.

We contributed to the initial definition and subsequent structural or functional characterization of several new integrins or integrin subunits from carcinoma cells.[2-7] Most of our efforts have focused on one integrin, α6β4, because of its potentially dominant role in regulating a process important to carcinogenesis, i.e., cell adhesion to basement membranes[8] (see below). In this chapter, we highlight published information from our laboratory and comment on its significance in the context of some of the work published by others.

II. PATTERNS OF EXPRESSION OF THE INTEGRIN, α6β4
A. INITIAL DEFINITION OF α6β4

We have characterized several structural and functional aspects of the integrin, α6β4. The α6β4 integrin was originally identified in our laboratory by the use of a monoclonal antibody, S3-41, raised against the pancreatic carcinoma cell line, FG.[9] When a panel of human tumor cell lines was screened with S3-41, it became apparent that this antibody preferentially reacted with carcinoma cell types. Several tumor cell lines of nonepithelial derivation were unreactive. For instance, no reactivity was observed with ten melanoma lines, three glioblastoma, three neuroblastoma, seven T or B leukemia, and two fibroblast cell lines. Not all carcinoma cell lines, however, were positive with S3-41. Oat cell carcinoma cells were consistently negative, and heterogeneity was found within breast cancer and lung and bladder carcinoma cell lines.[9]

With human tumor tissue specimens, similar patterns were observed. For instance, melanoma or sarcoma tumors were generally S3-41 negative, while strong positivity was consistently found in carcinomas of the pancreas, gastro-intestinal tract, bladder, prostate, and upper respiratory tract. Breast cancer specimens showed heterogeneity of staining and tended to be negative.[9]

B. EXPRESSION IN NORMAL TISSUES

From these results, the integrin α6β4 could be considered a carcinoma-associated antigen, and, given its nature of adhesion receptor, a role in carcinoma cell invasion and metastasis appears likely. To put this idea in a better perspective, it becomes important to consider the distribution of α6β4 in normal tissues.[9] Interestingly, α6β4 expression is confined to few sites. Several stratified epithelia, including epidermis, esophagus, and cervix, contain α6β4. However, in these tissues, antibodies to α6β4 stained exclusively the ventral surface of cells that are located at a basal position within the epithelium. That is, the staining was concentrated in the areas of contact between plasma membranes and basement membrane (this restricted topographical localization has important functional consequences, as discussed below). Other simple epithelia, including small and large intestine, breast, and cervix, were also intensely stained by S3-41, also at the region of cell contact with the basement membrane. These results suggest that the integrin α6β4 serves a role as adhesion receptor for some basement membrane components in various types of normal epithelia. It should be stressed, however, that there are many normal epithelia in which α6β4 is not found, e.g., pancreas, kidney, ovary, and lung. In these epithelia, the role of basement membrane receptor is presumably taken over by some other surface molecule, perhaps another integrin.

Table 1 **Trends in changes of α6β4 expression according to the tissue derivation of transformed epithelial cells**

α6β4 Neg → Pos	Pancreas Lung Kidney Ovary
α6β4 Pos → Pos	Esophagus Stomach Colon Cervix
α6β4 Pos → Neg	Breast

C. EXPRESSION IN CARCINOMAS
1. Overview
The limited distribution of α6β4 in normal tissues contrasts its consistent expression in carcinomas from a variety of sources. In Table 1, expression of α6β4 in normal vs. transformed epithelia is compared. The general tendency is that α6β4 is expressed in carcinoma cells, whether or not the normal tissue of origin is α6β4 positive. A notable exception is the breast gland, whose normal epithelium is α6β4 positive, while breast cancer cells tend to be α6β4 negative. Taken together, our observations support the concept of an association between α6β4 expression and carcinogenesis.

Results from other laboratories are in agreement with and extend this conclusion. Kennel and collaborators described TSP180, a cell surface protein complex found in murine spontaneous lung carcinomas, as a tumor-associated antigen.[10] TSP180 was later found to be equivalent to α6β4,[11] indirectly confirming the fact that, in humans, this integrin is not expressed on normal lung epithelium but is sporadically expressed in lung carcinomas.[12] Liebert and co-workers described in some detail the expression of α6β4 in the transitional epithelium of the bladder, as well as in urothelia in general.[13] They reported that transformed cells of this origin consistently express α6β4, and that disturbances in the polarized topography of this integrin are a frequent occurrence, perhaps related to clinical outcome. Virtanen, Gould, and co-workers studied the expression of several integrins in solid tumors, including breast[14] and colon,[15] and indicated that α6β4 is often topographically altered.

2. Breast Cancer
Expression patterns of α6β4 in breast cancer are worthy of some detailed discussion. Recently, a cluster of publications has appeared on the expression of integrins in human breast cancer.[16-28] Several integrins are expressed in the normal mammary epithelium, including collagen/laminin receptors (α1β1 and α2β1), fibronectin receptors (α5β1), and the basement membrane receptor α6β4. No staining patterns clearly distinguish myoepithelial from luminal epithelial cells.[23] The distribution of these integrins is substantially maintained in benign mammary lesions.[14] In contrast, the predominant trend in malignant lesions is the underexpression of the basement membrane receptor α6β4, regardless of the histological type. By compiling results from various laboratories,[14,19,20,23] approximately 30% of the lesions are α6β4 positive, but of these more than half have lost the typical polarized distribution of α6β4. Another 20% is α6 positive, but β4 negative. The remaining 40% is α6 negative, β4 negative. It should be stressed, however, that within the same lesion or between lesions in the same patient, the expression of α6β4 may vary. Natali et al.,[23] for instance, report one case in which the primary tumor is α6β4 negative, while a nodal metastatic site is positive. Jones et al.[20] concluded that

loss of α6β4 appears to be an early indicator of malignant transformation, since it appeared to be downregulated in normal portions of cancerous mammary glands. For these reasons, breast cancer may be an informative model for the role of α6β4 integrin in invasion and metastasis.

3. Head and Neck Cancer

Another type of cancer potentially revealing as to the role of α6β4 is represented by head and neck squamous carcinomas. Carey and co-workers[29] showed that a monoclonal antibody, UM-A9, reacts with a molecule identical to α6β4, by immunochemical criteria.[30] Positive staining with this antibody can be used as a prognostic criteria for recurrence of head and neck squamous carcinomas.[29] High expression of α6β4 in primary tumor specimens was associated with poor prognosis, characterized by metastasis and relapse. The combination of α6β4 and blood group expression on tumors was the variable that most accurately predicted disease-free survival, even when adjusted for other conventional prognostic factors such as tumor site, stage, and classification.[29] Loss or incorrect expression of α6β4 was also apparent in squamous or basal carcinomas of another stratified epithelium, the epidermis.[31] Models for keratinocyte transformation may therefore be excellent systems for studying α6β4 integrins in cancer.

4. Considerations

Potentially informative is also the behavior of α6β4 in psoriatic keratinocytes.[32,33] Psoriasis vulgaris is a relatively common skin disease, characterized by hyperproliferation of keratinocytes. In involved skin from these patients, the expression of α6β4 is altered in two respects: first, it is no longer limited to the basal layer of keratinocytes, but it extends to cells in suprabasal position; second, it is no longer polarized, but it is distributed over the entire surface of cells. Although psoriatic keratinocytes are clearly nonneoplastic cells, their proliferation is not properly controlled. This could, therefore, be considered another example of association between overexpression of α6β4 and uncontrolled proliferative state.

In summary, in spite of compelling evidence, there is little or no understanding of the mechanisms by which α6β4 integrins may promote malignant behavior. It seems logical that, as an integrin, α6β4 should act at the level of carcinoma cell adhesion and/ or migration. Unfortunately, the role of α6β4 in adhesion and migration of even normal cells is not properly understood.[34] In the following sections, we will summarize the available knowledge on structure and function of α6β4, and will speculate on the possible connections with the transformed state. We reiterate that most of the emphasis will be on results from our own laboratory.

III. STRUCTURE OF α6β4
A. OVERALL PROTEIN STRUCTURE
1. The α6β4 Heterodimer

In analogy to other integrins, α6β4 is a heterodimer comprised of structurally unrelated subunits, both glycosylated.[35] We purified α6β4 on a preparative scale from human placenta, from cultured pancreatic carcinoma cells FG and from lung carcinoma cells UCLA-P3, by using S3-41 antibody affinity columns. We determined the amino terminal sequence of both subunits by gas-phase automated microsequencing.[2] The amino terminal sequence of α6 shared unequivocal homologies with other integrin α chains. The amino terminus of β4 was identical at several positions with one or more of the other integrin β chains, and the spacing of three cysteine residues, a landmark of integrin β chains, was exactly conserved. Neither α6 nor β4, however, matched precisely the sequence of any other integrins, indicating that this epithelial heterodimer was composed of novel subunits.

2. Structure of β4 Subunit

A peculiar feature of the β4 chain is its molecular weight which is significantly higher than other integrin β chains.[2] Virtually all other integrin β chains have an apparent molecular weight of approximately 120,000 Da, under reducing conditions in SDS-PAGE. Under nonreducing conditions, this apparent size is smaller by 20 to 30,000 Da, due to intrachain disulfide bridges that keep the molecule in a more compact conformation. The apparent molecular weight of β4 is instead 200,000 under reducing, and 190,000 under nonreducing conditions in SDS-PAGE. Initially, this increase in size appeared to be due to high levels of posttranslational glycosylation, because the molecular weight of β4 decreased by about 50% upon treatment with neuraminidase.[2] However, it later turned out that those preparations of neuraminidase were contaminated with proteases (V. Quaranta, unpublished). The real explanation came from the cloning of the β4 cDNA, which revealed the existence of a large cytoplasmic domain, accounting for about 100,000 Da in molecular weight [5,36,37] (see below).

Another interesting feature of β4 is that two proteolytic fragments are generally found in purified preparation of α6β4, particularly from tissues such as placenta.[2] We unequivocally showed that such proteolytic fragments are due to deletions of sequences at the carboxy terminus of β4, by sequencing such fragments from a purified α6β4 preparation from placenta, and showing that their amino terminus has a sequence identical to that of the intact β4 protein.[2] In the case of placental preparations, it was clear that proteolysis of β4 occurred during purification, since it was enhanced by repeat freeze-thaw cycles.[2] It has been proposed, however, that such fragments may also be generated intracellularly and may have a physiological role.[38]

3. Molecular Clones of β4 Subunit

cDNAs for the β4 subunit were cloned in several laboratories at approximately the same time.[5,36,37] In our laboratory, we screened a λgt11 cDNA expression library from the carcinoma cell line FG, with polyclonal antisera raised against α6β4 purified from placenta. A series of overlapping cDNAs contained an opening reading frame that, by several independent criteria, encoded the β4 protein.[5] The structure of the β4 protein, however, contained a surprise. An extracellular portion homologous to that of other integrin β chains was followed by a cytoplasmic domain unique and unusually large. The β4 cytoplasmic domain encompasses more than 1000 residues, while all other integrin β chains contain cytoplasmic domains about 50 amino acids long. The function of the β4 cytoplasmic tail remains unsolved. It could represent a link protein to the cytoskeleton that has become incorporated in the sequence of the integrin receptor itself. In this regard, it is interesting to note that α6β4 is the only integrin known to be linked to the intermediate filament network, rather than to the microfilament cytoskeleton[34] (see below). Another possibility is that the β4 cytoplasmic domain may have some enzymatic function, promoting the polymerization-depolymerization of the local cytoskeleton, or engaging in a signaling pathway. At this point, these are speculations that need to be investigated by appropriate experimental schemes.

In extensive searches against databases of available sequences, the sequence of the β4 cytoplasmic domain did not show any overall significant identity with proteins that may shed light on its function. Three regions encompassing about 140 residues are homologous to type III fibronectin repeats[39] (Figure 1A). The tridimensional structure of this protein module was recently solved by two groups. One solved the structure of a type III repeat from fibronectin itself,[40] the other of a repeat from tenascin.[41] This information should be useful to guide structural studies on the β4 cytoplasmic domain, though the solved structures are from extracellular matrix proteins, and it is not clear whether the structure of the type III repeats is conserved in the intracellular portion of a molecule.

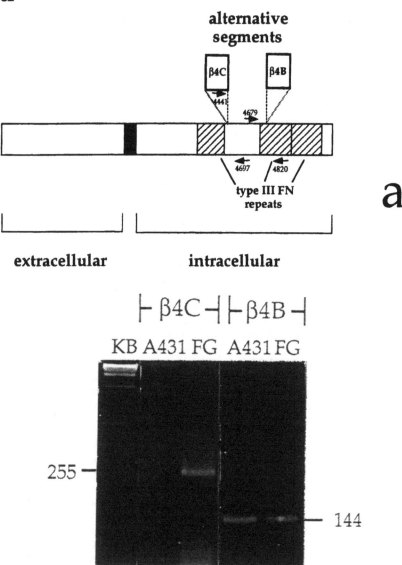

Figure 1 (A) Schematic diagram of the alternative segments in the cytoplasmic domain of the β4 subunit. Location with respect to the type III fibronectin repeats is shown. Arrows indicate the position of the oligonucleotide primer pairs used to amplify these regions by RT-PCR,[5] identified by the numbering of the first nucleotide: 4441 (40-mer) and 4697 (19-mer) for β4C; 4679 (19-mer) and 4820 (16-mer) for β4B. (B) Representative results obtained with primers specified in Figure 1A, with templates from the pancreatic carcinoma cell line FG, and the epidermoid carcinoma A431. Both β4B and β4C are expressed in FG cells, while only β4B is found in A431 cells.

4. Structure of α6 Subunit

Thus far, the only partner β4 has been recognized to form heterodimers with is α6 (previously referred to with the abandoned name αE[2,42]). The α6 subunit was cloned in our laboratory[5] from a pancreatic carcinoma cDNA library (FG), by screening with degenerate oligonucleotide probes based on 29 residues of the α6 amino-terminal sequence that we previously determined.[2] Two overlapping inserts, spanning 6 kb, contained an open reading frame encoding the entire α6 protein. The primary structure of α6 was deduced from the complete nucleotide sequence of these clones. The α6 protein was made of 1,073 amino acids and shared about 25% identity with most other integrin α chains. It does, however, share a higher level of identity (over 40%), as well as other properties (see below), with the subunits α3 and α7, so that these 3 α chains are usually grouped together in current integrin nomenclature.[7,43] In spite of the high homology, there is no evidence that α3 or α7 subunits, which normally form heterodimers with β1, may also associate with β4. Instead, in addition to forming α6β4 heterodimers, α6 can also form α6β1 heterodimers. Antisera to the carboxy-terminal region of α6, reacting with both native and heat/SDS denatured α6 in immunoprecipitation and Western blotting, were used to formally demonstrate that α6β4 heterodimers are found in cells of epithelial origin, whereas in most other cell types, which do not express β4, α6 is associated with β1.[5] While the precise rules that govern formation of one heterodimer over the other are not understood, it appears that in the presence of both β1 and β4 in the biosynthetic pathway, α6β4 heterodimers are formed preferentially.[38] This may be due to a higher affinity of α6 for β4 than β1. Alternatively, other biosynthetic mechanisms may be operational, such as chaperone proteins or pool accessibility. Competition experiments with purified subunits *in vitro* may solve this potentially critical issue.

5. Relationship to Other Integrins

As expected, no obvious serologic cross-reactivities have been detected between α6β4 and other integrins, by using a battery of monoclonal and polyclonal antisera.[2] An exception, though, is represented by antibodies directed to epitopes on the α6 subunit, such as GOH3[44] or BQ16.[13] The epitopes recognized by these antibodies are generally maintained both on α6β1 and α6β4 heterodimers, such that cross-reactivity between these heterodimers is observed.[45]

B. STRUCTURAL VARIANTS OF α6β4

An interesting possibility is that the functions of α6β4 may vary from one cell type to the other according to minor variations in the structure of either α6 or β4. We have identified variants in either subunit, which probably arise by alternative mRNA splicing. In addition, we investigated point mutations in the α6 subunits from pancreatic carcinoma cell lines. The evidence for these structural variations is described in the next few paragraphs, again, limited to our own results.

1. Isoforms of β4 Subunit

Variant mRNAs encode β4 proteins with slightly different cytoplasmic domains (β4A, β4B, and β4C) (Figure 1A). The β4B form contains an insertion of 53 amino acids at position 1519, which falls in the middle of the second type III fibronectin repeat (Figure 1A). The β4C contains a 70-amino acid insert at position 1369, between the first and second type III repeats (Figure 1A). Form β4A contains neither insert (Figure 1A). These β4 variants are likely to be encoded by alternatively spliced β4 mRNAs, though formal evidence to this fact is lacking. By reverse transcription-polymerase chain reaction (RT-PCR), cultured carcinoma cells (Figure 1B) often contain form A and C, while certain normal epithelial tissues contain A and B (Table 2). However, this correlation does not always hold, as is the case in the cervix. Normal cervix tissue contains all

Table 2 **Distribution of β4 cytoplasmic domain variants detected by RT-PCR in human cell lines and tissue specimens**

Cell Lines		β4A	β4B	β4C
FG-2	Pancreatic CA	+	−	+
1320 Met	Pancreatic CA	+	−	+
Panc I	Pancreatic CA	+	−	−
SG-R	Pancreatic CA	+	−	+
CoLo	Pancreatic CA	+	−	+
LoVo	Pancreatic CA	+	−	+
CaCo-2	Colon CA	+	−	+
HT-29	Colon CA	+	−	−
A431	Epidermoid CA	+	−	−
Jar	Choriocarcinoma	+	−	+
Jeg-3	Choriocarcinoma	+	−	+
BeWo	Choriocarcinoma	+	−	−
Hela	Cervix CA	+	−	−
M 21	Melanoma	−	−	−
Tissues				
Placenta		+	+	−
Normal cervix 1		+	+	+
Normal cervix 2		+	+	+
Normal cervix 3		+	+	+
Normal cervix 4		+	+	+
Squamous cell CA, cervix	(well-differentiated)	+	+	+
Squamous cell CA, cervix	(poorly differentiated)	+	+	−
Adeno carcinoma, Cervix		+	+	−

three β4 isoforms, while cervical carcinomas sometimes lose expression of the C isoform (Table 2). More precise quantitative assays and better resolution at the single cell level are needed to evaluate the siginificance of the β4 isoform distribution.

2. Isoforms of α6 Subunit

We also uncovered[6,46] the existence of two splice variants of α6, termed A and B, which differ in the size and sequence of their cytoplasmic domains. By cloning and sequencing of PCR fragments, we determined the structure of both cytoplasmic domains. The A cytoplasmic domain was identical to that of the original α6 cDNA clones. The B cytoplasmic domain represented a novel one. We made antipeptide antisera which distinguished between α6A and α6B. With these antisera, as well as by RT-PCR, we showed that the expression of α6A vs. α6B was regulated according to development and differentiation.[46] Teratocarcinomas or embryonic stem cells expressed exclusively α6B. Differentiated cell types, such as keratinocytes or fibroblasts, expressed exclusively α6A. However, most carcinoma cells we tested expressed both α6A and α6B, suggesting that the splicing mechanism was deregulated by transformation (Figure 2). Sequential immunoprecipitations showed that α6A and α6B could form heterodimers with either β1 or β4.[46] Functionally, α6Aβ1 and α6Bβ1 were both laminin receptors.[6] However, it is not clear whether the alternative $\alpha6$ cytoplasmic domains transmit distinct signals to the cell interior. Subsequently, we found two other integrin subunits, $\alpha3$ and $\alpha7$, that undergo alternative splicing of their cytoplasmic domains in a cell-type dependent fashion.[7,46] This may represent a widespread mechanism by which certain integrins modulate cellular responses to matrices according to differentiation, and widens the similarities among the subunits $\alpha3$, $\alpha6$, and $\alpha7$.

In normal keratinocytes, $\alpha6A$, not $\alpha6B$, is found in exclusive association with $\beta4$ (even though $\beta1$ is synthesized in these cells and forms heterodimers with other α subunits).[42] Thus, the $\alpha6$ integrin phenotype of these normal epithelial cells specifies $\alpha6A\beta4$ exclusively. Interestingly, such a narrowly defined phenotype is almost invariably lost in transformed epithelial cells (Figure 2). In such cells, expression of both $\alpha6A$ and $\alpha6B$ is essentially a constant, with varying relative ratios. Whether simultaneous expression of both $\alpha6$ isoforms is compatible with a normal epithelial phenotype is an important issue, thus far unresolved. Perspective to this issue will derive from an understanding of the functional value of the two $\alpha6$ isoforms.[47]

3. Point Mutations in $\alpha6$ Subunit

Other structural changes in $\alpha6\beta4$ may be caused by accumulation of mutations in the corresponding genes during carcinogenesis. While speculative, this possibility has many precedents in other genes,[48] and would be a powerful explanation for integrin-dependent invasive or metastatic behavior of cancer cells. We recently showed[49] that single base substitutions occur in the $\alpha6$ subunit gene in cultured carcinoma cells. The $\alpha6$ cDNA we sequenced derived from libraries of a pancreatic carcinoma cell line, FG. We then resequenced $\alpha6$ cDNAs from a library of normal human epidermal keratinocytes (Clontech, San Francisco). Three base changes occurred in the coding region. Two gave rise to amino acid substitutions; the third was silent. At bp 1114, a G$\rightarrow$A transition in the carcinoma sequence resulted in a methionine instead of valine (codon 323). At bp 2442, a G$\rightarrow$T transvection resulted in tyrosine instead of aspartic (codon 766). The methionine mutation (codon 323) immediately precedes a cation binding site of $\alpha6$, thought to be of functional importance. The tyrosine mutation results in a drastic change in the character of the amino acid at that position, such that both of these changes may affect function and/or expression of the $\alpha6$ chain. We excluded artifacts (e.g., sequencing, reverse transcription) by sequencing five times in both orientations at least two independent cDNA clones from the same library. These substitutions may still be unrelated to carcinogenesis, e.g., they may be polymorphisms in the human population. However, in the $\alpha6$ sequence from chicken,[50] both positions 1114 and 2442 agree with the normal keratinocyte human sequence. Conservation over a large evolutionary distance makes these unlikely sites for polymorphic variations. The mutations may have arisen in tissue culture, rather than being present in the original tumor from which FG was derived. Unfortunately, the original tumor or very early passages of FG cells is no longer available. The critical issue now is whether these or other such mutations can be found in other tumor specimens and cell lines.[49]

4. Significance of $\alpha6\beta4$ Variants in Carcinomas

The significance of these structural variants of $\alpha6$ and $\beta4$ in the context of epithelial cell transformation is not known. Recently, several gene products that confer a metastatic phenotype were identified. A splice variant of CD44, a nonintegrin adhesion molecule, was found in metastasizing, but not in nonmetastasizing pancreatic and breast carcinoma cell lines.[51] Transfection of such splice variants sufficed to transfer the metastatic phenotype.[51] Low expression of nm23, a nucleoside diphosphate kinase, correlated with metastasis in melanoma cells.[52] Matrix-degrading proteases, such as transin, collagenase, or urokinase, were found to be upregulated in invading cell lines.[8] These findings establish

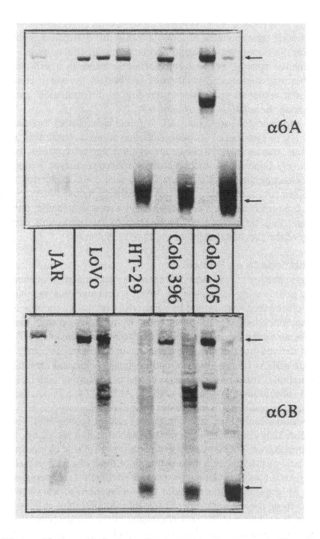

Figure 2 Western blotting of indicated cell detergent lysates, reacted with antibodies to the cytoplasmic domains of the α6 isoforms A and B.[46] Human cell lines are as follows: Jar, choriocarcinoma; LoVo, HT-29, Colo 396, and Colo 205, colon carcinomas. Each lysate was electrophoresed under nonreducing (left lane) or reducing (right lane) conditions. Under nonreducing conditions, the two proteolytically processed fragments of the α6 subunit are joined together by disulfide bridges, and run at an apparent molecular weight of 150,000 Da (upper arrow). Under reducing conditions, the two fragments separate, and the antisera recognize epitopes on the smaller fragments, running at about 30,000 Da (lower arrow). Note that in the cell line LoVo, proteolytic processing of α6 does not occur (H.M. Cooper and V. Quaranta, unpublished).

the important precedent that it is possible to use experimental assays to clearly correlate single gene products with metastasis. The CD44 results also indicate that arguably minor differences in gene expression, such as alternative splicing in inappropriate cells,[53] can have a major impact on metastatic properties. On the other hand, missense and other mutations in integrins can cause genetically transmitted diseases such as Glanzmann's thrombasthenia[54] and leukocyte adhesion deficiency[55] as well as loss or gain of function in several other integrin subunits.[56] Together, this evidence provides a rationale for studies aimed at correlating structural variations in α6β4 with invasion and metastasis. A critical complement to these studies, however, is an understanding of the functional properties of α6β4, i.e., a knowledge of its ligand(s). The next section deals with this issue.

IV. POSSIBLE LIGANDS FOR α6β4

A. LAMININ

By immunofluorescence staining with specific antibodies, α6β4 appears to be polarized to the basal plasmamembrane domain in a variety of epithelial cell types,[2,5] including cultured keratinocytes adhering to basement membrane substrates.[42] Extensive codistribution between α6β4 and laminin is also seen by immunostaining of human epidermis. Adhesion inhibition studies with specific antisera, however, have been thus far inconclusive in determining whether one of the known basement membrane components is the ligand for α6β4. In our hands, keratinocyte adhesion required 12 h or longer, making it difficult to assign all of the observed cell adhesion to the coating substrate used in a particular assay. That is, over such period of time, cells have the ability to synthesize and deposit their own matrix.[42] Unfortunately, shorter incubation times (e.g., 1 to 3 h) were insufficient to measure reproducibly and reliably normal keratinocyte adhesion.

In a different approach for identifying the α6β4 ligand, we prepared affinity columns with purified EHS laminin,[57,58] a major component of basement membranes. We then passed over these columns surface-radioiodinated cell lysates from carcinoma cells, in order to determine whether α6β4 would bind. These lysates were prepared in the detergent octyl-glucoside, previously shown to be compatible with integrin binding to ligands.[59] In several attempts, no α6β4 could be recovered from the laminin columns. Importantly, at least one β1 integrin, presumably one of the β1 laminin receptors,[43] was eluted, suggesting that, under our conditions, the interaction of at least some integrins with their ligand is possible (R. N. Tamura, unpublished). From these experiments, we concluded that laminin (at least its EHS isoform[57]) is not the ligand for α6β4. This conclusion is in agreement with results from another laboratory,[60] but is the exact opposite of the message from a recent report by Mercurio and co-workers,[61] indicating that laminin (EHS isoform) is a ligand for α6β4. This conflict in the data is unresolved as yet, but could have several explanations. One important detail is that laminin binding of α6β4 was observed only with detergent extracts from one human cell line, Clone A, derived from a poorly differentiated adenocarcinoma of the colon, and said to "adhere avidly" to laminin.[61] Besides trivial differences in affinity chromatography methods and ligand purification protocols, it could be that the state of activation of α6β4 in the particular cell line Clone A is such that laminin binding becomes detectable. State of activation may be the result of mutations in any of the subunits, or unknown intracellular or membrane factors, as seen for other integrins.[62] The physiological significance of this result for other cell types remains to be proven.

B. KALININ/NICEIN/EPILIGRIN

Another possible ligand for α6β4 is the recently described basement membrane component, kalinin.[63] This molecule is specifically found in some, but not all, basement membranes and is apparently related or identical to two other basement membrane components known as nicein[64] and epiligrin.[65] Kalinin/nicein/epiligrin is a ligand for the integrin α3β1.[65] However, a suggestion that this molecule might also be a ligand for α6β4 comes from the reported co-distribution of kalinin with hemidesmosomes, adhesive complexes of epithelial cells containing α6β4 (see below). To our knowledge, however, no biochemical data are available for or against this possibility.

C. 804G MATRIX

A ligand for α6β4 may be contained in the matrix deposited by the cell line, 804G. This cell derives from a rat bladder carcinoma, and has the unique ability to form hemidesmosomes *in vitro*.[66] In collaboration with the laboratory of J.C.R. Jones, we showed that the ability of 804G to form hemidesmosomes can be inhibited by anti-β4 antibodies.[67] Very recently, Jones and co-workers[68] showed that certain epithelial cell lines can themselves form hemidesmosomes when grown on 804G-deposited matrix. Matrix-coated glass coverslips were produced[68] by growing 804G cells to confluency, and then removing the cell layer by lysis in ammonium hydroxide. The α6β4 fluorescence appeared as coarsely granular, characteristically arranged as "Swiss cheese," in SSC12 human squamous carcinoma cells plated on 804G matrix. In contrast, on control matrix, α6β4 fluorescence was finely distributed in streaks. The components of 804G matrix presumably interacting with α6β4 were not characterized directly. However, an antiserum recognizing a pattern in the 804G matrix identical to that of the overlaying α6β4 was used to screen an 804G cDNA library. This resulted in the isolation of a cDNA encoding a protein with a high degree of identity to the laminin subunit isoform B2t.[68,69] It is therefore possible that the 804G cell line deposits a laminin-like molecule that both interacts with α6β4 and promotes hemidesmosome formation.

V. α6β4 AND HEMIDESMOSOMES
A. THE HEMIDESMOSOME ADHESION DEVICE

Hemidesmosomes represent the main anchor of epithelia to the basement membrane. They are considered a part of the basement membrane zone and regulate adhesive interactions with the basement membrane. The hemidesmosome[70] is a specialized device through which several types of epithelial cells adhere to the underlying connective tissue. Hemidesmosome function is comparable to that of focal adhesions,[71,72] at least superficially. Both hemidesmosomes and focal adhesions act as a plasma membrane site of attachment for cytoskeletal components. A major difference, however, is that whereas focal adhesions are the site of interaction for actin microfilaments, hemidesmosomes are the site of interaction of the intermediate filament network.[73] The hemidesmosome was originally described at the morphological level.[70] Several of its components have been characterized at the molecular level, including collagen type VII,[74] BP180 and BP230, two high molecular weight components identified with the aid of bullous pemphigoid autoantisera.[75]

B. ASSOCIATION OF α6β4 WITH HEMIDESMOSOMES

Several laboratories, at approximately the same time, produced data indicating that the integrin α6β4 is physically associated with hemidesmosomes.[67,76–78] In collaboration with J.C.R. Jones (Northwestern University) we found that α6β4 is part of the hemidesmosome[67] by immunofluorescence and immunoelectron microscopy with antibodies to both extra- and intracytoplasmic domains of α6β4. In normal stratified epithelia, such as bovine tongue, the staining obtained with monoclonal antibodies to β4 completely overlapped with that of antibodies to BP230, a cytoplasmic marker of hemidesmosomes. Jones and co-workers had previously shown that the rat bladder carcinoma cell line 804G was capable of forming structures *in vitro* that resembled ultrastructurally hemidesmosomes of epithelia.[66] In this cell line, colocalization of α6β4 with hemidesmosomal markers was also observed. We were, therefore, in a position to test the functional involvement of the α6β4 integrin during assembly of hemidesmosomes. A critical finding was that when 804G cells were plated in the presence of the anti-β4 antiserum 5710, no hemidesmosomes were detectable at the ultrastructural level, compared to control cells.[67] Identical results were obtained with normal corneal keratinocytes plated on denuded connective tissue.[79] These results suggested that α6β4 and its putative ligand played a role in assembly of hemidesmosomes.

C. ROLE OF α6β4 IN HEMIDESMOSOMES

It is important to recall that correct reformation of hemidesmosome adhesion complexes is a prerequisite for proper wound healing.[80] We showed[79] that α6β4 was diffusely distributed throughout the cytoplasm and the cell surface in corneal epithelial cells stimulated to migrate in an *in vitro* model for wound healing. As soon as migration halted, α6β4 became quickly polarized to the basal surface, in areas of contact with the basement membrane. Colocalization of other hemidesmosomal components and assembly of fully mature hemidesmosomes ensued. These results suggested that α6β4 had a role in catalyzing reorganization of hemidesmosomal components, perhaps after recognition of a ligand in the basement membrane.[79] We also showed that antibodies to α6 or to β4 can interrupt this process. Surprisingly, anti-β4 antisera prevented hemidesmosome assembly, but did not interfere with adhesion of cells. In contrast, anti-α6 monoclonal antibodies caused the detachment of cells, even those cells already possessing fully mature hemidesmosomes. Thus, the integrin α6β4 may have a dual role in hemidesmosomes, i.e., to trigger their assembly *and* to signal their disassembly. To accommodate the dichotomy of antibody inhibition data, we speculated[34,79] that the assembly signal may be transmitted through the β4 subunit, while the disassembly signal may be transmitted by the α6 subunit, concomitantly to a cell detachment signal.

D. α6β4, HEMIDESMOSOMES, AND CARCINOGENESIS

These functional results raise once again the issue of the role of α6β4 in carcinoma cells. Its recurrent expression may be viewed as an apparent contradiction with its postulated role as a receptor for the hemidesmosome, a strong anchoring device of epithelial cells. A receptor for "strong" adhesion to the basement membrane may indeed immobilize carcinoma cells, rather than favor their invasive and metastasizing properties. Note, however, that while carcinoma cell lines possess hemidesmosome components, they are almost always completely incapable of assembling them.[66] Thus, in carcinoma cells, α6β4 may be frozen in a "signaling" capacity that forces carcinoma cells *not* to assemble hemidesmosomes and therefore reduces their chances of becoming trapped by basement membranes. Furthermore, the puzzling cell detachment response induced by anti-α6 antibodies may also be relevant to carcinoma cells. Learning the nature of these signals and how to manipulate them is a challenge for the future.

VI. FINAL CONSIDERATIONS ON α6β4 AND CARCINOGENESIS

The information summarized in the preceding sections cannot be fit easily into a coherent picture for a role of α6β4 in carcinoma cells. There is no obvious, single feature that is shared by all cells expressing α6β4, whether or not transformed. In some instances, α6β4 is found on nonproliferating, basement membrane bound cells; in other cases, on proliferating, suprabasal cells. Sometimes it is basally polarized, sometimes all around the plasmamembrane of carcinoma cells.

It would then be easy to conclude that there is no direct correlation between expression of α6β4 and aggressiveness of cancer cells. However, the high incidence of deregulation in essentially any type of carcinoma (Table 1) makes the α6β4 integrin a good candidate for a "closely linked variable, more likely to be causative than associative", and perhaps "more likely to contribute directly to disease pathogenesis" (quoted from Reference 81). The problem then becomes to clarify the functional value of expressing α6β4 for transformed epithelial cells.

Postulating a dual functional role for α6β4 in epithelial cells could be one way out of the conflicting evidence that in some carcinomas α6β4 is overexpressed, while in others it is lost (Table 1). The first role would be of a receptor for true adhesion, i.e., dedicated to establishing mechanical connections between the basement membrane and the cytoskeleton. The second role would be of a signaling receptor, which could modulate proliferation or migration. Such a dual role has been shown for several integrins,[82] and is expected to be a general feature of the family. If this turns out to be true also for α6β4, then this integrin is in an ideal position to regulate mechanochemical transduction in epithelial cells, by interpreting signals originating in the basement membrane. For example, α6β4 may repress or relax cell proliferation and/or migration, depending on the local needs of the epithelium self-renewal operations. The status of the α6β4 ligands in the basement membrane may be a trigger for modulating such signals. Thus, wherever there occurs a loss of basement membrane structural integrity, or of the continuity of the epithelial sheet, α6β4 would be poised to consequently regulate local behavior of epithelial cells.

Mechanochemical transduction via α6β4 immediately suggests a hypothesis for its role in carcinomas, which takes into account both upregulation and downregulation in transforming epithelial cells. Thus, the important step in carcinogenesis would be the loss of α6β4 functions, which would free carcinoma cells of several constraints imposed by basement membranes on proliferation and migration. One obvious way to eliminate α6β4 functions is by extinguishing its expression. However, several other ways may achieve the same effect. Crippling mutations in either subunit of α6β4, mutations at locations upstream or downstream in the putative α6β4 signaling pathway, are examples of carcinoma-associated changes that may freeze the receptor in a state of incoherence between its adhesive and signaling capacities. Under these circumstances, overexpression of α6β4 may become an advantageous trait of a transforming cell, subsequently fixed in overt carcinoma.

In essence, this interpretation suggests that α6β4 may in fact be a tumor suppressor protein or be a part of a tumor suppressor pathway. It should not be impossible to address this issue in experimental terms.

ACKNOWLEDGMENTS

This work was supported by NIH grants CA47858 and GM46902 to VQ.

REFERENCES

1. Ruoslahti, E. and Giancotti, F. G., Integrins and tumor cell dissemination, *Cancer Cells,*, 1, 119, 1989.
2. Kajiji, S., Tamura, R. N., and Quaranta, V., A novel integrin ($\alpha_E\beta_4$) from human epithelial cells suggests a fourth family of integrin adhesion receptors, *EMBO J.*, 3, 673, 1989.
3. Cheresh, D. A., Smith, J. W., Cooper, H. M., and Quaranta, V., A novel vitronectin receptor integrin ($\alpha_V\beta_X$) is responsible for distinctive adhesive properties of carcinoma cells, *Cell*, 57, 59, 1989.
4. Sheppard, D., Rozzo, C., Starr, L., Quaranta, V., Erle, D. J., and Pytela, R., Complete amino acid sequence of a novel integrin β subunit (β_6) identified in epithelial cells using the polymerase chain reaction, *J. Biol. Chem.*, 265, 11502, 1990.
5. Tamura, R. N., Rozzo, C., Starr, L., et al., Epithelial integrin $\alpha_6\beta_4$: Complete primary structure of α_6 and variant forms of β_4, *J. Cell Biol.*, 111, 1593, 1990.
6. Cooper, H. M., Tamura, R. N., and Quaranta, V., The major laminin receptor of

mouse embryonic stem cells is a novel isoform of the β6β1 integrin, *J. Cell Biol.*, 115, 843, 1991.

7. Collo, G., Starr, L., and Quaranta, V., A new isoform of the laminin receptor integrin alpha7beta1 is developmentally regulated in skeletal muscle, *J. Biol. Chem.*, 268, 19019, 1993.

8. Liotta, L. A., Steeg, P. S., and Stetler-Stevenson, W. G., Cancer metastasis and angiogenesis: an imbalance of positive and negative regulation, *Cell*, 64, 327, 1991.

9. Kajiji, S., Davceva, B., and Quaranta, V., Six monoclonal antibodies to human pancreatic cancer antigens, *Cancer Res.*, 47, 1367, 1987.

10. Kennel, S. J., Foote, L. J., and Flynn, K. M., Tumor antigen on benign adenomas and on murine lung carcinomas quantitated by a two-site monoclonal antibody assay, *Cancer Res.*, 46, 707, 1986.

11. Kennel, S. J., Foote, L. J., Falcioni, R., Sonnenberg, A., Crouse, C., and Hemler, M. E., Analysis of the tumor-associated antigen TSP-180. Identity with alpha 6-beta 4 in the integrin superfamily, *J. Biol. Chem.*, 264, 15515, 1989.

12. Costantini, R. M., Falcioni, R., Battista, P., et al., Integrin (alpha 6/beta 4) expression in human lung cancer as monitored by specific monoclonal antibodies, *Cancer Res.*, 50, 6107, 1990.

13. Liebert, M., Wedemeyer, G., Stein, J. A., Washington, R. W., Jr., Carey, T. E., and Grossman, H. B., The monoclonal antibody BQ16 identifies the alpha 6 beta 4 on bladder cancer, *Hybridoma*, 12, 67, 1993.

14. Koukoulis, G. K., Virtanen, I., Korhonen, M., Laitinen, L., Quaranta, V., and Gould, V. E., Immunohistochemical localization of integrins in the normal, hyperplastic and neoplastic breast: correlation with their functions as receptors and cell adhesion molecules, *Am. J. Pathol.*, 139, 787, 1991.

15. Koukoulis, G. K., Virtanen, I., Moll, R., Quaranta, V., and Gould, V. E., Immunolocalization of integrins in the normal and neoplastic epithelium, *Virchows Arch. B Cell Pathol.*, 63, 373, 1993.

16. Brodt, P., Fallavollita, L., Sawka, R. J., et al., Tumor cell adhesion to frozen lymph node sections—a correlate of lymphatic metastasis in breast carcinoma models of human and rat origin, *Breast Cancer Res. Treat.*, 17, 109, 1990.

17. Clezardin, P., Frappart, L., Clerget, M., Pechoux, C., and Delmas, P. D., Expression of thrombospondin (TSP1) and its receptors (CD36 and CD51) in normal, hyperplastic, and neoplastic human breast, *Cancer Res.*, 53, 1421, 1993.

18. Coopman, P., Verhasselt, B., Bracke, M., et al., Arrest of MCF-7 cell migration by laminin in vitro: possible mechanisms, *Clin. Exp. Metast.*, 9, 469, 1991.

19. DArdenne, A. J., Richman, P. I., Horton, M. A., Mcaulay, A. E., and Jordan, S., Co-ordinate expression of the alpha-6 integrin laminin receptor sub-unit and laminin in breast cancer, *J. Pathol.*, 165, 213, 1991.

20. Jones, J. L., Critchley, D. R., and Walker, R. A., Alteration of stromal protein and integrin expression in breast—a marker of premalignant change, *J. Pathol.*, 167, 399, 1992.

21. Lazard, D., Sastre, X., Frid, M. G., Glukhova, M. A., Thiery, J. P., and Koteliansky, V. E., Expression of smooth muscle-specific proteins in myoepithelium and stromal myofibroblasts of normal and malignant human breast tissue, *Proc. Natl. Acad. Sci. U.S.A.*, 90, 999, 1993.

22. Mechtersheimer, G., Munk, M., Barth, T., Koretz, K., and Moller, P., Expression of beta 1 integrins in non-neoplastic mammary epithelium, fibroadenoma and carcinoma of the breast, *Virchows Arch. A Pathol. Anat. Histopathol.*, 422, 203, 1993.

23. Natali, P. G., Nicotra, M. R., Botti, C., Mottolese, M., Bigotti, A., and Segatto, O., Changes in expression of alpha 6/beta 4 integrin heterodimer in primary and metastatic breast cancer, *Br. J. Cancer*, 66, 318, 1992.

24. Miettinen, M., Castello, R., Wayner, E., and Schwarting, R., Distribution of VLA integrins in solid tumors. Emergence of tumor-type-related expression. Patterns in carcinomas and sarcomas, *Am. J. Pathol.*, 142, 1009, 1993.

25. Pignatelli, M., Cardillo, M. R., Hanby, A., and Stamp, G. W., Integrins and their accessory adhesion molecules in mammary carcinomas: loss of polarization in poorly differentiated tumors, *Hum. Pathol.*, 23, 1159, 1992.

26. Virtanen, I., Korhonen, M., Kariniemi, A. L., Gould, V. E., Laitinen, L., and Ylanne, J., Integrins in human cells and tumors, *Cell Differ. Dev.*, 32, 215, 1990.

27. Zutter, M. M., Krigman, H. R., and Santoro, S. A., Altered integrin expression in adenocarcinoma of the breast. Analysis by in situ hybridization, *Am. J. Pathol.*, 142, 1439, 1993.

28. Zutter, M. M., Mazoujian, G., and Santoro, S. A., Decreased expression of integrin adhesive protein receptors in adenocarcinoma of the breast, *Am. J. Pathol.*, 137, 863, 1990.

29. Wolf, G. T., Carey, T. E., Schmaltz, S. P., et al., Altered antigen expression predicts outcome in squamous cell carcinoma of the head and neck, *J. Natl. Cancer Inst.*, 82, 1566, 1990.

30. Van Waes, C., Kozarsky, K. F., Warren, A. B., et al., The A9 antigen associated with aggressive human squamous structurally and functionally similar to the newly defined alpha 6 beta 4, *Cancer Res.*, 51, 2395, 1991.

31. Peltonen, J., Larjava, H., Jaakkola, S., et al., Localization of integrin receptors for fibronectin, collagen, and laminin in human skin. Variable expression in basal and squamous cell carcinomas, *J. Clin. Invest.*, 84, 1916, 1989.

32. Pellegrini, G., De Luca, M., Orecchia, G., et al., Expression, topography, and function of integrin receptors are severely altered in keratinocytes from involved and uninvolved psoriatic skin, *J. Clin. Invest.*, 89, 1783, 1992.

33. Hertle, M. D., Kubler, M. D., Leigh, I. M., and Watt, F. M., Aberrant integrin expression during epidermal wound healing and in psoriatic epidermis, *J. Clin. Invest.*, 89, 1892, 1992.

34. Quaranta, V. and Jones, J. C. R., The internal affairs of an integrin, *Trends Cell Biol.*, 1, 2, 1991.

35. Sonnenberg, A., Modderman, P. W., and Hogervorst, F., Laminin receptor on platelets is the integrin VLA-6, *Nature*, 336, 487, 1988.

36. Hogervorst, F., Kuikman, I., vondemBorne, A. E. G. Kr., and Sonnenberg, A., Cloning and sequence analysis of β_4 cDNA: an integrin subunit that contains a unique 118kDa cytoplasmic domain, *EMBO J.*, 9, 765, 1990.

37. Suzuki, S. and Naitoh, Y., Amino acid sequence of a novel integrin β_4 subunit and primary expressio of the mRNA in epithelial cells, *EMBO J.*, 9, 757, 1990.

38. Giancotti, F. G., Stepp, M. A., Suzuki, S., Engvall, E., and Ruoslahti, E., Proteolytic processing of endogenous and recombinant beta 4 integrin subunit, *J. Cell Biol.*, 118, 951, 1992.

39. Quaranta, V., Epithelial integrins, *Cell Differ. Dev.*, 32, 361, 1991.

40. Main, A. L., Harvey, T. S., Baron, M., Boyd, J., and Campbell, I. D., The three-dimensional structure of the tenth type III module of fibronectin: an insight into RGD-mediated interactions, *Cell*, 71, 671, 1992.

41. Leahy, D. J., Hendrickson, W. A., Aukhil, I., and Erickson, H. P., Structure of a fibronectin type III domain from tenascin phased by MAD analysis of the selenomethionyl protein, *Science*, 258, 987, 1992.

42. DeLuca, M., Tamura, R. N., Kajiji, S., et al, Polarized integrin mediates keratinocyte adhesion to basal lamina, *Proc. Natl. Acad. Sci. U.S.A.*, 87, 6888, 1990.

43. Hynes, R. O., Integrins: versatility, modulation and signaling in cell adhesion, *Cell*, 69, 11, 1992.

44. Sonnenberg, A., Daams, H., Van Der Valk, M. A., Hilkens, J., and Hilgers, J., Development of mouse mammary gland: identification of stages in differentiation of luminal and myoepithelial cells using monoclonal antibodies and polyvalent antiserum against keratin, *J. Histochem. Cytochem.*, 34, 1037, 1986.

45. Hemler, M. E., Crouse, C., and Sonnenberg, A., Association of the VLA subunit α_6 with a novel protein. A possible alternative to the common VLA β_1 subunit on certain cell lines, *J. Biol. Chem.*, 264, 6529, 1989.

46. Tamura, R. N., Cooper, H. M., Collo, G., and Quaranta, V., Cell-type specific integrin variants with alternative α chain cytoplasmic domains, *Proc. Natl. Acad. Sci. U.S.A.*, 88, 10183, 1991.

47. Hogervorst, F., Admiraal, L. G., Niessen, C., et al., Biochemical characterization and tissue distribution of the A variants of the integrin alpha 6 subunit, *J. Cell Biol.*, 121, 179, 1993.

48. Weinberg, R., Oncogenes, tumor suppressor genes, and cell transormation: trying to put it all together, in *Origins of Human Cancer. A Comprehensive Review*, Brugge, J., Curran, T., Harlow, E., and McCormick, F., Eds., CSHL Press, New York, 1991, 1.

49. Starr, L. and Quaranta, V., An efficient and reliable method for cloning PCR-amplification products: a survey of point mutations in integrin cDNA, *Biotechniques*, 13, 612, 1992.

50. deCurtis, I., Quaranta, V., Tamura, R. N., and Reichardt, L. F., Laminin receptors in the retina: sequence analysis of the chick integrin α_6 subunit; evidence for transcriptional and posttranslational regulation, *J. Cell Biol.*, 113, 405, 1991.

51. Gunthert, U., Hofmann, M., Rudy, W., et al., A new variant of glycoprotein CD44 confers metastatic potential to rat carcinoma cells, *Cell*, 65, 13, 1991.

52. Leone, A., Flatow, U., King, C. R., et al., Reduced tumor incidence, metastatic potential, and cytokine responsiveness of nm23-transfected melanoma cells, *Cell*, 65, 25, 1991.

53. Arch, R., Wirth, K., Hofmann, M., et al., Participation in normal immune responses of a splice variant of CD44 [see comments], *Science*, 257, 682, 1992.

54. Newman, P. J., Seligsohn, U., Lyman, S., and Coller, B. S., The molecular genetic basis of Glanzmann thrombasthenia in the Iraqi-Jewish and Arab populations in Israel, *Proc. Natl. Acad. Sci. U.S.A.*, 88, 3160, 1991.

55. Yong, K. and Khwaja, A., Leucocyte cellular adhesion molecules, *Blood Rev.*, 4, 211, 1990.

56. O'Toole, T. E., Mandelman, D., Forsyth, J., Shattil, S. J., Plow, E. F., and Ginsberg, M. H., Modulation of the affinity of integrin αIIbβ3 (GPIIb-IIIa) by the cytoplasmic domain of αIIb, *Science*, 254, 845, 1991.

57. Kleinman, H. K., McGarvey, M. L., Liotta, L. A., Gehron-Robey, R., Tryggvason, K., and Martin, G. R., Isolation and characterization of type IV procollagen, laminin, and heparan sulfate proteoglycan from EHS sarcoma, *Biochemistry*, 21, 6188, 1982.

58. Timpl, R., Rohde, H., Robey, P. G., Rennard, S. I., Foidart, J.-M., and Martin, G. R., Laminin—A glycoprotein from basement membranes, *J. Biol. Chem.*, 254, 9933, 1979.

59. Pytela, R., Pierschbacher, M. D., Argraves, W. S., Suzuki, S., and Ruoslahti, E., Arginine-glycine-aspartic acid adhesion receptors, *Methods Enzymol.*, 144, 475, 1987.

60. Sonnenberg, A., Linders, C. J., Modderman, P. W., Damsky, C. H., Aumailley, M., and Timpl, R., Integrin recognition of different cell-binding fragments of laminin (P1, E3, E8) and evidence that alpha 6 beta 1 but not alpha 6 beta 4 functions as a major receptor for fragment E8, *J. Cell Biol.*, 110, 2145, 1990.

61. Lee, E. C., Lotz, M. M., Steele, G. D., Jr. and Mercurio, A. M., The integrin alpha 6 beta 4 is a laminin receptor, *J. Cell Biol.*, 117, 671, 1992.

62. Masumoto, A. and Hemler, M. E., Multiple activation states of VLA-4. Mechanistic differences adhesion to CS1/fibronectin and to vascular cell adhesion, *J. Biol. Chem.*, 268, 228, 1993.

63. Rousselle, P., Lunstrum, G. P., Keene, D. R., and Burgeson, R. E., Kalinin: an epithelium-specific basement membrane adhesion molecule that is a component of anchoring filaments, *J. Cell. Biol.*, 114, 567, 1991.

64. Verrando, P., Hsi, B.-L., Yeh, C.-J., Pisani, A., Serieys, N., and Ortonne, J.-P., Monoclonal antibody GB3, a new probe for the study of human basement membranes and hemidesmosomes, *Exp. Cell Res.*, 170, 116, 1987.

65. Carter, W. G., Ryan, M. C., and Gahr, P. J., Epiligrin, a new cell adhesion ligand for integrin α3β1 in epithelial basement membranes, *Cell*, 65, 599, 1991.

66. Riddelle, K. S., Green, K. J., and Jones, J. C. R., Formation of hemidesmosomes *in vitro* by a transformed rat bladder cell line, *J. Cell Biol.*, 112, 159, 1991.

67. Jones, J. C. R., Kurpakus, M. A., Cooper, H. M., and Quaranta, V., A function for the integrin α6β4 in the hemidesmosome, *Cell Regul.*, 2, 427, 1991.

68. Langhofer, M., Hopkinson, S. B., and Jones, J. C. R., The matrix secreted by 804G cells contains laminin related components that participate in hemidesmosome assembly in vitro, *J. Cell Sci.*, in press.

69. Kallunki, P., Sainio, K., Eddy, R., et al., A truncated laminin chain homologous to the B2 chain: structure, spatial expression, and chromosomal assignment, *J. Cell Biol.*, 119, 679, 1992.

70. Staehelin, L. A., Structure and function of intercellular junctions, *Int. Rev. Cytol.*, 39, 191, 1974.

71. Geiger, B., Involvement of vinculin in contact-induced cytoskeletal interactions, *Cold Spring Harbor Symp. Quant. Biol.*, XLVI(Abstr.), 671, 1981.

72. Burridge, K., Molony, L., and Kelly, T., Adhesion plaques: sites of transmembrane interaction between the extracellular matrix and the actin cytoskeleton, *J. Cell Sci. Suppl.*, 8, 211, 1987.

73. Gipson, I. K., Spurr-Michaud, S., Tisdale, A., and Keough, M., Reassembly of the anchoring structures of the corneal epithelium during wound repair in the rabbit, *Invest. Ophthal. Vis. Sci.*, 30(Abstr.), 425, 1989.

74. Sakai, L. Y., Keene, D. R., Morris, N. P., and Burgeson, R. E., Type VII collagen is a major structural component of anchoring fibrils, *J. Cell Biol.*, 103, 1577, 1986.

75. Klatte, D. H., Kurpakus, M. A., Grelling, K. A., and Jones, J. C. R., Immunochemical characterization of three components of the hemidesmosome and their expression in cultured epithelial cells, *J. Cell Biol.*, 109, 3377, 1989.

76. Sonnenberg, A., Calafat, J., Janssen, H., et al., Integrin alpha 6/beta 4 complex is located in hemidesmosomes, suggesting a major role in epidermal cell-basement membrane adhesion, *J. Cell Biol.*, 113, 907, 1991.

77. Stepp, M. A., Spurrmichaud, S., Tisdale, A., Elwell, J., and Gipson, I. K., Alpha-6-beta-4 integrin heterodimer is a component of hemidesmosomes, *Proc. Natl. Acad. Sci. U.S.A.*, 87, 8970, 1990.

78. Carter, W. G., Kaur, P., Gil, S. G., Gahr, P. J., and Wayner, E. A., Distinct functions for integrins α3β1 in focal adhesions and α6β4/bullous pemphigoid antigen in a new stable anchoring contact (SAC) of keratinocytes: relation to hemidesmosomes, *J. Cell Biol.*, 111, 3141, 1990.

79. Kurpakus, M. A., Quaranta, V., and Jones, J. C. R., Surface relocation of α6β4 integrins and assembly of hemidesmosomes in an *in vitro* model of wound healing, *J. Cell Biol.*, 115, 1737, 1991.

80. Kurpakus, M. A., Stock, E. L., and Jones, J. C. R., Analysis of wound healing in

an *in vitro* model: early appearance of laminin and a $125 \times 10^3 M_r$ polypeptide during adhesion complex formation, *J. Cell Sci.*, 96(Abstr.), 651, 1990.

81. Lippman, M. E., The development of biological therapies for breast cancer, *Science*, 259, 631, 1993.

82. Schwarz, M. A., Signaling by integrins: implications for tumorigenicity, *Cancer Res.*, 53, 1503, 1993.

Chapter 9

Internalization of Microbial Pathogens by Integrin Receptors and the Binding of the *Yersinia pseudotuberculosis* Invasin Protein

Ralph R. Isberg

CONTENTS

I. INTRODUCTION

During the course of an infectious disease, most microbial pathogens interact with host tissues in some fashion. This interaction can be of a direct nature, by binding receptors found on host cells, or by binding secreted components that may either immobilize the microorganism at an extracellular site or promote adhesion to host cells. The ultimate site of localization during the infection of the microorganism is most probably determined by the particular array of surface molecules found on the microbial cell surface.[1] An internal or external localization of the microorganism relative to host cells is in large part determined by the particular receptors to which microbial surface components bind.[2] Furthermore, there is some evidence that the survival of a microorganism within a particular cell type may depend on the particular receptor bound.[3,4]

The interaction of microbial pathogens with integrin receptors will be analyzed, with particular focus on the *Yersinia pseudotuberculosis* invasin protein. As will be seen, the encounter between the microorganism and integrin leads to a variety of results, many of which are advantageous to the invading microbe. These include the binding of the microbe on the external surface of mammalian cells, internalization by normally nonphagocytic cells, as well as survival of the microorganism in the normally hostile environment of the macrophage. In some of these cases, the integrin receptor seems well suited to allowing the pathogen to establish a niche. In other instances, the interaction with the particular integrins chosen by the microbe seems difficult to understand, or counterproductive to the infection. It is from analyses of these latter events in particular that we may get the most insight, since they uncover novel features of the infectious process that are not readily apparent on casual observation.

II. MICROBIAL BINDING OF EXTRACELLULAR MATRIX COMPONENTS

The earliest evidence that microorganisms could interact at an indirect level with integrin receptors came from studies demonstrating that a variety of microbial pathogens bind extracellular matrix components, such as fibronectin,[5] which in turn bind integrins. The *in vivo* significance of this interaction is not clear, although the fact that a wide variety of microorganisms possess fibronectin receptors makes it seem likely that they play important roles in the infection process.[6-12] Thus far, there are three proposed roles for binding to extracellular matrix components. The first potential role is for the microorganism to actually bind and colonize within the extracellular matrix or collagenous tissues such as cartiledge.[6,8-12] In this case, the interaction with host cells is quite indirect, and the goal of the microorganism may be to establish a niche that is poorly accessible to phagocytic cells. The second potential role is that the coating of the microorganism by components such as fibronectin is a tactic that facilitates adhesion to host cell integrins.[13] This event may allow the microorganism to adhere to phagocytic cells in a fashion that favors cellular adhesion of the parasite without subsequent efficient phagocytosis by the host cell. Finally, the binding of integrin receptors by extracellular matrix components could enhance the efficiency of phagocytosis[14,15] in an event that is not the result of direct opsonization of the microorganism.

A number of microorganisms, including the yeast *Candida albicans*, *Treponema pallidum* (the causative agent of syphilis), pathogenic *Escherichia coli*, *Streptococcus pyogenes*, and *Staphylococcus aureus* have receptors for fibronectin.[16-20] The latter organism is particularly interesting in this regard, because, in addition to binding fibronectin, *S. aureus* apparently has high affinity receptors for collagen, vitronectin, and fibrinogen, all extracellular matrix proteins that could serve as links to integrin receptors.[6] The interaction of fibronectin with these microorganisms offers some contrast in styles as well as in the nature of the factors promoting binding.

Staphylococcus aureus binds with a high affinity to the 29-kDa amino terminal portion of fibronectin,[21] a region of this ECM protein that appears to be involved in binding a number of other bacterial species.[8] This fibronectin binding protein has a structure typical of other membrane proteins found in Gram-positive bacteria, although it has a distinct 38-amino acid repeat that is apparently necessary and sufficient to bind the amino terminal fragment of fibronectin.[22] Synthetic peptides corresponding to one of these repeats are able to competitively inhibit binding to the fibronectin receptor,[23] and mutant derivatives lacking the repeats are defective for binding.[22] It is of interest that the involvement of repeated sequence in binding ECM components appears to be a common theme among Gram-positive bacterial species. For instance, *Streptococcal* species encode a fibronectin receptor that contains a series of repeats, although there is little apparent homology between these repeats and those found in the *Staphylococcal* fibronectin binding protein.[20]

The amino terminal heparin-binding domain of fibronectin is also able to adhere to a variety of *E. coli* strains,[17,18,24] but the bacterial structure involved appears to be quite different from that found in the Gram-positive cocci. Binding to *E. coli* occurs via bundles of distinct surface structures that bear some resemblance to bacterial pili.[18] The fine and twisted appearance of these organelles in the electron microscope has earned them the name of *curli*, which are made of repeated monomers of the curlin subunit.[18] The expression of these filamentous structures is under very tight thermoregulatory control, and since most *E. coli* isolates have the ability to express curli, it appears that there is some selective pressure to maintain fibronectin binding *in vivo*. The potential *in vivo* roles will be discussed below.

Unlike the examples discussed above, the spirochete *Treponema pallidum* appears to coat itself with fibronectin by binding to the cell adhesion domain of this ECM protein.[16] This was demonstrated by showing that proteolytic fragments encompassing the cell adhesion domain bind the bacterium, and that synthetic peptides containing the tripeptide sequence arg-gly-asp (RGD) inhibit adhesion to fibronectin. The multimeric nature of soluble fibronectin allows the spirochete bacterium to coat itself via this domain and still maintain the ability to bind mammalian cells.[25,26] It has been reported that several spirochete-encoded proteins are able to bind fibronectin, although recent data on the nature of the bacterial cell surface call these results into question. It is now believed that there are no bacterial-encoded proteins exposed on the spirochete surface; so, fibronectin must be binding some lipid or carbohydrate component. This result is even more unusual given the fact that RGD peptides seem to inhibit the interaction. Perhaps the charge of the peptide interferes with binding, or else there is some nonproteinaceous structure that is able to mimic the site recognized by fibronectin in integrin receptors.

The bacteria cited above utilize adhesion to ECM components for somewhat unrelated reasons. For Gram-positive cocci, such as *S. aureus* and *Streptococci*, it can be argued that ECM binding is irrelevant to microbial adhesion to host cells and, furthermore, is actually a tactic that allows the microbe to avoid undesired contact with cells such as phagocytes. Invasive diseases by *S. aureus* are associated with abscess formation and colonization of fibrin-rich deposits in the host. Within these sites of heavy ECM deposition, the bacteria can become entrapped and find a niche protected from immune-reactive cells. An additional site of colonization of these microorganisms is within joint tissue, especially in the presence of prosthetic devices.[6] The microorganisms could colonize within cartiledge, in such cases, by adhering to ECM components in these sites. A role for the *S. aureus* collagen receptor has been proposed in this type of colonization, and experimental support exists for this hypothesis.[27]

It seems likely that the pathogenic *S. pyogenes* also employs receptors for extracellular matrix proteins to facilitate invasive disease in disparate tissue sites, as well as for the colonization of in-dwelling medical devices. In the case of oral *Streptococci*, however, it is more likely that adhesion to ECM components is critical to maintain cellular adherence in the mouth, the normal site of colonization. The oral mucous layer contains considerable amounts of fibronectin, and adhesion to this ECM component would be expected to promote indirect cellular adhesion of the bacteria to integrin receptors. This is supported by studies of a *Streptococcus* derivative that has been mutated by insertion of a drug-resistant element in the gene for fibronectin receptor. In such strains, adhesion to host cells is totally eliminated, indicating that the *Streptococcal* fibronectin receptor is the primary strategy used for cellular adhesion by the microorganism (M. Caparon, personal communication).

The role for the treponemal fibronectin receptor also seems to be to use fibronectin as a bridge between host cell integrin receptor and the microorganism. *T. pallidum* appears to colonize extracellularly in subendothelial regions. In order to move to these sites, the microorganism must penetrate across a layer of cells, either by transcytosis or by moving in between host cells. Evidence exists that the movement of *T. pallidum* to subendothelial sites is analogous to the migration of neutrophils between endothelial cells.[28] Using polarized human umbilical vein endothelial cells (HUVEC) cultured on porous filters, it was shown that the microorganism was able to attach to lateral regions of these cells and migrate through the filter after moving in between the polarized cells.[29]

The role of fibronectin receptors in colonization of *E. coli* is somewhat more difficult to explain than for the other above-described microorganisms. If the organism is using curlin to colonize the intestine, it suggests that there may be some fibronectin secreted on the luminal side of the colonic mucous layer. The problem then arises in determining

Table 1 **Examples of binding of microorganisms to extracellular matrix components**

Organism	Extracellular Matrix Protein
S. aureus	Fibronectin, collagen, vitronectin
	Fibrinogen
Streptococal species	Fibronectin
E. coli	Fibronectin
Y. enterocolitica	Collagen, fibronectin
T. palladum	Fibronectin
C. albicans	Fibronectin

whether host cells are immobilizing the fibronectin, or if the protein is merely present in a mucous meshwork that entraps the bacterium. The fact that integrin receptors are, by and large, located at the basal region of intestinal epithelial cells suggests that the latter possibility is more likely. This model is not entirely satisfactory, however, since not all intestinal epithelial cells have identical integrin localization properties;[30] a recently described integrin appears to have apical orientation, and the localization of integrins in certain cell types in the intestine is still unknown (see below). Perhaps more critical to this discussion is whether, in fact, the bacterial fibronectin receptor plays a role in *E. coli* extraintestinal infections rather than intestinal colonization. Infection of mice by encapsulated *E. coli* strains indicates that *oral* colonization of the bacterium is a critical prerequisite for infantile meningitis caused by this bacterium.[31] Perhaps the fibronectin receptor plays a role in *E. coli* meningitis that is identical to that described above for the streptococcal fibronectin receptor. Similarly, uropathogenic *E. Coli* must colonize bladder and kidney tissue to establish disease.[32]

The ability to bind fibronectin may be important for invasion of the microorganism into these tissue sites.

III. MICROBIAL INTERACTION WITH INTEGRINS FOUND ON PHAGOCYTIC CELLS

Integrins are associated with three major aspects of microbial uptake by phagocytic cells. First, host cell proteins that are ligands for integrins may coat a microorganism, resulting in opsonization. Second, many organisms encode surface-localized factors that apparently mimic host-encoded ligands, and these are directly recognized by integrin receptors found on phagocytic cells prior to uptake. Finally, integrin-mediated adherence of either monocytes or polymorphonuclear leukocytes (PMN) to surfaces coated with ECM proteins causes activation of the phagocyte, resulting in enhanced levels of binding and phagocytosis of microorganisms, as well as stimulation of a variety of pathways involved in antimicrobial tactics.[33,34]

A. SURFACE COATING

Multiple members of the β2 chain family of integrins are able to bind microorganisms coated with complement component C3bi. Opsonic coating by C3bi can take place either via the alternative and classical pathways of complement fixation on the microbial surface or via the presentation of a specific macromolecule on the surface of the microorganism that binds directly to C3bi.[35] The former tactic is an important pathway for clearing the host of invading microorganism and will not be covered here. The latter strategy is used by a variety of pathogenic microorganisms and is usually associated with a phagocytosis pathway associated with intracellular survival of the microorganism.

It is apparent that coating of the microorganism with C3b components is an important survival strategy for many microorganisms, because it allows an important cell-mediated killing pathway to be bypassed. The initial step after the phagocyte encounters the microbe is for the phagocyte to release toxic oxygen radicals which are generated during the respiratory burst and fusion of primary granules with the phagocytic membrane.[36]

The release of these toxic compounds only requires adhesion of the microorganism to the phagocyte surface and is not necessarily associated with microbial internalization. Most microorganisms are highly susceptible to this killing pathway, so using a strategy that bypasses the release of compounds causing oxygen-dependent killing is critical for surviving the encounter with a phagocyte. Receptors that bind C3b and C3bi somehow allow this pathway to be bypassed. Wright and Silverstein showed that erythrocyte surfaces coated by C3b (E-C3b) or C3bi (E-C3bi) could bind monocytes or neutrophils, but were unable to induce the release of H_2O_2.[4] In contrast, erythrocytes coated by IgG (E-IgG) generated a vigorous release of H_2O_2. The primary receptors involved in binding the EC3b and EC3bi particles are complement receptors CR1 and CR3 (Mac-1; $\alpha mac\beta_2$), the latter of which is an integrin.[37] Therefore, binding to an integrin allows the microorganism to uncouple uptake from release of toxic compounds.

One such organism that takes advantage of this pathway is *Leishmania mexicana*. *Leishmania* promastigates are internalized by macrophages shortly after inoculation of the host by a sandfly.[38] The microorganism grows within a phagolysosome in these cells, where it is apparently resistant to killing by lysosomal components.[39] The organism, however, is sensitive to oxidative killing pathways.[39] Attachment of the microorganism to phagocytes results in a vigorous respiratory burst and loss of viability of the microorganism. In contrast, parasites coated with C3b are internalized primarily via $\alpha m\beta 2$,[40-42] with the result that no such burst takes place, and efficient intracellular growth of the parasite follows. This demonstrates that the particular receptor chosen by an intracellular parasite is an important factor in determining the ultimate fate of the microorganism after encounter with a phagocytic cell, and that the $\alpha m\beta 2$ integrin belongs to a set of receptors which allows adhesion of the microorganism to be uncoupled from the respiratory burst.

The alternative pathway of complement activation allows deposition of C3b on the surface of *Leishmania*, and purified soluble C3 is able to bind the promastigote directly.[43] Co-immunoprecipitation experiments in which detergent-solubilized parasites exposed to purified C3 are incubated in the presence of anti-C3 serum revealed that the predominant associated *Leishmania* protein was gp63. gp63 is a multifunctional glycoprotein that is the most abundant protein found on the surface of the parasite. In addition to binding C3b, it is an endoprotease having considerable sequence-specificity and, as will be discussed below, is also able to bind directly the $\alpha m\beta 2$.[44] There is some disagreement as to whether the C3b-binding activity or the integrin-binding activity is the relevant property for pathogenesis of this microorganism, but it is agreed that the normal route for leishmanial uptake involves binding of complement receptors one way or the other by gp63.

Bacterial pathogens such as *Legionella pneumophila* are also capable of taking advantage of the strategy of complement coating. Serum coating of the bacterium stimulates phagocytic uptake, and this effect is apparently due to coating by complement component C3b.[45] *L. pneumophila* is unable to bind phagocytes plated on coverslips coated with C3b or with a combination of antibodies directed against the complement receptors CR1 and $\alpha m\beta 2$ integrin, presumably because macrophage adhesion to the coverslips under these conditions results in a net deficit of receptors available for binding bacteria.[45] Binding of the microorganism to these receptors is required for intracellular growth, since antibody directed against CR1 and αmβ2 integrin prevents phagocytosis and growth within monocytes. As is the case with *Leishmania, L. pneumophila* apparently has a specific protein on its surface that is able to bind C3b and fix it to its surface. This was demonstrated by showing a direct physical interaction between the major outer membrane protein (MOMP) and C3b.[46]

It should be noted that although *Leishmania* and *L. pneumophila* uptake appears to occur via the same receptors, the two organisms cause different diseases, and they have contrasting styles of replication within the phagocytic cell. *L. pneumophila* causes a pneumonic disease as the result of specific growth within the alveolar macrophages of the host.[47,48] After uptake into the cell via a process aptly described as "coiling phagocytosis", the organism establishes a sequestered phagosome that bypasses fusion with the lysosomal network.[49,50] Once replication is established, a variety of organelles and smooth vesicles are recruited to this site, referred to as a "replicative phagosome".[49] This site of replication is very different from the fused phagolysosome seen with *Leishmania* parasites. The fact that the two microorganisms use different growth strategies does not diminish the fact that at early stages in the phagocytic process, before their respective sites of replication are established, they encounter similar problems. As stated above, the phagocytic uptake process is potentially antimicrobial, and complement coating seems to be a common mechanism for bypassing this important host defense, no matter what strategy the intracellular pathogen ultimately uses to grow within the cell.

B. MICROBIAL SURFACE PROTEINS BINDING PHAGOCYTE INTEGRINS

Complement coating of microorganisms clearly has its disadvantages as well as its advantages. Deposition of C3 on the surface of the microorganism is a first step in complement fixation, with potential for release of a variety of modulators that draw immune-reactive cells to the site of infection, as well as for the initiation of the pore-forming cascade. A more efficient route would be for the invading organism to actually synthesize and localize on its surface a factor that directly binds receptors that might bypass host killing mechanisms. The first realization that microorganisms could directly bind integrin receptors on phagocytes was the work of Lachmann and co-workers, who demonstrated that yeast could bind directly to the αmβ2 integrin without opsonization.[51,52] Significantly, this interaction was inhibited by N-acetyl glucosamine, and could be reproduced by incubating macrophages with cell wall material containing mostly polysaccharide and no protein. This indicates that αmβ2 integrin has a lectin-like activity, which is an often overlooked property of this receptor.

Evidence from *Leishmania mexicana* and the pathogenic yeast *Histoplasma capsulatum* indicates that the lectin-like activity may be shared by more than one integrin containing the β2 chain. In the case of *Leishmania*, macrophages and monocytes are able to bind to surfaces coated with abundant promastigote glycolipid lipophosphoglycan (LPG). Macrophages isolated from patients suffering from leukocyte adhesion deficiency (LAD), who are defective for synthesis of functional β2 chain integrins, are unable to bind LPG.[53] Monoclonal antibodies directed specifically against either αmβ2, α150β2, or β2 chain alone were able to inhibit this binding, implying that multiple receptors are able to recognize the glycolipid. Peptides containing the tripeptide sequence arg-gly-asp, which normally inhibit the binding of complement to αmβ2, had no effect on binding LPG, indicating that the lectin site on the integrin receptor is distinct from the complement binding site.[53] Similar results have been obtained for both the binding of *Histoplasma capsulatum* and *Escherichia coli*.[54-56]

In the latter case, the surface molecule recognized by β2 chain integrins appears to be the core saccharide moiety of lipopolysaccharide.[57,58]

Binding of a microorganism to the entire family of β2 chain integrins seems to facilitate a strategy of evasion similar to that described above for coating by C3b. This was illustrated by analyzing the interaction of *Histoplasma capsulatum* with human neutrophils.[54,59,60] Antibodies against any single β2 chain integrin member caused partial blocking of yeast adhesion to human macrophages. On the other hand, much more significant blocking of adhesion takes place if an antibody directed to the entire β2 chain integrin family is used. Uptake of *Histoplasma* into macrophages via this pathway is associated with a total lack of secretion of superoxide anion (O_2^-).[55] This is reminiscent of the lack of H_2O_2 production after adhesion of C3-coated particles to monocytes,[4] and seems to be an important property for an organism that is known to grow within alveolar macrophages in infected hosts.

In addition to the lectin activity of the leukocyte integrins, at least one β2 integrin is able to specifically recognize microbial surface proteins. *Bordetella pertussis*, the causative agent of the respiratory tract disease whooping cough, is able to bind phagocytic cells.[61,62] This binding requires the presence of filamentous hemagglutinin, a protein encoded by the bacterium that forms a flocculent structure on its surface. It is possible to partially block this adhesion by galactose, but, unlike the above examples, it appears that it is the bacterially encoded protein responsible for lectin activity.[63] Much more striking blocking of adhesion takes place, however, when an antibody directed against the αmβ2 integrin is included during the incubation of bacteria with phagocytic cells. Two pieces of evidence indicate that the site on this integrin recognized by FHA is identical to that bound by arg-gly-asp (RGD). FHA contains two RGD sequences, one of which is critical for binding the integrin. This was demonstrated by showing that site-specific mutations in one of the RGD sequences could prevent attachment of the organism to human macrophages. Secondly, peptides corresponding to this sequence in FHA could block adhesion of the microorganism.[63] The role of binding αmβ2 in the disease process is totally unclear, however. In animal models, mutants defective in FHA are unable to associate with macrophages, but this has no consequences on the disease process.[64] It is possible that this binding is a tactic that allows targeting of the microorganism to neutrophils. This is advantageous to the microorganism, since the bacterium secretes a number of potent toxins, and binding to αmacβ2 could allow a concentrated attack on an important cell involved in the first line of host defense.

As is true of *B. pertussis*, *Leishmania mexicana* encodes a protein that binds αmβ2, and, surprisingly, it is the multifunctional gp63 surface protein that also has C3 binding activity.[65] Beads coated with gp63 bind phagocytes in the absence of serum, and this binding is blocked by RGD-containing peptides.[44] the adhesion activity is distinct from that seen with C3, because blocking antibody directed against αmβ2 integrin is sufficient to prevent binding of the gp63-coated particles, whereas such antibodies are not sufficient to block C3-mediated adhesion to its multiple receptors. Interestingly, gp63 seems to have great structural similarity to C3, since an antibody directed against a synthetic peptide of the *Leishmanial* surface protein cross-reacts with the α chain of the human protein.[66]

C. INTEGRIN-MEDIATED ADHESION TO ECM AND ACTIVATION OF PHAGOCYTOSIS

Resting monocytes and unstimulated PMNs are relatively inefficient at phagocytosis of particles via the αmβ2 integrin, Fc and CR1 receptors.[33] Uptake is stimulated, however, by a variety of agonists such as phorbol myristate acetate[67,68] and fmet-leu-phe. Activation occurs rapidly and without *de novo* protein synthesis.[67] The result of this activation is that the avidity of the receptors for substrate is increased, and this causes an increase in the efficiency of phagocytosis. Thus, phagocytosis is a highly regulated process that requires external signals for successful internalization of particles.

The activation of receptors involved in phagocytosis can occur following adhesion of the phagocyte to any of a number of common integrin substrates, such as fibronectin, vitronectin, fibrinogen, and collagen.[34,68,69] Consistent with this observation, activation by adhesion is blocked by the presence of soluble peptides containing the RGD sequence, while activation can also occur if cells are allowed to adhere to a solid matrix coated with proteins modified to contain multiple copies of the RGD sequence.[70-72] It has been proposed that a particular β3 chain integrin is the adhesion molecule involved in causing the signal to be sent in the activation process.[71] The myriad of adhesive substrates capable of activating phagocytosis indicates either that this is a rather promiscuous receptor, or that multiple adhesion receptors must be capable of mediating this activation process. A recent report suggests that integrins may be physically associated with other proteins exposed on the cell surface, and that coupling between two factors is necessary for activation to occur.[69] The evidence for this is based on the fact that a MAb directed against a 50-kDa cell surface protein is capable of blocking activation of phagocytes after cellular adhesion to a variety of integrin substrates.[69]

From the point of view of the host, utilizing cell adhesion as an activation signal for phagocytosis is a high efficient strategy for dealing with infectious agents. In response to infection, a circulating PMN will migrate through the endothelium to the site of microbial invasion.[73] In this subendothelial region, the phagocyte is exposed to ECM components. Amplification of this signal is facilitated because there amy be deposition of ECM components, particularly fibrinogen and fibronectin, in response to microbial replication. In this fashion, the phagocyte has received a signal informing it of its arrival at a site of infection, causing the leukocyte to be retained at this location and activated to engulf microorganisms. The actual signal pathway involved in this process is unclear, but a recent tantalizing report indicates that production of a novel lipid occurs after activation, and this factor directly binds leukocyte integrins, causing an increase in avidity for substrates, and presumably increased phagocytosis.[74]

IV. MICROBIAL INTERACTION WITH INTEGRINS FOUND ON NONPHAGOCYTIC CELLS

The association of cell adhesion molecules and integrins with phagocytosis by neutrophils and macrophages has been known for a number of years. It is now apparent that intracellular microorganisms may associate with integrins found on a variety of other cells, including lymphocytes and epithelial cells, by synthesizing ligands that directly bind the host cell receptor. So far, three organisms have been shown to adhere to such integrins, two of them bacterial in origin and the third viral. In cultured cells, binding of the microorganism to the integrin can lead to uptake into a membrane-bound phagosome, similar to what is seen with phagocytic cells, and recent studies indicate that the mechanism of phagocytosis may be very similar among the various cell types.

A. THE YERSINIA INVASIN PROTEIN AND UPTAKE INTO NONPHAGOCYTIC CELLS

The enteropathogenic *Yersinia* initiate disease after ingestion of contaminated foodstuffs. Starting from the lumen of the terminal ileum, the microorganism assumes the intracellular phase of the infection process by entering into cells found in the intestinal epithelial layer of the host, followed by uptake into lymphocytes and phagocytes found in submucosal regions.[75] The organisms then drain to regional lymph nodes, the liver, and the spleen, all of which support growth of the bacterium, predominantly in an extracellular mode.[76] The ability to convert from an intracellular to an extracellular mode of replication appears to be tightly regulated and is controlled at the level of synthesis of a variety of bacterial-encoded factors that either stimulate or inhibit uptake into host cells.[77,78]

Three different factors encoded by enteropathogenic *Yersinia* are able to promote internalization of the bacterium into normally nonphagocytic cultured cells,[78] one of which is able to directly bind integrin receptors.[79] The first factor identified that was able to promote bacterial uptake was the protein invasin, the 986-amino acid product of the *Y. pseudotuberculosis inv* gene, localized in the bacterial outer membrane.[80] This factor was originally identified by selecting for molecular clones, derived from *Y. pseudotuberculosis* chromosomal DNA, that were able to confer on the normally innocuous *E. coli* K12 strain the ability to be internalized by cultured cells.[81] Invasin is able to bind mammalian cells. This was shown by demonstrating that mammalian cells were able to adhere and spread on solid supports containing the immobilized protein. Monoclonal antibodies directed against invasin that block this binding event are also able to block uptake of bacteria into mammalian cells, indicating that binding of the bacterium via this protein is required for internalization.[82]

Several lines of evidence indicate that invasin is a relatively passive partner in the bacterial uptake process, with its role primarily being to bind a mammalian cell receptor prior to internalization of the microorganism. Analysis of hybrid proteins shows that the carboxyl terminal 192 amino acids of invasin are necessary and sufficient for the protein to bind mammalian cells.[83] This binding domain also is sufficient to promote uptake of particles. Latex beads coated with invasin are able to be internalized, and *Staphylococcus aureus* coated with a hybrid protein containing the carboxyl-terminal 192 amino acids of invasin is efficiently internalized by cultured cells.[84] Therefore, it appears that mere binding of invasin to the appropriate receptor is sufficient to promote uptake, and that the mammalian cell is the active partner in the internalization process.

Uptake of *E. coli* strains expressing invasin on their surface requires binding to an integrin receptor, since a variety of anti-integrin monoclonal antibodies that block the binding of invasin also block uptake of the microorganism.[79] The ability of invasin to bind multiple members of the integrin superfamily of cell adhesion molecules appears consistent with the hypothesis that invasin need not play an active role in uptake. Attachment of a mammalian cell to a surface coated with an integrin substrate is often coordinated with spreading of the cell on the solid surface.[85] This spreading is topologically similar to the movement of the mammalian cell surface about the circumference of the bacterium during phagocytosis.[86] At a first approximation, thus, uptake of the microorganism can be viewed as a consequence of simply taking advantage of the normal host cell processes of spreading and migration, with the bacterium, rather than the extracellular matrix, becoming the substratum. This useful model is somewhat of an oversimplification, since important physical factors play a role in determining whether binding of a microorganism to an integrin results in simple surface adhesion or internalization by the mammalian cell (see Section V, below).

Invasin is able to bind five integrins, each of which contain the β1 chain.[79] This was shown by performing affinity chromatography on a variety of cell lines and tissue types, using a carboxyl-terminal fragment of invasin as the affinity matrix.[79] The particular integrins identified in this fashion primarily function as receptors for extracellular matrix components such as fibronectin, laminin, and collagen, although one of the receptors, the α4β1 integrin, is involved in binding of T lymphocytes to inflamed endothelial cells that express the surface protein VCAM-1.[87,88] This distribution of receptors bound by invasin does not suggest a particular tissue tropism for the microorganism, nor is it obvious whether the choice of receptors directs the microorganism toward a favored route in the host, since almost all organs contain cells that have at least one invasin receptor. On the other hand, the localization of the identified invasin receptors on polarized cells, such as those found in epithelial layers, does appear to have a significant role in determining the pathway of infection of *Y. pseudotuberculosis*. Localization of β1 chain integrins on epithelial cells is usually limited to the basal and lateral surfaces,[30]

Table 2 Binding of microorganisms to integrins encoded by phagocytic cells. Shown are organisms known to bind specific integrins on phagocytic cells and the putative ligands that allow attachment. CR1, although not an integrin, is included to illustrate that particular ligands can bind to more than one receptor

Organism	Receptor	Ligand	Reference
Bordetella pertussis	αmβ2	FHA	63
E. coli	αmβ2, αLβ2, α150β2	LPS	57
H. capsulatum	αmβ2, αLFA1β2, α150β2	?	56
L. pneumophila	αmβ2, CR1	C3b,C3bi	45,46
L. mexicana	αmβ2, CR1	C3b,C3bi	35
L. mexicana	αmβ2	gp63	65
L. mexicana	αmβ2, αLFA1β2, α150β2	LPG	53
Yersinia species	α4β1, α5β1	Invasin	79

as would be expected for receptors involved in adherence to the extracellular matrix. This localization is probably responsible for the choice of cells involved in the penetration of the microorganism into intestinal submucosal regions, as well as the localization of the organism after this penetration occurs. By and large, the bacterium resident in the terminal ileum is unable to bind the lumenal surface of most epithelial cells, choosing instead to interact with a small population of cells that couple the epithelial layer with lymphoid tissue (see Section VI).

From the diversity of receptors that bind invasin, it might be supposed that this protein is relatively promiscuous in its integrin binding. Several lines of evidence suggest that this is not the case. As far as can be discerned, invasin only binds integrins containing the β1 chain. Cell lines that fail to express the β1 chain do not bind invasin,[90] and affinity purification of receptors from cell lines that encode integrins with the β2, β3, and β5 chains never allows isolation of heterodimers having one of these chains. Furthermore, comparison of the behavior of heterodimers having the identical α chain, but different β chains, is consistent with the importance of the β1 chain in binding

Table 3 Human β1 chain integrins and biochemically identified invasin receptors. Displayed are different cell types used to identify invasin receptors by affinity chromatography, the known receptor spectrum from each line, and the receptors isolated. List includes only receptors that have been purified by affinity chromatography on resins linked to invasin

Cell	β1 Chain Integrins	Invasin Receptors
HEp-2 cells	α2β1, α3β1, α5β1	α3β1, α5β1
EJ cells	α2β1, α3β1, α5β1	α3β1, α5β1
K562 cells	α5β1	α5β1
JAR cells	α5β1, others?	α5β1
HPB-MLT	α4β1	α4β1
Platelets	α2β1, α5β1, α6β1	α5β1, α6β1
Placenta	α2β1, α3β1, α5β1	α5β1
IMR32	α1β1, αvβ1	αvβ1

invasin. For instance, the $\alpha4\beta1$ and $\alpha6\beta1$ heterodimers bind invasin, whereas the $\alpha4\beta7$ and $\alpha6\beta4$ integrins do not.[90] On the other hand, it does not appear that invasin is binding directly to the $\beta1$ chain or indiscriminately binding to any heterodimer containing this chain. The isolated $\beta1$ chain is unable to bind invasin, and some $\alpha\beta1$ heterodimers are similarly unable to bind invasin.[79,91] It is significant that the two $\alpha\beta1$ heterodimers known not to bind invasin, $\alpha1\beta1$ and $\alpha2\beta1$, have an added sequence determinant inserted in the α chain called an I-domain, which is located near the divalent cation binding repeat sequence.[92] This domain, which is found in a number of other integrin α chains as well as a variety of serum proteins, may cause structural changes that interfere with the ability of invasin to recognize a site on the heterodimer.[93]

B. OTHER PATHOGENS THAT INTERACT WITH INTEGRINS ON NONPHAGOCYTIC CELLS

At least two pathogens in addition to enteropathogenic *Yersinia* encode proteins that directly bind integrins found on normally nonphagocytic cells. *Bordetella pertussis* encodes a protein called pertactin that allows binding to unidentified integrins found on cultured nonphagocytic cells.[94] This 65-kDa protein is a major outer membrane porin that contains an RGD sequence that is surface exposed. Mammalian cells can adhere to solid substrates coated with purified pertactin, and this binding is inhibited by peptides corresponding to the region of FHA containing the RGD sequence. Bacteria coated with pertactin can be internalized by cultured cells, although the efficiency of uptake is considerably lower than that found with invasin.[95] The reason for this difference will be discussed in the following section.

The most recent pathogen shown to bind an integrin is the small RNA genome echovirus. Monoclonal antibodies were raised against a cell line that was sensitive to echovirus, and hybridomas were selected that secreted antibodies able to block echovirus uptake. The antibodies raised in this fashion were able to immuneprecipitate a heterodimeric surface protein with physical properties of a member of the integrin family.[96] Immune depletion experiments indicated that the receptor was the $\alpha2\beta1$ integrin. Consistent with this result, cell lines unable to bind echovirus do not express the $\alpha2\beta1$ receptor, although a variety of other $\beta1$ integrins may be expressed in such lines. Most importantly, transfection of the gene encoding the $\alpha2$ chain into resistant cell lines encoding the $\beta1$ chain results in conferral of the ability to bind the virus and support viral growth. Interestingly, the site of binding of the virus on the integrin appears to be different from the site recognized by the most common ligands for $\alpha2\beta1$, since monoclonal antibodies that block binding of ligands such as collagen and laminin have no effect on viral binding, while the MAbs isolated in this study that block echovirus growth have no effect on the binding the ECM proteins.[96]

V. MECHANISM OF INTEGRIN-MEDIATED UPTAKE OF MICROORGANISMS

The fact that both phagocytic and nonphagocytic cells use members of the identical receptor family to internalize microorganisms is consistent with the notion that they share a common mechanism for uptake of particles. One problem posed by the internalization process, true for all host cell types, is that the uptake of relatively large particles (≥ 1 μm) requires cytoskeletal rearrangement.[97] For instance, cytochalasin D uniformly blocks phagocytic events,[98] and fluorescence microscopy of microorganisms undergoing internalization reveals that uptake is often accompanied by localization of actin aggregates around the phagocytic process.[99,100] Therefore, binding of the microorganism must be coordinated with transduction of a signal that directs these cytoskeletal rearrangements to occur.

Adhesion of the microorganism to an integrin receptor may facilitate cytoskeletal rearrangements in multiple ways. First, the cytoplasmic domain of the β chain seems certain to directly interact with the cell cytoskeleton, by binding either talin or α-actinin,[101–103] allowing a simple route of communication between the external surface of the host cell and its cytoplasm. Second, a growing number of studies indicate that integrin-mediated adhesion of mammalian cells to solid surfaces results in a variety of signals being transduced, presumably as a result of multivalent receptor ligation to the substrate.[104–106] One such signal that has been characterized results in the phosphorylation of pp125[FAK], a 125-kDa species that shows high sequence similarity to members of the protein kinase family.[107] Other signal pathways are activated which are apparently independent of this phosphorylation event, and in at least one instance this can be measured as a change in the cytoplasmic pH of the host cell.[104] These results indicate that a variety of signal cascades could occur after binding of a microorganism to integrin receptors, and sorting out which specific signals actually play a role in the uptake process will be a significant challenge. For a more detailed discussion of the role of integrins in signal transduction, see a recent review by Hynes.[108]

The fact that binding of an integrin to a polyvalent substrate results in signal transduction suggests that mere adhesion of an organism to integrin receptors should promote efficient internalization. By and large this does not appear to be the case. Fibronectin-coated particles adhere to the surface of cells, but are poorly internalized by a variety of cell types, while the *B. pertussis* protein pertactin is inefficient relative to invasin at promoting uptake into cultured cells.[95] In all probability, these three substrates bind to similar sites on their integrin receptors; so, the site of interaction does not appear to be an important determinant in distinguishing between extracellular adhesion and internalization of a microorganism.[91] The most compelling explanation for the differences in these responses is that, under conditions in which there is a limiting amount of receptor available to bind the microorganism, substrates with the highest affinity for the integrin receptor are the most efficiently internalized.[109] This is either because the phagocytic process requires a series of high affinity interactions to stably drive the uptake process, or else the particular signal sent for phagocytosis is generated by a polyvalent ligation of receptors that requires a high affinity interaction. Further analysis of how these fates are determined is treated in detail elsewhere.[2]

VI. ROUTE OF CELLULAR PENETRATION *IN VIVO* AND THE PROBLEM OF BASOLATERAL LOCALIZATION

The use of cultured cell models has been extremely useful in identifying factors encoded by both the microorganism and the host cell involved in microbial internalization. There are several difficulties, however, in applying the results obtained in these model systems to events that occur *in vivo*. First, continued growth of mammalian cells in culture may result in selection for variants with altered integrin expression, so that some receptors are overproduced relative to their *in vivo* situation, whereas others are synthesized at aberrantly low levels or not at all. Second, particularly in the case of epithelial cells, an integrin may have distinct localization properties that are lost on continued culture. This problem is partially solved by analyzing polarized cells in culture,[110] but even this model generates some unique problems, since the cell lines used may have little in common with the cells normally encountered by the microorganism. Finally, the actual cells that are the targets for the uptake of the pathogen may be in very small quantities in the host tissue, may turn over rapidly, or be impossible to culture. These problems are related to one another and worth considering further, especially regarding the role of integrin receptors.

Continued growth of adherent cell lines potentially selects for variants best able to bind and spread on cultureware. The receptors responsible for this attachment are, by and large, members of the integrin receptor family. Since growth in culture may favor excess expression of integrins, the usual localization properties of integrins becomes disrupted. In the case of epithelial cells, β1 integrins are usually found localized on the basal and lateral surfaces of cells.[111] Growth of cells in a nonpolarized fashion encourages a more random distribution of integrins in cultured cells than found *in vivo*.[112] Receptors that would normally be found exclusively localized on the basal surface of epithelial cells, and not usually found in contact with microorganisms, are now available for binding and internalization of bacteria in culture. This raises the possibility that cultured cells may actually internalize a microorganism more efficiently than do host cells *in vivo*.

Growth of cultured epithelial cells in a polarized fashion may mimic more precisely the general localization properties found *in vivo*. On the other hand, use of these cells, with their predictable distribution of receptors, does not take into account the general nonhomogeneity of cell populations found *in vivo*. First, the localization of receptors *in vivo* may be somewhat less predictable than in culture. Second, *in vivo*, there is constant loss and regeneration of epithelium, so that freshly dividing cells have more uniform receptor distribution than more mature cells. Third, localized cellular damage in the epithelium may be common, so that microorganisms may gain access to integrins found on either the lateral surfaces of the epithelium, or on freshly divided cells at the sites of damage which do not occur in an intact polarized monolayer.[110] Finally, *in vivo*, some pathogens generally avoid utilizing the majority of polarized cells found in an epithelial layer, and simply do not enter the apical surface of these cells. So, the use of polarized cells may be of questionable relevance in these particular cases.

As alternatives to entry via the apical end of epithelial cells, it now appears that several, but not all, bacterial pathogens gain access to deeper tissues by two other tactics. In the first tactic, which appears to be employed by *Y. pseudotuberculosis* and *S. flexneri*, internalization occurs via M cells that are dispersed within the epithelial layer.[113–115] These cells, which are involved in antigen sampling, are usually localized over patches of lymphoid tissue, such as those found in the Peyer's patches in the small intestine. Although evidence argues that M cells are more discriminate in internalization of microorganisms than are professional phagocytes, they do appear to be particularly active in internalizing normally intracellular pathogens.[114–116] This may be because M cells are relatively free of the mucous layer that protects epithelial cells from microorganisms, or because the distribution of receptors is less stringently localized in M cells than in other cells found in the epithelial layer (as in cultured cells grown in a nonpolarized fashion). The second tactic is for the microorganism to produce localized damage in the epithelium, via cytotoxins, or for the microorganism to take advantage of damage induced by accessory factors, such as other infectious agents. These factors then allow exposure of lateral surfaces of the epithelium, where integrins active in internalization of microorganisms can now be available for binding microorganisms. There is some evidence that *S. flexneri* takes advantage of this tactic and binds lateral regions of the epithelial cell after some destruction of the tight junctions between cells;[110] so, there is no reason to believe that a variety of microorganisms may likewise pursue this path.

In summary, cultured cell models are important probes of cellular penetration by bacteria. Since no single cultured cell model totally reproduces the *in vivo* infection, a variety of strategies must be taken to analyze uptake of microorganisms, including the analysis of animal infection models using histochemical techniques.

VII. CONCLUSION

It is clear that a variety of pathogens bind integrins as a primary tactic for colonization of the host. In many cases, this adhesion is a prelude to the uptake of the microorganism into the host cell. It is significant that integrins are important receptors involved in clearing of organisms by phagocytes: the ability of this class of receptors to coordinate binding of extracellular ligands to cytoskeletal rearrangements is probably critical to successful completion of the phagocytic process. Since this same basic communication occurs with integrins found on normally nonphagocytic cells, it would appear that microorganisms have simply extended the tactic to bind members of this receptor family on all cell types. In the future, it will be important to determine which factors allow the integrin to communicate with the cytoskeleton and identify the steps in the signaling pathway that lead to phagocytosis. Presumably, the pathway of phagocytosis is very similar in all cell types, and the use of simple model systems will yield results that are generally applicable.

ACKNOWLEDGMENTS

I would like to thank John Leong, Guy Tran Van Nhieu, and Susannah Rankin for their work that contributed to our knowledge of invasin. Work on invasin was supported by NIH grant AI23538 and as well as by an NSF Presidential Young Investigator Award to R.I. R.I. was a Searle-Chicago Community Trust scholar during the course of the work reviewed.

REFERENCES

1. Sharpe, A. H. and Fields, B. N., Pathogenesis of viral infections. Basic concepts derived from the reovirus model, *N. Engl. J. Med.*, 312, 486, 1985.
2. Isberg, R. R., Discrimination between intracellular uptake and surface adhesion of bacterial pathogens, *Science*, 252, 934, 1991.
3. Joiner, K. A., Fuhrman, S. A., Miettinen, H. M., Kasper, L. H., and Mellman, I., *Toxoplasma gondii:* fusion competence of parasitophorous vacuoles in Fc receptor-transfected fibroblasts, *Science*, 249, 641, 1990.
4. Wright, S. D. and Silverstein, S. C., Receptors for C3b and C3bi promote phagocytosis but not the release of toxic oxygen from human phagocytes, *J. Exp. Med*, 158, 2016, 1983.
5. Hook, M., McGavin, M. J., Switalski, L. M., Raja, R., Raucci, G., Lindgren, P. E., Lindberg, M., and Signas, C., Interactions of bacteria with extracellular matrix proteins, *Cell Differ. Dev.*, 32, 433, 1990.
6. Hook, M., Switalski, L. M., Wadstrom, T., and Lindberg, M., in *Fibronectin*, Mosher, D., Ed., Academic Press, New York, 1989, 295.
7. Wadstrom T., Rubin, K., Ljungh, A., Hook, M., and Switalski, L. M., Fibronectin and toxic shock syndrome [letter], *JAMA*, 252, 343, 1984.
8. Hook, M., M. M. J., Switalski, L. M., Raja, R., Raucci, G., Lindgren, P. E., Lindberg, M., and Signas, C., Interactions of bacteria with extracellular matrix proteins, *Cell Differ. Dev.*, 32, 433, 1990.
9. Speziale, P., Raucci, G., Visai, L., Switalski, L. M., Timpl, R., and Hook, M., Binding of collagen to *Staphylococcus aureus* Cowan 1, *J. Bacteriol.*, 167, 77, 1986.
10. Voytek, A., Gristina, A. G., Barth, E., Myrvik, Q., Switalski, L., Hook, M., S. P., Staphylococcal adhesion to collagen in intra-articular sepsis, *Biomaterials*, 9, 107, 1988.
11. Switalski, L. M., Speziale, P., and Hook, M., Isolation and characterization of a putative collagen receptor from *Staphylococcus aureus* strain Cowan 1, *J. Biol. Chem.*, 264, 21080, 1989.
12. Faris, A., Froman, G., Switalski, L., and Hook, M., Adhesion of enterotoxigenic (ETEC) and bovine mastitis *Escherichia coli* strains to rat embryonic fibroblasts: role of amino-terminal domain of fibronectin in bacterial adhesion, *Microbiol. Immunol.*, 32, 1, 1988.

13. Verbrugh, H. A., Peterson, P. K., Smith, D. E., Nguyen, B. T., Hoidel, B. T., Wilkinson, J. R., Verhoef, J., and Furcht, L. T., *Infect. Immun.*, 33, 811, 1981.
14. Wright, S. D., Craigmyle, L. S., and Silverstein, S. C., Fibronectin and serum amyloid P component stimulate C3b- and C3bi-mediated phagocytosis in cultured human monocytes, *J. Exp. Med.*, 158, 1338, 1983.
15. Newman, S. L., and Tucci, M. A., Regulation of human monocyte/macrophage function by extracellular matrix. Adherence of monocytes to collagen matrices enhances phagocytosis of opsonized bacteria by activation of complement receptors and enhancement of Fc receptor function, *J. Clin. Invest.*, 86, 703, 1990.
16. Peterson, K. M., Baseman, J. B., and Alderete, J. F., *Treponema pallidum* receptor binding proteins interact with fibronectin, *J. Exp. Med.*, 157, 1958, 1983.
17. Froman, G., Switalski, L. M., Faris, A., Wadstrom, T., and Hook, M., Binding of *Escherichia coli* to fibronectin. A mechanism of tissue adherence, *J. Biol. Chem.*, 259, 14899, 1984.
18. Olsen, A., Jonsson, A., and Normark, S., Fibronectin binding mediated by a novel class of surface organelles on *Escherichia coli*, *Nature*, 338, 652, 1989.
19. Abraham, S. N., Beachey, E. H., and Simpson, W. A., Adherence of *streptococcus pyogenes*, *Escherichia coli*, and *Pseudomonas aeruginosa* to fibronectin-coated and uncoated epithelial cells, *Infect. Immun.*, 41, 1261, 1983.
20. Lindgren, P. E., Speziale, P., McGavin, M., Monstein, H. J., Hook, M., Visai, L., Kostiainen, T., Bozzini, S., and Lindberg, M., Cloning and expression of two different genes from *Streptococcus dysgalactiae* encoding fibronectin receptors, *J. Biol. Chem.*, 267, 1924, 1992.
21. Froman, G., Switalski, L. M., Speziale, P., and Hook, M., Isolation and characterization of a fibronectin receptor from *Staphylococcus aureus*, *J. Biol. Chem.*, 262, 6564, 1987.
22. Flock, J. I., Froman, G., Jonsson, K., Guss, B., Signas, C., Nilsson, B., Raucci, G., Hook, M., Wadstrom, T., and Lindberg, M., Cloning and expression of the gene for a fibronectin-binding protein from *Staphylococcus aureus*, *EMBO J.*, 6, 2351, 1987.
23. Raja, R. H., Raucci, G., and Hook, M., Peptide analogs to a fibronectin receptor inhibit attachment of *Staphylococcus aureus* to fibronectin-containing substrates, *Infect. Immun.*, 58, 2593, 1990.
24. Visai, L., Bozzini, S., Petersen, T. E., Speciale, L., and Speziale, P., Binding sites in fibronectin for an enterotoxigenic strain of *E. coli* B342289c, *FEBS Lett.*, 290, 111, 1991.
25. Thomas, D. D., Baseman, J. B., and Alderete, J. F., Fibronectin tetrapeptide is target for syphilis spirochete cytadherence, *J. Exp. Med.*, 162, 1715–9, 1985.
26. Thomas, D. D., Baseman, J. B., and Alderete, J. F., Fibronectin mediates Treponema pallidum cytadherence through recognition of fibronectin cell-binding domain, *J. Exp. Med.*, 161, 514, 1985.
27. Patti, J. M., Jonsson, H., Guss, B., Switalski, L. M., Wiberg, K., Lindberg, M., and Hook, M., Molecular characterization and expression of a gene encoding a *Staphylococcus aureus* collagen adhesin, *J. Biol. Chem.*, 267, 4766, 1992.
28. Lawrence, M. B. and Springer, T. A., Leukocytes roll on a selectin at physiologic flow rates—distinction from and prerequisite for adhesion through integrins, *Cell*, 65, 859, 1991.
29. Thomas, D. D., Navab, M., Haake, D. A., Fogelman, A. M., Miller, J. N., and Lovett, M. A., Treponema pallidum invades intercellular junctions of endothelial cell monolayers, *Proc. Natl. Acad. Sci. U. S. A.*, 85, 3608, 1988.
30. Choy, M. Y., Richman, P. I., Horton, M. A., and MacDonald, T. T., Expression of the VLA family of integrins in human intestine, *J. Pathol.*, 160, 35, 1990.
31. Bloch, C. A. and Orndorff, P. E., Impaired colonization by and full invasiveness of

Escherichia coli K1 bearing a site-directed mutation in the type 1 pilin gene, *Infect. Immun.*, 58, 275, 1990.

32. Svanborg-Eden, C., Andersson, B., Aniansson, G., Leffler, H., Lomberg, H., Mestecky, J., and Wold, A. E., Glycoconjugate receptors for bacteria attaching to mucosal sites: examples for *Escherichia coli* and *Streptococcus pneumoniae, Adv. Exp. Med. Biol.*, 1987.

33. Wright, S. D. and Griffin, F. M., Jr., Activation of phagocytic cells' C3 receptors for phagocytosis, *J. Leukocyte Biol.*, 38, 327, 1985.

34. Brown, E. J., Complement receptors and phagocytosis, *Curr. Opin. Immunol.*, 3, 76, 1991.

35. Russell, D. G., The macrophage-attachment glcoprotein gp63 is the predominant C3-acceptor site on *Leishmania mexicana* promastigotes, *J. Immunol.*, 1986.

36. Babior, B. M., Oxygen-dependent killing by phagocytes, *N. Engl. J. Med.*, 298, 659, 1978.

37. Corbi, A. L., Kishimoto, T. K., Miller, L. J., and Springer, T. A., The human leukocyte adhesion glycoprotein Mac-1 (complement receptor type 3, CD11b) alpha subunit. Cloning, primary structure, and relation to the integrins, von Willebrand factor and factor B, *J. Biol. Chem.*, 263, 12403, 1988.

38. Wilson, M. E., and Pearson, R. D., Roles of CR3 and mannose receptors in the attachment and ingestion of *Leishmania donovani* by human mononuclear phagocytes, *Infect. Immun.*, 56, 363, 1988.

39. Mosser, D. M. and Edelson, P. J., The third component of complement (C3) is responsible for the intracellular survival of *Leishmania major, Nature*, 327, 329, 1987.

40. Mosser, D. M. and Edelson, P. J., The mouse macrophage receptor for C3bi (CR3) is a major mechanism in the phagocytosis of *Leishmania promastigotes, J. Immunol.*, 135, 2785, 1985.

41. Mosser, D. M., Vlassara, H., Edelson, P. J., and Cerami, A., *Leishmania promastigotes* are recognized by the macrophage receptor for advanced glycosylation endproducts, *J. Exp. Med.*, 165, 140, 1987.

42. Mosser, D. M., Vlassara, H., Edelson, P. J., and Cerami, A., *Leishmania promastigotes* are recognized by the macrophage receptor for advanced glycosylation endproducts, *J. Exp. Med.*, 165, 140, 1987.

43. Russell, D. G., The macrophage-attachment glcoprotein gp63 is the predominant C3-acceptor site on *Leishmania mexicana* promastigotes, *J. Immunol.*, 1986.

44. Russell, D. G. and Wright, S. D., Complement recept type 3 (CR3) binds to an Arg-Gly-Asp-Containing region of the major surface glycoprotein, gp63, of *Leishmania promastigotes*, 144, 4817, 1990.

45. Payne, N. R. and Horwitz, M. A., Phagocytosis of *Legionella pneumophila* is mediated by human monocyte complement receptors, *J. Exp. Med.*, 166, 1377, 1987.

46. Bellinger-Kawahara, C. and Horwitz, M. A., Complement component C3 fixes selectively to the major outer membrane protein (MOMP) of *Legionella pneumophila* and mediates phagocytosis of liposome-MOMP complexes by human monocytes, *J. Exp. Med.*, 172, 1201, 10, 1990.

47. Horwitz, M. A. and Silverstein, S. C., Interaction of the Legionnaires' disease bacterium (*Legionella pneumophila*) with human phagocytes. I. *L. pneumophila* resists killing by polymorphonuclear leukocytes, antibody, and complement, *J. Exp. Med.*, 153, 386, 1981.

48. Horwitz, M. A. and Silverstein, S. C., Interaction of the legionnaires' disease bacterium (*Legionella pneumophila*) with human phagocytes. II. Antibody promotes binding of *L. pneumophila* to monocytes but does not inhibit intracellular multiplication, *J. Exp. Med.*, 153, 398, 1981.

49. Horwitz, M. A., Formation of a novel phagosome by the Legionnaires' disease

bacterium (*Legionella pneumophila*) in human monocytes, *J. Exp. Med.*, 158, 1319, 1983.

50. Horwitz, M. A., Phagocytosis of the Legionnaires' disease bacterium (*Legionella pneumophila*) occurs by a novel mechanism: engulfment within a pseudopod coil, *Cell*, 36, 27, 1984.

51. Ross, G. D., Cain, J. A., Myones, B. L., Newman, S. L., and Lachmann, P. J., Specificity of membrane complement receptor type three (CR3) for beta-glucans, *Complement*, 4, 61, 1987.

52. Ross, G. D., Cain, J. A., and Lachmann, P. J., Membrane complement receptor type three (CR3) has lectin-like properties analogous to bovine conglutinin as it functions as a receptor for zymosan and rabbit erythrocytes as well as a receptor for iC3b, *J. Immunol.*, 134, 3307, 1985.

53. Talamas-Rohana, P., Wright, S. D., Lennartz, M. R., and Russell, D. G., Lipophospho-glyan from *Leishmania mexicana* promastigotes binds to members of the CR3, p150, 95 and LFA-1 family of leukocyte integrins, *J. Immunol.*, 144, 4817, 1990.

54. Bullock, W. E. and Wright, S. D., Role of the adherence-promoting receptors, CR3, LFA-1, and p150, 95, in binding of Histoplasma capsulatum by human macrophages, *J. Exp. Med.*, 165, 195, 1987.

55. Schnur, R. A. and Newman, S. L., The respiratory burst response to *Histoplasma capsulatum* by human neutrophils. Evidence for intracellular trapping of superoide anion, 144, 4765, 1990.

56. Newman, S. L., Bucer, C., Rhodes, J., and Bullock, W. E., Phagocytosis of *His-toplasma caulatum* yeasts and microconidia by human cultured macrophages and aleolar macrophages. Cellular, *J. Clin. Invest.*, 85, 223,

57. Wright, S. D. and Jong, M. T., Adhesion-promoting receptors on human macrophages recognize *Escherichia coli* by binding to lipopolysaccharide, *J. Exp. Med.*, 164, 1876, 1986.

58. Wright, S. D., Levin, S. M., Jong, M. T., Chad, Z., and Kabbash, L. G., CR3 (CD11b/CD18) expresses one binding site for Arg-Gly-Asp-containing peptides and a second site for bacterial lipopolysaccharide, *J. Exp. Med.*, 169, 175, 1989.

59. Cain, J. A., Newman, S. L., and Ross, G. D., Role of complement receptor type three and serum opsonins in the neutrophil response to yeast, *Complement*, 4, 75, 1987.

60. Newman, S. L., Bucer, C., Rhodes, J., and Bullock, W. E., Phagocytosis of His-toplasma capsulatum yeasts and microconidia by human cultured macrophages and aleolar macrophages, Cellular, *J. Clin. Invest.*, 85, 223,

61. Tuomanen, E., Weiss, A., Rich, R., Zak, F., and Zak, O., Filamentous hemagglutinin and pertussis toxin promote adherence of Bordetella pertussis to cilia, *Dev. Biol. Stand.*, 61, 197, 1985.

62. Relman, D. A., Domenighini, M., Tuomanen, E., Rappuoli, R., and Falkow, S., Filamentous hemagglutinin of Bordetella pertussis: nucleotide sequence and crucial role in adherence, *Proc. Natl. Acad. Sci. U.S.A.*, 86, 2637, 1989.

63. Relman, D., Tuomanen, E., Falkow, S., Golenbock, D. T., Saukkonen, K., and Wright, S. D., Recognition of a bacterial adhesion by an integrin: macrophage CR3 (αMβ2, CD11b/CD18) binds filamentous hemagglutinin of Bordetella pertussis, *Cell*, 61, 1375, 1990.

64. Saukkonen, K., Cabellos, C., Burroughs, M., Prasad, S., and Tuomanen, E., Integrin-mediated localization of Bordetella pertussis within macrophages: role in pulmonary colonization, *J. Exp. Med.*, 173, 1143, 1991.

65. Russell, D. G. and Wright, S. D., Complement receptor type 3 (CR3) binds to an Arg-Gly-Asp-containing region of the major surface glycoprotein, gp63, of Leishmania promastigotes, *J. Exp. Med.*, 168, 279, 1988.

66. Russell, D. G., Talamas-Rohana, P., and Zelechowski, J., Antibodies raised against

synthetic peptides from the Arg-Gly-Asp-containing region of the Leishmania surface protein gp63 cross-react with human C3 and interfere with gp63-mediated binding to macrophages, *Infect. Immun.*, 57, 630, 1989.

67. Wright, S. D., and Meyer, B. C., Phorbol esters cause sequential activation and deactivation of complement receptors on polymorphonuclear leukocytes, *J. Immunol.*, 136, 1759, 1986.

68. Wright, S. D., Licht, M. R., Craigmyle, L. S., and Silverstein, S. C., Communication between receptors for different ligands on a single cell: ligation of fibronectin receptors induces a reversible alteration in the function of complement receptors on cultured human monocytes, *J. Cell Biol.*, 1984.

69. Brown, E., Hooper, L., Ho, T., and Gresham, H., Integrin-associated protein: a 50-kD plasma membrane antigen physically and functionally associated with integrins, *J. Cell Biol.*, 1990.

70. Bohnsack, J. F., OShea, J. J., Takahashi, T., and Brown, E. J., Fibronectin-enhanced phagocytosis of an alternative pathway activator by human culture-derived macrophages is mediated by the C4b/C3b complement receptor (CR1), *J. Immunol.*, 135, 2680, 1985.

71. Gresham, H. D., Goodwin, J. L., Allen, P. M., Anderson, D. C., and Brown, E. J., A novel member of the integrin receptor family mediates Arg-Gly-Asp-stimulated neutrophil phagocytosis, *J. Cell Biol.*, 108, 1935, 1989.

72. Gresham, H. D., Zheleznyak, A., Mormol, J. S., and Brown, E. J., Studies on the molecular mechanisms of human neutrophil Fc receptor-mediated phagocytosis. Evidence that a distinct pathway for activation of the respiratory burst results in reactive oxygen metabolite-dependent amplification of ingestion, *J. Biol. Chem.*, 265, 7819, 1990.

73. Dustin, M. L. and Springer, T. A., Role of lymphocyte adhesion receptors in transient interactions and cell locomotion, 9, 27, 1991.

74. Hermanowski-Vosatka, A., Van Strijp, J. A., Swiggard, W. J., and Wright, S. D., Integrin modulating factor-1: a lipid that alters the function of leukocyte integrins, *Cell*, 68, 341, 1992.

75. Carter, P. B., Pathogenecity of Yersinia enterocolitica for mice, *Infect. Immun.*, 11, 164, 1975.

76. Simonet, M., Richard, S., and Berche, P., Electron microscopic evidence for in vivo extracellular localization of Yersinia pseudotuberculosis harboring the pYV plasmid, *Infect. Immun.*, 58, 841, 1990.

77. Rosqvist, R., Bolin, I., and Wolf, W. H., Inhibition of phagocytosis in Yersinia pseudotuberculosis: a virulence plasmid-encoded ability involving the Yop2b protein, *Infect. Immun.*, 56, 2139, 1988.

78. Isberg, R. R., Pathways for the penetration of enteropathogenic Yersinia into mammalian cells, *Mol. Biol. Med.*, 7, 73, 1990.

79. Isberg, R. R. and Leong, J. M., Multiple β1 chain integrins are receptors for invasin, a protein that promoted bacterial penetration into mammalian cells, *Cell*, 60, 861, 1990.

80. Isberg, R. R., Voorhis, D. L., and Falkow, S., Identification of invasin: a protein that allows enteric bacteria to penetrate cultured mammalian cells, *Cell*, 50, 769, 1987.

81. Isberg, R. R. and Falkow, S., A single genetic locus encoded by Yersinia pseudotuberculosis permits invasion of cultured animal cells by Escherichia coli K-12, *Nature*, 317, 262, 4, 1985.

82. Isberg, R. R. and Leong, J. L., Cultured mammalian cells attach to the invasin protein of Yersinia pseudotuberculosis, *Proc. Natl. Acad. Sci. U.S.A.*, 85, 6682, 1988.

83. Leong, J. M., Fournier, R., and Isberg, R. R., Identification of the integrin binding domain of the Yersinia pseudotuberculosis invasin protein, *EMBO J.*, 9, 1979, 1990.

84. Rankin, S., Isberg, R. R., and Leong, J. M., An integrin binding domain that is

sufficient to allow bacteria to be internalized by cultured mammalian cells, *Infect. Immun.*, in press.

85. Hynes, R. O., Integrins: a family of cell surface receptors, *Cell*, 48, 549, 1987.

86. Theriot, J. A. and Mitchison, T. J., Actin microfilament dynamics in locomoting cells, *Nature*, 352, 126, 1991.

87. Elices, M. J., Osborn, L., Takada, Y., Crouse, C., Luhowskyj, S., Hemler, M. E., and Lobb, R. R., VCAM-1 on activated endothelium interacts with the leukocyte integrin VLA-4 at a site distinct from the VLA-4/fibronectin binding site, *Cell*, 60, 577, 1990.

88. Shimizu, Y., Newman, W., Gopal, T. V., Horgan, K. J., Graber, N., Beall, L. D., and Vanseventer, G. A., Four molecular pathways of T-cell adhesion to endothelial cells—roles of LFA-1, VCAM-1, and ELAM-1 and changes in pathway hierarchy under different activation conditions, *J. Cell Biol.*, 113, 1203, 1991.

89. DeStrooper, B., Van, d. S. B., Jaspers, M., Saison, M., Spaepen, M., Van, L. F., Van der Schueren, B. H., and Cassiman, J. J., Distribution of the β1 subgroup of the integrins in human cells and tissues, *J. Histochem. Cytochem.*, 37, 299, 1989.

90. Krukonis, E. and Isberg, R. R., unpublished observations, 1992.

91. Tran Van Nhieu, G. and Isberg, R. R., The Yersinia pseudotuberculosis invasin protein and human fibronectin bind to mutually exclusive sites on the α5β1 integrin receptor, *J. Biol. Chem.*, 266, 24367, 1991.

92. Springer, T. A., Adhesion receptors of the immune system, *Nature*, 346, 425, 1990.

93. Dustin, M. L., Garcia, A. J., Hibbs, M. L., Larson, R. S., Stacker, S. A., Staunton, D. E., Wardlaw, A. J., and Springer, T. A., Structure and regulation of the leukocyte adhesion receptor LFA-1 and its counterreceptors, ICAM-1 and ICAM-2, *Cold Spring Harbor Symp. Quant. Biol.*, 2, 753, 1989.

94. Leininger, E., Roberts, M., Kenimer, J. G., Charles, I. G., Fairweather, N., Novotny, P., and Brennan, M. J., Pertactin, an Arg-Gly-Asp-containing Bordetella pertussis surface protein that promotes adherence of mammalian cells, *Proc. Natl. Acad. Sci. U.S.A.*, 88, 345, 1991.

95. Leininger, E., Ewanowich, C. A., Bhargava, A., Peppler, M. S., Kenimer, J. G., and Brennan, M. J., Comparative roles of the arg-gly-asp sequence present in the *Bordetella pertussis* adhesins pertactin and filamentous hemagglutinin, *Infect. Immun.*, 60, 2380, 1992.

96. Bergelson, J. M., Shepley, M. P., Chan, B. M., Hemler, M. E., and Finberg, R. W., Identification of the integrin VLA-2 as a receptor for echovirus 1 [see comments], *Science*, 255, 1718, 1992.

97. Michl, J., Pieczonka, M. M., Unkeless, J. C., Bell, G. I., and Silverstein, S. C., Fc receptor modulation in mononuclear phagocytes maintained on immobilized immune complexes occurs by diffusion of the receptor molecule, *J. Exp. Med.*, 157, 2121, 1983.

98. Finlay, B. B. and Falkow, S., Comparison of the invasion strategies used by Salmonella cholerae-suis, Shigella flexneri and Yersinia enterocolitica to enter cultured animal cells: endosome acidification is not required for bacterial invasion or intracellular replication, *Biochimie*, 70, 1089, 1988.

99. Greenberg, S., Elkhoury, J., Divirgilio, F., Kaplan, E. M., and Silverstein, S. C., Ca2+-independent F-actin assembly and dissembly during Fc receptor-mediated phagocytosis in mouse macrophages, *J. Cell Biol.*, 113, 757, 1991.

100. Young, V. B., Falkow, S., and Schoolnik, G. K., The invasin protein of Yersinia enterocolitica: internalization of invasin-bearing bacteria by eukaryotic cells is associated with reorganization of the cytoskeleton, *J. Cell Biol.*, 116, 197, 1992.

101. Horwitz, A. F., Duggan, K., Buck, C., Beckerle, M. C., and Burridge, K., Interaction of plasma membrane fibronectin receptor with talin—a transmembrane linkage, *Nature*, 320, 531, 1986.

102. Burridge, K., Fath, K., Kelly, T., Nuckolls, G., and Turner, C., Focal adhesions: transmembrane junctions between the extracellular matrix and the cytoskeleton, *Annu. Rev. Cell Biol.*, 4, 487, 1988.

103. Otey, C. A., Pavalko, F. M., and Burridge, K., An interaction between α-actinin and the β1 integrin subunit in vitro, *J. Cell Biol.*, 111, 721, 1990.

104. Schwartz, M. A., Lechene, C., and Ingber, D. E., Insoluble fibronectin activates the Na/H antiporter by clustering and immobilizing integrin α5β1, independent of cell shape, *Proc. Natl. Acad. Sci. U.S.A.*, 88, 7849, 1991.

105. Guan, J. L., Trevithick, J. E., and Hynes, R. O., Fibronectin/integrin interaction induces tyrosine phosphorylation of a 120-kDa protein, *Cell Regul.*, 2, 951, 1991.

106. Kornberg, L. J., Earp, H. S., Turner, C. E., Prockop, C., and Juliano, R. L., Signal transduction by integrins: increased protein tyrosine phosphorylation caused by clustering of β1 integrins, *Proc. Natl. Acad. Sci. U.S.A.*, 88, 8392, 1991.

107. Schaller, M. D., Borgman, C. A., Cobb, B. S., Vines, R. R., Reynolds, A. B., and Parsons, J. T., pp125[FAK], a structurally distinctive protein tyrosine kinase associated with focal adhesions, *Proc. Natl. Acad. Sci. U.S.A.*, 89, 5192, 1992.

108. Hynes, R. O., Integrins: versatility, modulation, and signaling in cell adhesion, *Cell*, 69, 11, 1992.

109. Tran Van Nhieu, G. and Isberg, R. R., Affinity and receptor density are primary determinants for integrin-mediated bacterial internalization, *Embo J.*, 12, 1887, 1993.

110. Mounier, J., Vasselon, T., Hellio, R., Lesourd, M., and Sansonetti, P. J., *Shigella flexneri* enters human colonic Caco-2 epithelial cells through the basolateral pole, *J. Bacteriol.*, 60, 237, 1992.

111. Burridge, K. and Fath, K., Focal contacts: transmembrane links between the extracellular matrix and the cytoskeleton, *Bioessays*, 10, 104, 1989.

112. Dedhar, S., Argraves, W. S., Suzuki, S., Ruoslahti, E., and Pierschbacher, M. D., Human osteosarcoma cells resistant to detachment by an Arg-Gly-Asp-containing peptide overproduce the fibronectin receptor, *J. Cell Biol.*, 105, 1175, 1987.

113. Amerongen, H., Neutra, M., and Isberg, R. R., Unpublished observations, 1992.

114. Wassef, J. S., Keren, D. F., and Mailloux, J. L., Role of M cells in initial bacterial uptake and in ulcer formation in the rabbit intestinal loop model in shigellosis, *Infect. Immun.*, 57, 858, 1989.

115. Sansonetti, P. J., Arondel, J., Fontaine, A., d'Hauteville, H., and Bernardini, M. L., *ompB* (osmo-regulation) and *icsA* (cell to cell spread) mutants of *Shigella flexneri*: vaccine candidates and probes to study the pathogenesis of shigellosis, *Vaccine*, 9, 416, 1991.

116. Bye, W. A., Allan, C. H., and Trier, J. S., Structure, distribution and origin of M cells in Peyer's patches of mouse ileum, *Gastroenterology*, 86, 789, 1984.

Chapter 10

Integrin Function in Early Vertebrate Development: Perspectives from Studies of Amphibian Embryos

Douglas W. DeSimone

CONTENTS

I. INTRODUCTION

Embryogenesis is characterized by a highly orchestrated series of changes in adhesion that mediate the interactions of cells with one another and with the extracellular matrix (ECM). The cellular rearrangements of the gastrula, the migration of neural crest and primordial germ cells, and the formation of neural connections are but a few of the more spectacular examples of the precision with which embryonic cells and tissues use adhesive mechanisms to play out their morphogenetic programs. These events are likely to require the diverse functional activities of a large number of adhesion molecules with differing specificities and affinities. The integrins are an example of such a family of receptors thought to play important functional roles during development.

Interest in integrins as participants in morphogenetic processes has increased in recent years because it is now clear that the integrins represent a large family of structurally related receptors with multiple adhesive functions.[1] Collectively, these receptors display considerable functional diversity, but it is important to note that integrins represent only

0-8493-4711-4/94/$0.00+$.50
© 1994 by CRC Press, Inc.

a subset of the multiple adhesive interactions that are likely to operate during embryonic development.[2,3] A major challenge to the developmental biologist, therefore, is to distinguish among these interactions in order to elucidate the functional contributions of individual adhesion molecules.

Although interest in integrins is high among developmental biologists, progress in this area has been slow relative to advances in the cell biology of these receptors[1,4,5] (see also other chapters, this volume). This is due, in part, to a lack of reagents needed to study integrins in vertebrate systems suitable for embryologic analyses. Considerable progress has been made, however, with functional analyses of integrins in *Drosophila* and the identity of specific integrin α and β subunits expressed in rodent, avian, and amphibian embryos.[6-8]

This review will focus on studies of integrin expression and function in amphibian embryos. Recent studies have revealed considerable integrin diversity throughout amphibian development. These data suggest roles for integrins in some of the earliest morphogenetic events that occur during vertebrate embryogenesis. Work on amphibian integrins will be considered in light of recent advances in our understanding of embryonic induction, morphogenesis, and differentiation. The suitability of amphibian systems for analyses of integrin function in development is also discussed.

A. A BRIEF OVERVIEW OF THE AMPHIBIAN SYSTEM

Amphibian eggs are released and fertilized in an aqueous environment where they develop into larval and adult forms. The zygote contains all of the informational macromolecules and nutrients needed to support embryogenesis without further input from the parents. This reproductive strategy, although highly successful, comes with a price; large numbers of embryos must be produced in order to counter the pressures of predation and other natural hazards associated with parental abandonment. It is precisely these characteristics, however, that make the amphibian embryo particularly attractive to students of vertebrate embryology. Impressive numbers of embryos can be obtained easily and reared in the laboratory, making the entire developmental sequence accessible for experimental analyses. Their large size facilitates a variety of microsurgical procedures, and embryo fragments or individual cells obtained from dissociated embryos can be cultured in simple salt solutions lacking serum and other supplements. Although not suited to the powerful genetic approaches developed for mammalian systems, recent advances make it possible to apply "reverse-genetic" techniques to studies of early amphibian embryogenesis.[9] These features have made the amphibian embryo a particularly useful system for investigating cellular mechanisms of morphogenesis (e.g., gastrulation, neurulation, cell migration) and have provided a valuable means with which to test the importance of ECM molecules such as fibronectin (FN) in these processes.[10-12]

B. AMPHIBIAN GASTRULATION AS A MODEL FOR CELL-CELL AND CELL-ECM INTERACTIONS

Gastrulation is a dramatic morphogenetic event that results in the formation of the three embryonic germ layers, which in turn provide the foundation for future differentiation and organogenesis. Amphibian gastrulation has received considerable attention from developmental biologists through the years, and, as a result, a great deal is now known about the cellular mechanisms involved in this process.[10-17] Unlike the situation in other developmental systems, where only correlative evidence is available to implicate various adhesive interactions, studies of amphibian gastrulation have made it possible to test, *in vivo*, the importance of specific adhesion molecules in supporting gastrulation movements.[12] The following discussion will present a brief description of the embryology of amphibian gastrulation sufficient to provide the necessary background information for a thorough review of integrin-ECM studies in this system.

When considering mechanisms of amphibian gastrulation, it is necessary to bear in mind that some important details differ between species, and generalizations must be made with caution. This review will concentrate on studies undertaken with urodeles (salamanders; e.g., *Ambystoma* and *Pleurodeles* sp.) and anurans (frogs; e.g., *Rana* and *Xenopus* sp.), which represent the 2 major orders of amphibia favored by embryologists. The following descriptions focus primarily on analyses of *Xenopus laevis* and *Pleurodeles waltl* embryos because most of what we know about amphibian integrins comes from studies of these species. Significant exceptions or differences in experimental systems are noted where relevant.

1. Embryology and Cellular Considerations

Mechanistically, the process of gastrulation may be considered on several levels that include: (1) the "behavior" of the individual cells that collectively serve to "drive" morphogenesis at gastrulation, (2) the molecules mediating the shape changes and adhesive properties of these cells, and (3) the control of the timing and patterning of the cellular movements involved. Following the progression through early cleavage and blastula stages, gastrulation begins with the appearance of a slit-like invagination of bottle cells, termed the blastopore, on the dorsal side of the embryo as illustrated in Figure 1. The involuting mesoderm subsequently comes in contact with the blastocoel roof and travels along it in the direction of the animal pole. The zone of involution initiated at the dorsal lip of the blastopore spreads laterally and ventrally to enclose the endoderm, which remains visible as a yolk plug through late gastrulation. The superficial cells of the animal pole and equatorial marginal zone spread by epiboly during this process, thus covering the entire outer surface of the embryo. Inside the embryo, the endodermally derived archenteron forms as the mesoderm advances, resulting in the displacement of the blastocoel (Figure 1).

The cellular movements that accompany gastrulation in *Xenopus* have been described in considerable detail by Keller and colleagues.[13,16,17] For the most part, these movements arise from a coordinated series of cellular rearrangements termed "convergence" and "extension". The overall result is the narrowing and lengthening of dorsal tissues at the site of involution, thereby providing the mechanical force that "drives" mesoderm involution and leads to the elongation of the body axis. Cells undergoing these movements actively change position with respect to one another by local intercalation, resulting in a thinner but more elongated mass of cells. This involuting mass of cells converges upon the dorsal lip and extends along the anterioposterior axis, giving rise to the notochord and somitic mesoderm.

Convergence and extension appear to play a minor role in driving gastrulation in *Pleurodeles*, suggesting a basic mechanistic difference between anuran and urodele embryos.[18] In *Pleurodeles*, the involuting mesodermal cells actively migrate across the inner surface of the blastocoel roof, in contrast to the situation for *Xenopus*, where a cohesive mass of mesodermal cells undergoes coordinated movement that is independent of migration. As discussed below, this is not meant to imply that the migration of individual cells plays no role in *Xenopus* gastrulation. In fact, a small population of cells at the leading edge of the involuting mesoderm actively migrates across the blastocoel roof in *Xenopus*.[11,19] The trailing presomitic and prenotochordal mesoderm will also migrate as a cohesive mass when placed *in vitro* on artificial substrates.[19] The latter observation suggests that although these cells travel across the blastocoel roof using convergence and extension movements they are competent to behave as a migratory population.

2. The Role of Fibronectin in Gastrulation

Whatever the relative contributions of cell migration and convergent extension to amphibian gastrulation, cell adhesion is likely to play a part in both types of cellular behaviors.

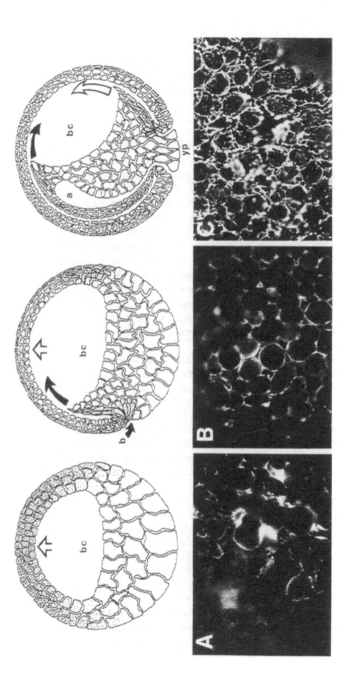

Figure 1 Cellular movements and fibronectin synthesis during *Xenopus* gastrulation. Top panels illustrate the major cellular movements associated with gastrulation in *Xenopus*. Drawings represent sections through the midline of blastula, early, and late gastrula stage embryos. Each is oriented with their dorsal sides to the left and animal poles at the top. Gastrulation begins with the invagination of bottle cells (b, curved arrow) at the dorsal side of the embryo. This primary invagination initiates the involution of the mesoderm. The mesoderm moves inward and upward (direction indicated by the curved, solid arrow), coming into contact with the inner surface of the blastocoel (bc) "roof" and migrating toward the animal pole. Although involution begins at the dorsal side of the embryo (curved, solid arrow), it proceeds laterally and then ventrally in later gastrulae (curved, open arrow), resulting in the formation of the circular yolk plug (yp). Eventually, the bc is displaced by the involuting mesodermal mass and the forming archenteron (a), or primitive gut.

Whole mount immunofluorescence[71] is used to localize fibronectin during gastrulation of *Xenopus*. FN synthesis begins at the mid-blastula stage (A) where staining is first noted at the periphery of cells that line the inner surface of the blastocoel roof at the animal pole. The level of staining increases as gastrulation proceeds (B). Eventually, a dense network of FN fibrils is formed (c) coincident with mesodermal cell migration. Immunostaining was accomplished using the 4H2 monoclonal antibody, which is directed against *Xenopus* FN (Ramos and DeSimone, unpublished). Open arrows in illustrations correspond to the approximate location of the FN fibrils depicted in the accompanying photomicrographs.

The identification and characterization of adhesion molecules present during gastrulation is, therefore, of considerable importance.[12,49,51] Fibronectin was one of the first adhesive proteins to be described and implicated in amphibian gastrulation. It is largely because of the FN work that attention has now shifted to integrins and the roles of these receptors. The following sections will summarize the evidence in favor of FN and integrin involvement in supporting gastrulation and later morphogenetic events.

a. Fibronectin Expression in Early Embryos

Several lines of circumstantial evidence indicate that FN is involved in supporting morphogenetic movements at gastrulation. Early studies demonstrated that FN is specifically localized along the inner surface of the blastocoel roof during this process[20,21] (see Figure 1). FN is synthesized and secreted throughout the embryo but is preferentially localized to the blastocoel roof[21] where it forms an elaborate fibrillar matrix as gastrulation begins (Figure 1). The mechanism by which FN becomes preferentially associated with the blastocoel roof is unclear. One possibility is that integrin receptors for FN (e.g., $\alpha5\beta1$) are expressed by cells in the blastocoel roof and that these receptors help promote the assembly of the FN matrix. In any event, the localization of FN along the pathway of the involuting mesoderm provided an early clue that FN might be influencing the migratory behavior of these cells.

Another indication that FN may be important in gastrulation comes from analyses of the timing of its expression. FN synthesis commences around the time of the mid-blastula transition[21] (MBT), which marks the onset of zygotic transcription in *Xenopus*[22] and other amphibian embryos. However, the FN synthesized at this point and during gastrulation is encoded not by newly transcribed mRNAs, but by maternal transcripts stored in the unfertilized egg.[21,23,24] It is not until late in gastrulation that accumulation of zygotic FN transcripts is detected.[23] An intriguing question, therefore, is what controls the translational activation of FN maternal mRNAs at the MBT? Experiments performed by Lee et al.[21] suggest that FN synthesis is independent of the MBT because FN is still translated "on time" in parthenogenetically activated eggs that do not divide, do not undergo the MBT, and, therefore, do not transcribe RNA. These studies indicate that translation of FN maternal mRNAs is under the control of a "cytoplasmic clock", which is independent of the MBT.

FN cDNAs have now been cloned from both *Xenopus laevis*[23] and *Pleurodeles waltl*.[24] Deduced protein sequences of these cDNAs reveal a high degree of similarity (approximately 70% identity overall) with their avian and mammalian counterparts. The integrin-binding sites, arg-gly-asp (RGD) in the tenth type III repeat and the leu-asp-val (LDV) present in the V-region, are conserved in both *Pleurodeles* and *Xenopus* FNs. The three known regions of alternative splicing (e.g., EIIIA, EIIIB, and V-region) are also conserved. All three alternatively spliced exons are fully included in FN transcripts in the early embryo, indicating that both the RGD and LDV binding sites are present during gastrulation.

b. Fibronectin Function in Gastrulation

The expression of FN at both the right place and time suggests a role for this protein in supporting cell adhesion and migration at gastrulation. The most convincing evidence, however, comes from a series of experiments done with urodele embryos by Boucaut, Johnson, and their colleagues.[10,12] These investigators microsurgically inverted a small "patch" of the blastocoel roof so that the FN-rich ECM faced outside and the apical surface inside. As gastrulation proceeds, mesodermal cells are unable to migrate across the patches and, instead, stream around either side.[25] Although this suggests that the ECM might be necessary for promoting mesoderm migration, it does not specifically address the importance of FN. Microinjection of anti-FN Fab' fragments into the blastocoel provided the first direct evidence that FN is likely to be the molecule responsible.[25] Similar results were obtained by injecting RGD peptides into the blastocoel.[26] In both cases, involuting mesodermal cells are unable to migrate, with many cells detaching from the roof and rounding-up on the floor of the blastocoel. These experimental perturbations appear specific and nontoxic in that other morphogenetic movements, such as the epibolic spreading of overlying cells, continue normally.

In *Xenopus*[27] and other anurans,[12] injection of anti-FN antibodies has a similar effect on gastrulation, in that bottle cell invagination and the formation of the dorsal lip of the blastopore will occur while mesoderm involution does not. However, Keller[28,29] has shown that removal of the blastocoel roof has little affect on gastrulation (i.e., convergence-extension and involution movements) in *Xenopus*. Gastrulation is also reported to be unaffected in *Xenopus* embryos injected with RGD peptides even though this treatment results in the elimination of FN fibrils,[30] presumably by interfering with integrin-mediated matrix assembly. An obvious interpretation of these latter studies is that cell adhesion to the blastocoel roof is relatively unimportant for gastrulation in *Xenopus*.

How can these conflicting pieces of information be reconciled? One possible explanation is that adhesion and migration of mesodermal cells at the leading edge are required only to initiate and perhaps guide early involution in *Xenopus*. In this case, convergence and extension movements would provide the force necessary to complete gastrulation even in the absence of additional directional clues, or the adhesive "traction" provided by an overlying ECM. This would explain why gastrulation is perturbed following anti-FN antibody injections in *Xenopus*[27] but not by removal of the blastocoel roof.[28,29] It is more difficult to explain why gastrulation is apparently normal in embryos lacking a FN-matrix as a result of RGD peptide injection into the blastocoel.[30] A recent reinvestigation of these experiments confirms that FN fibrils are indeed lost at the animal pole following RGD injection (Ramos and DeSimone, unpublished observations). However, some FN-fibrils remain associated with the blastocoel roof at the marginal zone, which is the site of mesoderm involution. Whether the remaining FN is both sufficient and necessary to initiate mesoderm involution remains unclear.

One approach likely to prove useful in addressing the importance of FN in *Xenopus* gastrulation is the cytoplasmic injection of antisense oligodeoxynucleotides in order to eliminate specifically FN maternal mRNAs from the egg.[23] In preliminary experiments, these "FN-minus" embryos develop normally until the early gastrula stage when a dorsal lip forms, but mesoderm involution does not occur and FN fibrils are not detected (DeSimone, Heasman, and Wylie, unpublished observations).

c. Mesodermal Cell Adhesion and Migration In Vitro

The behavior of embryonic cells and tissues on ECM substrates *in vitro* has been studied in detail by a number of investigators.[10-12] The observation that the filopodial and lamellipodial extensions of migrating mesodermal cells is often associated with ECM fibrils on the blastocoel roof of urodele embryos[31] prompted Nakatsuji and Johnson[32] to study the behavior of isolated cells on conditioned substrates *in vitro*. This was accomplished by explanting fragments of the blastocoel roof, ECM side down, onto culture dishes. After a short incubation period the tissue fragment is removed, leaving behind an extracellular network of fibrils attached to the substrate. When individual mesodermal cells are seeded onto the conditioned substrate, they will adhere and migrate in a persistent fashion. In both *Xenopus*[19,33] and *Pleurodeles*,[34] explanted nondissociated-mesoderm migrates on conditioned substrates as a coherent mass toward the original location of the animal pole. This behavior is dependent upon FN because conditioned substrates will not support directional migration in the presence of RGD peptides and anti-FN antibodies.[34-36] In addition, artificial substrates coated with exogenous FN will support adhesion and locomotion but not directional migration. These data suggest that the organization of FN-fibrils *in vivo* is critical in providing guidance cues to the advancing mesoderm. The ability to migrate directionally also appears to be dependent upon the intrinsic cohesive properties of the mesodermal mass, which in turn appears to be dependent on the cell-cell adhesion molecule U-cadherin.[33] This is further demonstrated by the fact that individual mesodermal cells obtained from dissociated *Xenopus* explants migrate randomly on blastocoel roof-conditioned substrates.

More recently, a thorough structure-function analysis of *Xenopus* FN has been undertaken using bacterial fusion proteins (Ramos and DeSimone, in preparation[37]). This approach makes it possible to analyze the individual contributions of multiple cell-interactive sites on the FN molecule. Results suggest that mesodermal cell adhesion to FN is a complex process involving several different structural domains and cellular receptors. For example, RGD-containing fusion proteins support mesodermal cell adhesion and spreading in cooperation with an upstream synergistic region.[38] The V-region, which contains the LDV sequence recognized by integrin $\alpha 4\beta 1$,[39,40] does not support cell attachment. Interestingly, the C-terminal heparin binding domain promotes mesoderm attachment but not cell spreading. When both the heparin binding domain and the V-region are present, however, the cells attach and actively migrate. In contrast, little migration is observed on fusion proteins containing either the RGD or heparin binding sites alone.

d. Integrins and Fibronectins in Development

The investigation of FN function in amphibian gastrulation has inevitably led to the study of integrins and, therefore, now serves as a prototypical case study of cell-matrix interactions *in vivo*. The relatively "simple" process of gastrulation coupled with several potentially powerful experimental approaches available in the amphibian system may help sort out the multiple adhesive interactions that are operating in the embryo to elicit morphogenetic change.

There are several practical but important issues raised by the FN studies, many of which are likely to be answered as more information becomes available about integrins that are coexpressed in the early embryo (see below). The assembly of the FN matrix along the blastocoel roof, the adhesion and migration of involuting mesodermal cells, and the subsequent patterning of mesodermally derived tissues are all events in which integrins are likely to participate. One obvious prediction, therefore, is that multiple integrin receptors are required to mediate these functions and that integrin expression is tightly regulated both spatially and temporally, even at the earliest stages of development. Preliminary indications are that this is indeed the case, as discussed in the second half of this review.

II. AMPHIBIAN INTEGRINS
A. IDENTIFICATION OF INTEGRINS EXPRESSED IN EARLY EMBRYOS

The amphibian gastrulation studies clearly indicate that a more complete understanding of cell-ECM interactions in development depends upon a thorough analysis of integrin receptors. As mentioned earlier, this has been difficult to accomplish in experimentally suitable developmental systems due to the lack of reagents needed to identify and evaluate the functions of specific integrins *in vivo*. The initial emphasis, therefore, has been to obtain cDNAs and antibodies that can be used to investigate integrin expression and function in development. We are only now at the stage where enough new information is available to make a review of this important topic both timely and appropriate.[8] What has emerged thus far is the somewhat surprising indication that a large number of

Table 1   Expression of Integrin mRNAs in amphibian embryos

Subunit	First Detected	Temporal and/or Spatial Distribution
β1[a]	Oocyte	Expressed at high levels throughout embryogenesis, widespread distribution[41,44,45,49,55]
β2	Stage 35	Levels constant from tailbud to tadpole[49]
β3	Egg	Neural groove, neural folds, ventral blood-forming regions, epidermis[49]
β6	Egg	Increased expression by late tailbud[49]
α2	Stage 11	Late gastrula, increasing at neurulation[51]
α3	Egg	Increased expression at gastrulation in dorsal meso-derm, notochord, branchial arches[51]
α4	Stage 10	Increased levels of expression from late gastrula-tion onward[51]
α5	Egg	Expressed at high levels throughout embryogene-sis,[51] widespread distribution[b]
α6	Stage 10	Neural ectoderm,[51] CNS, pronephros[c]
αIIb	Stage 17	Increased expression in tailbud/tadpole[51]
αV	Egg	Expressed at high levels throughout embryogenesis[d]

[a]Both mRNA and protein data available for *Xenopus* and *Pleurodeles*. All other information based on *Xenopus* except for the αV, which is only available for *Pleurodeles*. Work unpublished or in preparation: [b](Meng, Whittaker, and DeSimone); [c](Lallier, Whittaker, and DeSimone); [d](Alfandari, Whittaker, DeSimone, and Darribere). Stages 10 and 11, gastrula; stage 17, neurula; stage 35, tailbud.

different integrins are expressed at very early stages of development (Table 1). These results suggest that a high degree of integrin diversity and functional redundancy are necessary in order to mediate many of the complex and often subtle morphogenetic events that characterize embryogenesis.

1. Characterization of β Subunits

Initial characterizations of amphibian integrins were done in *Pleurodeles*[41] and *Xeno-pus*,[42,43] using cross-reactive polyclonal antibodies directed against avian β1 integrins. The electrophoretic properties of amphibian integrins are similar to those described for avian and mammalian cells with respect to mobility, subunit profile, and behavior on reduction. Confirmation of the presence of the β1 subunit in amphibians was obtained by cloning the cDNA for this subunit from *Xenopus*[44,45] and more recently from *Pleuro-deles* (Shi, Riou, Darribere, DeSimone, and Boucaut, unpublished data). The amphibian β1 sequences share 81 to 86% amino acid identity with other vertebrate β1 integrins, including the presence of 56 cysteines in the extracellular domain and complete conserva-tion of C-terminal cytoplasmic domain sequences. In contrast, amino acid identities of different β subunits within a single species (e.g., human) range from approximately 30 to 50%. The high degree of identity noted among β1 subunits from several different vertebrate species suggests that integrin β subunits may have diverged from a common ancestral gene more than 500 million years ago prior to the appearance of the vertebrates.[44] This point is further supported by direct comparisons of β subunit sequences deduced from cloned cDNAs obtained from *Drosophila*, avian, mammalian, and amphibian sources.[44,46-49]

Polymerase chain reaction (PCR) methods and degenerate oligodeoxynucleotide prim-ers have recently been used to identify a number of additional β subunits expressed in *Xenopus* embryos[49] (Table 1). These efforts have resulted in the cloning of integrin cDNAs with homology to the human β2, β3, and β6 subunits. The deduced protein sequences of the PCR amplified regions and their human homologues range in identity from 70 to 74%, which is slightly lower than that for β1 across the same region (85%). Complete sequence information is also now available for the *Xenopus* β3 subunit, which is 76% identical overall to human β3.[49] *Xenopus* integrin sequence analysis is useful from the standpoint of evolutionary comparisons, which have already aided in the identification of conserved structural features with functional significance.[50] Although not all of the known vertebrate β subunits have yet been identified in *Xenopus*, the available information suggests that amphibia are likely to share the same number and variety of integrins as found in mammals.

2. Characterization of α Subunits

Homology PCR has been used to identify integrin α subunits expressed in neurula stage *Xenopus* embryos[51] (Table 1). Integrins α2, α3, α4, α5, α6, and αIIb were each identified in this way, providing evidence that a large number of different integrins are indeed present at a very early stage in development. Sequence comparisons of amphibian and human α subunit homologues reveal identities from 58 to 76% across the amplified region, which is generally lower than that observed for *Xenopus* β subunits (see above) or for specific α subunits from different mammalian species. These data suggest that integrin α subunits diverged from one another earlier in evolution than have the β subunits. This is perhaps not surprising given the larger number of α subunits and the fact that functional variability (i.e., ligand binding specificity) is largely determined by α subunit composition.[1] On the other hand, β subunit diversity may have been more constrained by conservation of specific functional features such as ligand binding sites in both the extracellular and cytoplasmic domains.[50,52-54]

3. Multiple Integrins are Coexpressed in Early Embryos

The following sections summarize information on the spatial and temporal patterns of expression of integrin α and β subunits during amphibian development. These studies represent the most complete data currently available concerning the expression and function of integrins in early vertebrate embryogenesis.[8]

a. Expression of Integrin β Subunits

The integrin β1 subunit is maternally encoded by transcripts synthesized during oogenesis.[43] Levels of β1 mRNA remain constant and at relatively high levels throughout development.[45,49] Not unexpectedly, β1 mRNA is widely distributed in the embryo[49] in agreement with immunolocalization studies, which demonstrate that the protein is expressed in most, if not all, cells as development proceeds.[41,42,45,55] However, *in situ* hybridization studies suggest that the amount of β1 mRNA expressed by different cells and tissues may vary considerably depending on the stage of development.[49]

Studies of β3 mRNA expression in *Xenopus* reveal that this subunit is also expressed as a maternal mRNA but at very low levels relative to the β1 subunit.[49] Expression increases by neurula stages where transcripts are first detected in the sensorial layer of the ectoderm, the neural folds, and in the bottle cells of the neural groove. As the neural folds come together, β3 expression is progressively lost at the site of neural tube closure in both the folds and the underlying neural groove. The synthesis of β3 mRNA decreases following neurulation until tailbud stages, where it is then localized in a small group of cells in the ventral blood-forming region. As the vascular system develops and the heart begins to beat, the β3-positive cells are dispersed throughout the embryo.[49] Although the identity of these cells is uncertain, it is likely that they represent amphibian thrombocytes, which are functionally homologous to mammalian platelets.

Expression of β2 subunit mRNAs is not detected until late in development[49] at a time coincident with hematopoietic differentiation, in keeping with the role of β2 integrins as leukocyte adhesion molecules.[1] In contrast, the β6 subunit mRNA is detected at very low levels in the egg and throughout early embryogenesis. Levels of β6 mRNA first increase at late tailbud stages.

b. Integrin β1 Subunit Synthesis and Processing

β1 integrins are synthesized during oogenesis and are present at the oocyte plasma membrane until maturation, at which time the receptor is cleared from the surface and stored in membrane vesicles.[56] Integrin β1 synthesis and processing continues during this period adding to a maternal store of mature heterodimer. Following fertilization, β1 receptors are then inserted into all newly formed membranes, presumably from these intracellular stores.[45] Although new synthesis of β1 subunits also begins sometime after fertilization,[42] it is not until neurulation that appreciable amounts of precursor become processed to mature αβ heterodimer and appear at the cell surface.[27,45] The reasons for this apparently complex regulation of expression are unclear. One obvious explanation for the storage of β1 subunits is to provide the cleaving egg with integrins sufficient for insertion into rapidly forming, new plasma membrane.[45] However, what is the signal for later processing and mobilization of β1 to the cell surface? Perhaps the synthesis of α subunits provides a regulatory step in the processing of mature heterodimer (see also below). Additional experiments are needed to determine whether β1 processing to the cell surface is controlled, in part, simply by the transcriptional or translational regulation of α subunit synthesis.

c. Expression of Integrin α Subunits

At least five different α subunit mRNAs are expressed by gastrulation in *Xenopus*[51] (refer to Table 1). Each of the α subunits encoded by these mRNAs is known to form dimers with the β1 subunit. Each subunit mRNA differs with respect to the timing and/ or level of expression as quantified by RNase protection analysis. For example, α5 is present as an abundant maternal mRNA that is maintained at relatively constant levels throughout development. The α3 is a low level maternal transcript with increased synthesis first noted by early gastrulation. Transcripts encoding α2, α4, and α6 are first detected during gastrulation. None of these mRNAs are synthesized during the onset of zygotic transcription at the MBT. Furthermore, several of these subunits are sequentially expressed over a short time period at gastrulation. These data are consistent, therefore, with a role for multiple integrins at gastrulation. Notably three of these integrins (i.e., α3β1, α4β1, and α5β1) are reported to bind FN.[1]

One prediction from the FN studies discussed earlier is that multiple FN receptors are likely to be synthesized during gastrulation and possibly expressed on distinct populations of cells. For example, the cells lining the roof of the blastocoel might be expected to express a "high affinity" receptor (e.g., α5β1) involved in assembling a FN matrix, whereas involuting mesodermal cells might have associated with their cell surfaces a lower affinity FN receptor (e.g., α3β1, α4β1) in order to promote migration. In this capacity, integrins could conceivably be providing "position specific" information necessary for defining morphogenetic fields of distinct cellular behaviors (e.g., adhesive, migratory).

Interestingly, *in situ* hybridization results confirm that α3 mRNAs are expressed in the dorsal involuting mesoderm, which comes in contact with the FN-rich blastocoel roof during gastrulation[51] (Figure 2). As development proceeds, these cells actively involute through the blastopore and extend along the anterioposterior axis as a narrow band of prenotochordal mesoderm. The expression of α3 is limited to the notochord as development continues and begins to decrease in this tissue at later stages. The α3

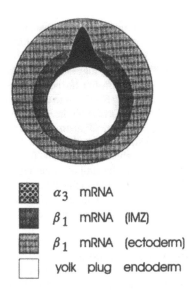

α_3 mRNA

β_1 mRNA (IMZ)

β_1 mRNA (ectoderm)

yolk plug endoderm

Figure 2 Integrin mRNAs are differentially expressed in the early amphibian embryo. Diagram summarizes *in situ* hybridization data demonstrating that integrin α_3 mRNAs are localized to the dorsal involuting mesoderm of *Xenopus* gastrulae.[51] The β_1 subunit is present as a maternal mRNA expressed throughout the early embryo.[42,45] However, β_1 expression is most abundant in the mesodermal cells of the involuting marginal zone (IMZ).[49] The expression of α_3 is restricted to cells of the dorsal involuting mesoderm. As discussed in Figure 1 and the text, these cells undergo active convergence and extension movements at gastrulation and are destined to give rise to the notochord.

subunit then becomes localized to the developing branchial arches and prosencephalon at the tailbud stage. Future experiments will address the role played by integrin $\alpha_3\beta_1$ in supporting the morphogenetic movements of these cells.

In contrast, α_6 transcripts are specifically localized to the neural ectoderm during gastrulation (Lallier, Whittaker, and DeSimone, unpublished observations). Again, the nonoverlapping patterns of the α_3 and α_6 subunits suggest that integrin expression may provide important positional information to the early embryo prior to tissue differentiation and segregation. Later in development, α_6 mRNAs are localized to discrete segments of the brain, the cranial nerves, neural tube, and the pronephros. The spatial localization of the remaining α subunits expressed at these stages awaits the availabilty of additional cDNAs suitable for the *in situ* hybridization procedure.

B. INTEGRIN FUNCTION IN DEVELOPMENT
1. Antibody Perturbation Experiments
The FN studies reviewed earlier suggested that integrins mediate many of the morphogenetic events associated with gastrulation. Darribere et al.[41] were the first to demonstrate that polyclonal Fab' fragments directed against avian β_1 integrins would perturb gastrulation when microinjected into the blastocoels of *Pleurodeles* embryos. The antibody arrested gastrulation by causing the detachment of migrating cells from the fibrillar FN-matrix on the blastocoel roof. These effects are strikingly similar to those obtained by the same investigators using anti-FN Fabs or RGD peptides,[25,26] further strengthening the case that integrin-FN interactions play a significant functional role in promoting the attachment and migration of mesodermal cells at gastrulation. More recently, Smith and colleagues[27,57] took the same approach in analyzing β_1 function during *Xenopus*

gastrulation. Injection of a more crudely defined (but primarily β1-specific) anti-β1 antisera resulted in a similar disruption of mesoderm involution, indicating that integrins are also involved in supporting gastrulation movements in anuran embryos.

The blastocoel injection experiments suggest that β1 integrins are necessary for mesodermal cell adhesion and migration but do not address the question of whether these receptors are involved in localizing FN fibrils to the blastocoel roof. Darribere et al.[58] investigated the question of FN matrix assembly in *Pleurodeles* by injecting polyclonal antibodies directed against the cytoplasmic domain of the β1 subunit into the cytoplasm of blastomeres at the 1- or 2-cell stage. Injected embryos develop normally through the blastula stage, but gastrulation is abnormal and FN fibrils are not assembled. One interpretation of this experiment is that the antibody interferes with normal β1-cytoskeletal interactions, which, in turn, affect the organization of FN-fibrils. An alternative explanation is that the antibody interferes with integrin transit to the plasma membrane; however; it was not reported whether or not receptors could be detected on the surfaces of these cells.

2. Integrin β3 Function in Development?

Little information is currently available concerning the identity of α subunits that are likely to associate with β3 in the early *Xenopus* embryo.[49] However, the identification of the αIIb subunit in tailbud stages[51] suggests that the "platelet" αIIbβ3 receptor may also be expressed in amphibian "thrombocytes". The cloning of αV cDNAs from *Pleurodeles* embryos (Alfandari, Whittaker, DeSimone, and Darribere, unpublished observation) indicates that αVβ3 may be the predominant β3 integrin expressed prior to the tailbud stage. However, the αV subunit has not yet been identified in *Xenopus*, and, similarly, there is no evidence available for β3 in *Pleurodeles* embryos. In any event, the importance of β3 during neural development is somewhat questionable in view of the fact that the human disease Glanzmann's Thrombasthenia results from the absence of functional β3 protein.[59] Glanzmann's patients suffer from bleeding disorders but presumably have proceeded through early development normally without serious defects. It is feasible, of course, that β3 expression is critical to *Xenopus* but not human development, and future experiments will therefore be needed to address its possible functions.

3. Mesoderm Induction and Activation of Integrin Function

Considerable progress has been made in recent years in our understanding of the molecular mechanisms involved in patterning the early embryo. Much of this work has been done using *Xenopus* as a model system, and several recent reviews address major advances.[60-62] In the early blastula stage, signals from the vegetal pole induce cells in the overlying equatorial region to form mesoderm.[63] This can be demonstrated experimentally by coculturing animal cap ectoderm (which is not normally in contact with the vegetal pole and does not form mesoderm) with vegetal pole endoderm. Under these *in vitro* conditions the animal cap is induced to form mesodermal tissues. The inducing activity of the vegetal pole can also be mimicked by growth factors such as activin, a member of the TGFβ family, which induces isolated animal caps to form mesoderm *in vitro*.[64]

One of the earliest responses of activin-induced animal cap tissues is the ability to spread and migrate on FN substrates.[42,43,57] The acquisition of FN binding activity occurs within minutes following exposure to activin. The binding is inhibited in the presence of RGD-containing peptides, suggesting the involvement of integrins in this process.[42] Could the control of mesodermal cell migration at gastrulation be triggered by inductive signals? If so, one hypothesis is that integrin expression should be upregulated in response to the inducer. However, no significant increase in the synthesis of integrin β1 protein[42] or mRNA[51] is observed following exposure to activin. There are several possible explanations for these results. As discussed earlier, a number of investigators have reported that the majority of the β1 subunit is stored in the egg and early embryo as a precursor form,[27,43,45,56] similar to the situation observed for many cultured cells.[65] It is possible that the appearance of new integrins at the cell surface following induction, therefore, might be controlled by the synthesis or processing of α subunits.

Whittaker and DeSimone[51] reported that the synthesis of α3, α4, and α6 mRNAs increases in animal cap cells in response to induction with activin. However, the expression of these α subunits is a relatively "late" event in comparison to early response genes such as the mesoderm gene marker, brachyury.[66] Given the time course, increased synthesis of α subunits cannot account for the rapid acquisition of FN binding that is observed. Another possibility is that non-β1-containing integrins present in the egg are responsible for activin-induced adhesion to FN. This remains a formal possibility given that at least two other integrin subfamiles, β3 and β6, are expressed in eggs and early embryos.[49] In the case of β3 mRNA, however, synthesis is also unaffected in animal cap cells exposed to activin.[51] Further evaluation of these integrins awaits the availability of specific antibodies and the identification of α subunits that are known to form dimers with the β3 and β6 subunits and also function as FN receptors.[1] It is possible that increased expression of one or more of these receptors at the cell surface might account for the increase in FN binding observed.

In this regard, it is interesting to consider one example of how a growth factor could be involved in controlling integrin α subunit expression. Goosecoid (gsc) is a homeobox gene expressed in the dorsal lip region of the blastopore at gastrulation and is believed to be a critical component in the control of dorsal cell fate and axis specification in *Xenopus*.[67] Recent studies indicate that gsc expression is important for controlling cell migration and mesoderm involution.[68] Gsc is induced in isolated animal caps by concentrations of activin[69] that also increase the expression of integrin α3, α4, and α6 mRNAs[51] (Whittaker and DeSimone, unpublished observations). In addition, gsc and α3 are coexpressed in the dorsal lip region of the embryo. It is possible, therefore, that α3 expression is controlled by gsc or another inducible upstream gene product. The study of integrin gene expression is certain to receive increased attention in the next few years as investigators begin to focus more on the specific morphogenetic events that are triggered by induction.

One additional possibility is that adhesion to FN following activin induction could arise via activation of preexisting integrins already present at the embryonic cell surface. This would be analogous to the situation observed for platelets, where the activation of αIIbβIIIa leads to the rapid upregulation of adhesive response.[70] The posttranslational activation of preexisting integrins at various times in development might provide a positional mechanism whereby adhesive activity could be rapidly modulated by embryonic inducers.

III. CONCLUSIONS AND PERSPECTIVES

Integrins display surprisingly complex patterns of expression at very early stages of development. Furthermore, the timing of expression is tightly regulated and often dynamic. The identification of multiple integrins in amphibian embryos has provided evidence suggesting that these receptors participate in development by specifying position, directing cellular movements, and signaling changes in the extracellular environment.

This chapter has focused on integrins in amphibian embryos, but the basic biological problem area has widespread applicability to the study of these receptors in a variety of developmental systems. The amphibian offers the advantage of easy accessibility to early developmental stages, and the promise of reverse genetic techniques to analyze integrin-ECM interactions *in vivo*. Other systems (e.g., mice, *Drosophila*) provide powerful genetic approaches that can be used to evaluate the functions of individual subunits with great precision. Whatever the system, the challenge is likely to be the same: to define the strategies required for identifying the contribution of a single type of adhesion molecule against a backdrop of considerable functional diversity and redundancy. The technical problems are daunting but the rewards potentially enormous to our understanding of the forces that help drive morphogenetic change.

ACKNOWLEDGMENTS

I thank the many colleagues who provided reprints and data prior to publication for inclusion in this article. I am also grateful to Mary Kate Worden, Tom Lallier, Mark Hens, Joe Ramos, David Ransom, and Charles Whittaker for their critical review of the manuscript and for helpful suggestions. I also acknowledge the generous research support of the Pew Charitable Trusts, the American Cancer Society, and the National Institutes of Health (NICHHD26402).

REFERENCES

1. Hynes, R. O., Integrins: versatility, modulation, and signaling in cell adhesion, *Cell*, 69, 11, 1992.
2. Hynes, R. O. and Lander, A. D., Contact and adhesive specificities in the associations, migrations, and targeting of cells and axons, *Cell*, 68, 303, 1992.
3. Takeichi, M., Cadherin cell adhesion receptors as a morphogenetic regulator, *Science*, 251, 1451, 1991.
4. Albelda, S. M. and Buck, C. A., Integrins and other cell adhesion molecules, *FASEB J.*, 4, 2868, 1990.
5. Schwartz, M. A., Transmembrane signalling by integrins, *Trends Cell Biol.*, 2, 304, 1992.
6. Hortsch, M. and Goodman, C. S., Cell and substrate adhesion molecules in Drosophila, *Ann. Rev. Cell Biol.*, 7, 505, 1991.
7. Glukhova, M. A. and Thiery, J. P., Fibronectin and integrins in development in *Seminars in Cancer Biology*, Yamada, K. M. and Akiyama, S., Eds., Saunders Scientific Publications, Academic Press, New York, In press.
8. Lallier, T., and Hens, M. D., DeSimone, D. W., Integrins in developent in *Integrins: Molecular and Biological Response to the Extracellular Matrix*, Cheresh, D. A. and Mecham, R. P., Eds., Academic Press, New York, (in press)
9. Vize, P. D., Melton, D. A., Hemmati-Brivanlou, A., and Harland, R. M., Assays for gene function in developing Xenopus embryos, *Methods Cell Biol.*, 36, 367, 1991.
10. Boucaut, J. C., Darribere, T., Shi, D. L., Riou, J. F.,Johnson, K. E., and Delarue, M., Amphibian gastrulation: the molecular bases of mesodermal cell migration in urodeles embryos, in *Gastrulation Movements, Patterns, and Molecules*, Keller, R., Clark, W. H., and Griffin, F., Eds., Plenum Press, New York, 1991, 169.
11. Winklbauer, R., Selchow, A., Nagel, M., Stoltz, C., and Angres, B., Mesoderm cell migration in the Xenopus gastrula, in *Gastrulation Movements, Patterns, and*

Molecules, Keller, R., Clark, W. H., and Griffin, F., Eds., Plenum Press, New York, 1991, 147.

12. Johnson, K. E., Boucaut, J. C., and DeSimone, D. W., Role of the extracellular matrix in amphibian gastrulation, in *Current Topics in Developmental Biology*, Pedersen, R.A., Eds., Academic Press, San Diego, 1992, 91.

13. Keller, R., Shih, J., and Wilson, P., Cell motility, control and function of convergence and extension during gastrulaton in Xenopus in *Gastrulation Movements, Patterns, and Molecules*, Keller, R., Clark, W. H., and Griffin, F., Eds., Plenum Press, New York, 1991, 101.

14. Keller, R. and Winklbauer, R., The role of the extracellular matrix in amphibian gastrulation, *Semin. Dev. Biol.*, 1, 25, 1990.

15. Keller, R. and Winklbauer, R., Cellular basis of amphibian gastrulation, in *Current Topics in Developmental Biology*, Pedersen, R. A., Eds., Academic Press, San Diego, 1992, 39.

16. Keller, R., Shih, J., Wilson, P., and Sater, A., Pattern and function of cell motility and cell interactions during convergence and extension in Xenopus, in *Cell interaction in Early Development*, Gerhart, J., Eds., Wiley-Liss, New York, 1991, 31.

17. Keller, R., Early embryonic development of Xenopus laevis, *Methods Cell Biol.*, 36, 61, 1991.

18. Shi, D. L., Delarue, M., Darribere, T., Riou, J. F., and Boucaut, J. C., Experimental analysis of the extension of the dorsal marginal zone in *Pleurodeles waltl* gastrulae, *Development*, 100, 147, 1987.

19. Winklbauer, R., Mesodermal cell migration during Xenopus gastrulation, *Dev. Biol.*, 142, 155, 1990.

20. Boucaut, J. C. and Darribere, T., Fibronectin in early amphibian embryos. Migrating mesodermal cells contact fibronectin established prior to gastrulation, *Cell Tissue Res.*, 234, 135, 1983.

21. Lee, G., Hynes, R. O., and Kirschner, M., Temporal and spatial regulation of fibronectin in early Xenopus development, *Cell*, 36, 729, 1984.

22. Newport, J. and Kirschner, M., A major developmental transition in early Xenopus embryos I. Characterization and timing of cellular changes at the midblastula stage, *Cell*, 30, 675, 1982a.

23. DeSimone, D. W., Norton, P. A., and Hynes, R. O., Identification and characterization of alternatively spliced fibronectin mRNAs expressed in early Xenopus embryos, *Dev. Biol.*, 149, 357, 1992.

24. Clavilier, L., Riou, J. C., Shi, D. L., DeSimone, D. W., and Boucaut, J. C., Amphibian *Pleurodeles waltl* fibronectin: cDNA cloning and developmental expression of spliced variants, *Cell Adhesion Commun.*, 1, 83, 1993.

25. Boucaut, J. C., Darribere, T., Boulekbache, H., and Thiery, J. P., Prevention of gastrulation but not neurulation by antibodies to fibronectin in amphibian embryos, *Nature*, 307, 364, 1984.

26. Boucaut, J. C., Darribere, T., Poole, T. J., Aoyama, H., Yamada, K. M., and Thiery, J. P., Biologically active synthetic peptides as probes of embryonic development: a competitive peptide inhibitor of fibronectin function inhibits gastrulation in amphibian embryos and neural crest cell migration in avian embryos, *J. Cell Biol.*, 99, 1822, 1984.

27. Howard, J. E., Hirst, E. M., and Smith, J. C., Are beta 1 integrins involved in Xenopus gastrulation?, *Mech. Dev.*, 38, 109, 1992.

28. Keller, R. E., Danilchik, M., Gimlich, R., and Shih, J., The function and mechanism of convergent extension during gastrulation of *Xenopus laevis*, *J. Embryol. Exp. Morphol.*, 89, 185, 1985.

29. Keller, R. and Jansa, S., Xenopus gastrulation without a blastocoel roof, *Dev. Dyn.*, 195, 162, 1992.

30. Yost, J., Regulation of vertebrate left-right asymmetries by extracellular matrix, *Nature*, 357, 158, 1992.

31. Nakatsuji, N., Gould, A. C., and Johnson, K. E., Movement and guidance of migrating mesodermal cells in *Ambystoma maculatum* gastrulae, *J. Cell Sci.*, 56, 207, 1982.

32. Nakatsuji, N. and Johnson, K. E., Experimental manipulation of a contact guidance system in amphibian gastrulation by mechanical tension, *Nature*, 307, 453, 1984.

33. Winklbauer, R., Selchow, A., Nagel, M., and Angres, B., Cell interaction and its role in mesoderm cell migration during Xenopus gastrulation, *Dev. Dyn.*, 195, 290, 1992.

34. Shi, D. L., Darribere, T., Johnson, K. E., and Boucaut, J. C., Initiation of mesodermal cell migration and spreading relative to gastrulation in the urodele amphibian *Pleurodeles waltl*, *Development*, 105, 351, 1989.

35. Winklbauer, R. and Nagel, M., Directional mesoderm cell migration in the Xenopus gastrula, *Dev. Biol.*, 148, 573, 1991.

36. Winklbauer, R., Nagel, M., and Selchow, A., Factors controlling the directionality of mesoderm cell migration in the Xenopus gastrula, in *Biological Pattern Formation NATO ASI Series*, Maini, K. P., Eds., in press.

37. Ramos, J. W. and DeSimone, D. W., Xenopus embryonic cell adhesion and migration are supported by multilpe cell-interactive sites on fibronectin, *J. Cell Biol.*, in preparation.

38. Nagai, T., Yamakawa, N., Aota, S., Yamada, S.S., Akiyama, S. K., Olden, K., and Yamada, K. M., Monoclonal antibody characterization of two distant sites required for function of the central cell-binding domain of fibronectin in cell adhesion, cell migration and matrix assembly, *J. Cell Biol.*, 114, 1295, 1991.

39. Guan, J. L. and Hynes, R. O., Lymphoid cells recognize an alternatively spliced segment of fibronectin via the integrin receptor alpha 4 beta 1, *Cell*, 60, 53, 1990.

40. Wayner, E. A. and Kovach, N. L., Activation-dependent recognition by hematopoietic cells of the LDV sequence in the V region of fibronectin, *J. Cell Biol.*, 116, 489, 1992.

41. Darribere, T., Yamada, K. M., Johnson, K. E., and Boucaut, J. C., The 140-kDa fibronectin receptor complex is required for mesodermal cell adhesion during gastrulation in the amphibian Pleurodeles waltlii, *Dev. Biol.*, 126, 182, 1988.

42. Smith, J. C., Symes, K., Hynes, R. O., and DeSimone, D., Mesoderm induction and the control of gastrulation in Xenopus laevis: the roles of fibronectin and integrins, *Development*, 108, 229, 1990.

43. DeSimone, D. W., Smith, J. C., Howard, J. E., Ransom, D. G., and Symes, K., The expression of fibronectins and integrins during mesodermal induction and gastrulation in Xenopus, in *Gastrulation Movements, Patterns, and Molecules*, Keller, R., Clark, W. H., and Griffin, F., Eds., Plenum Press, New York, 1991, 185.

44. DeSimone, D. W. and Hynes, R. O., *Xenopus laevis* integrins. Structural conservation and evolutionary divergence of integrin beta subunits, *J. Biol. Chem.*, 263, 5333, 1988.

45. Gawantka, V., Ellinger-Ziegelbauer, H., and Hausen, P., Beta 1-integrin is a maternal protein that is inserted into all newly formed plasma membranes during early Xenopus embryogenesis, *Development*, 115, 595, 1992.

46. MacKrell, A. J., Blumberg, B., Haynes, S. R., and Fessler, J. H., The lethal myospheroid gene of Drosophila encodes a membrane protein homologous to vertebrate integrin beta subunits, *Proc. Natl. Acad. Sci. U.S.A.*, 85, 2633, 1988.

47. Erle, D. J., Sheppard, D., Breuss, J., Ruegg, C., and Pytela, R., Novel integrin alpha and beta subunit cDNAs identified in airway epithelial cells and lung leukocytes using the polymerase chain reaction, *Am. J. Respir. Cell Mol. Biol.*, 5, 170, 1991.

48. Yee, G. H. and Hynes, R. O., A novel, tissue specific integrin subunit, beta nu, expressed in the midgut of Drosophila melanogaster, *Development*, 118, 845, 1993.

49. Ransom, D. G., Hens, M. D., and DeSimone, D. W., Integrin expression in early amphibian embryos: cDNA cloning and characterization of beta1, beta2, beta3 and beta6 subunits, *Dev. Biol.*, 160, 265, 1993.
50. D'Souza, S. E., Ginsburg, M. H., Burke, T. A., Lam, S. C. T., and Plow, E. F., Localization of an RGD recognition site within an integrin adhesion receptor, *Science*, 242, 91, 1988.
51. Whittaker, C. A. and DeSimone, D. W., Integrin alpha subunit mRNAs are differentially expressed in early Xenopus embryos, *Development*, 117, 1239, 1993.
52. Marcantonio, E. E. and Hynes, R. O., Antibodies to the conserved cytoplasmic domain of the integrin beta 1 subunit react with proteins in vertebrates, invertebrates, and fungi, *J. Cell Biol.*, 106, 1765, 1988.
53. Solowska, J., Edelman, J. M., Albelda, S. M., and Buck, C. A., Cytoplasmic and transmembrane domains of integrin beta 1 and beta 3 subunits are functionally interchangeable, *J. Cell Biol.*, 114, 1079, 1991.
54. Reszka, A. A., Hayashi, Y., and Horwitz, A. F., Identification of amino acid sequences in the integrin beta 1 cytoplasmic domain implicated in cytoskeletal association, *J. Cell Biol.*, 117, 1321, 1992.
55. Krotoski, D. and Bronner-Fraser, M., Distribution of integrins and their ligands in the trunk of *Xenopus laevis* during neural crest cell migration, *J. Exp. Zool.*, 253, 139, 1990.
56. Muller, A. H. J, Gawantka, V., Ding, X., and Hausen, P., Maturation induced internalization of beta1-integrin by Xenopus oocytes and formation of the maternal integrin pool, *Mech. Dev.*, 42, 77, 1993.
57. Smith, J. C. and Howard, J. E., Mesoderm-inducing factors and the control of gastrulation, *Development (Suppl.)*, 127, 1992.
58. Darribere, T., Guida, K., Larjava, H., Jonston, K. E., Yamada, K. M., Thiery, J. P., and Boucaut, J. C., *In vivo* analyses of integrin beta1 subunit function in fibronectin matrix assembly, *J. Cell Biol.*, 110, 1813, 1990.
59. Kieffer, T. K. and Phillips, D. R., Platelet membrane glycoproteins: functions in cellular interactions, *Annu. Rev. Cell Biol.*, 6, 329, 1990.
60. New, H. V., Howes, G., and Smith, J. C., Inductive interactions in early embryonic development, *Curr. Opin. Genet. Dev.*, 1, 196, 1991.
61. Kimmelman, D., Christian, J. L., and Moon, R. T., Synergistic principles in development: overlapping patterning systems in Xenopus mesoderm induction, *Development*, 116, 1, 1992.
62. Sive, H. L., The frog prnce-ss: a molecular formula for dorsoventral patterning in Xenopus, *Genes Dev.*, 7, 1, 1993.
63. Nieuwkoop, P. D., The formation of mesoderm in urodelean amphibians. I. Induction by the endoderm, *Roux Arch. Dev. Biol.*, 162, 341, 1969.
64. Smith, J. C., Mesoderm induction and mesoderm-inducing factors in early amphibian development, *Development*, 105, 665, 1989.
65. Heino, J., Ignotz, R. A., Hemler, M. E., Crouse, C., and Massague, J., Regulation of cell adhesion receptors by transforming growth factor-beta. Concomitant regulation of integrins that share a common beta 1 subunit, *J. Biol. Chem.*, 264, 380, 1989.
66. Smith, J. C., Price, B. M., Green, J. B., Weigel, D., and Herrmann, B. G., Expression of a Xenopus homolog of Brachyury (T) is an immediate-early response to mesoderm induction, *Cell*, 67, 79, 1991.
67. Cho, K. W., Blumberg, B., Steinbeisser, H., and De Robertis, E. M., Molecular nature of Spemann's organizer: the role of the Xenopus homeobox gene goosecoid, *Cell*, 67, 1111, 1991.
68. Niehrs, C., Keller, R., Cho, K. W., and De Robertis, E. M., The homeobox gene goosecoid controls cell migration in Xenopus embryos, *Cell*, 72, 491, 1993.

69. Green, J. B., New, H. V., and Smith, J. C., Responses of embryonic Xenopus cells to activin and FGF are separated by multiple dose thresholds and correspond to distinct axes of the mesoderm, *Cell*, 71, 731, 1992.

70. Phillips, D. R., Charo, I. F., and Scarborough, R. M., GPIIb-IIIa: the responsive integrin, *Cell*, 65, 359, 1991.

71. DeSimone, D. W. and Johnson, K. E., The Xenopus embryo as a model system for the study of cell-extracellular matrix interactions, *Methods Cell Biol.*, 36, 527, 1991.

INDEX

T - #0520 - 101024 - C0 - 222/151/14 - PB - 9781138560444 - Gloss Lamination